Lecture Notes in Computer Science 5858

Commenced Publication in 1973
Founding and Former Series Editors:
Gerhard Goos, Juris Hartmanis, and Jan van Le

Guido Governatori   John Hall
Adrian Paschke (Eds.)

Rule Interchange and Applications

International Symposium, RuleML 2009
Las Vegas, Nevada, USA, November 5-7, 2009
Proceedings

 Springer

Volume Editors

Guido Governatori
National ICT Australia
Queensland Research Laboratory
PO Box 6020
St Lucia 4067, Queensland, Australia
E-mail: guido.governatori@nicta.com.au

John Hall
Model Systems
15-19 Cavendish Place
London W1G 0DD, UK
E-mail: johnhall@modelsys.com

Adrian Paschke
Freie Universität Berlin
FB Mathematik und Informatik
Institut für Informatik
AG Corporate Semantic Web
Königin-Luise-Str. 24/26
14495 Berlin, Germany
E-mail: paschke@inf.fu-berlin.de

Library of Congress Control Number: 2009936145

CR Subject Classification (1998): D.3.1, F.3.2, H.5.3

LNCS Sublibrary: SL 2 – Programming and Software Engineering

ISSN   0302-9743

ISBN   978-3-642-04984-2 Springer Berlin Heidelberg New York

springer.com

© Springer-Verlag Berlin Heidelberg 2009

Typesetting: Camera-ready by author, data conversion by Scientific Publishing Services, Chennai, India
Printed on acid-free paper   SPIN: 12776832   06/3180   5 4 3 2 1 0

Preface

The 2009 International Symposium on Rule Interchange and Applications (RuleML 2009), collocated in Las Vegas, Nevada, with the 12th International Business Rules Forum, was the premier place to meet and to exchange ideas from all fields of rules technologies. The aims of RuleML 2009 were both to present new and interesting research results and to show successfully deployed rule-based applications. This annual symposium is the flagship event of the Rule Markup and Modeling Initiative (RuleML).

The RuleML Initiative (www.ruleml.org) is a non-profit umbrella organization of several technical groups organized by representatives from academia, industry and public sectors working on rule technologies and applications. Its aim is to promote the study, research and application of rules in heterogeneous distributed environments such as the Web. RuleML maintains effective links with other major international societies and acts as intermediary between various 'specialized' rule vendors, applications, industrial and academic research groups, as well as standardization efforts from, for example, W3C, OMG, and OASIS. To emphasize the importance of rule standards RuleML 2009 featured, besides a number of tutorials on various rule aspects, a tutorial and a workshop dedicated to the newly released W3C Rule Interchange Format (RIF).

After a series of successful international RuleML workshops and conferences, the RuleML Symposium, held since 2007, constitutes a new kind of event where the Web rules and logic community joins the established, practically oriented business rules forum community (www.businessrulesforum.com) to help the cross-fertilization between Web and business logic technology. The symposium supports the idea that there is a successful path from high-quality research results to applied applications. It brings together rule system providers, representatives of, and participants in, rule standardization efforts and open source rule communities, practitioners and technical experts, developers, users, and researchers, to exchange new ideas, practical developments and experiences on issues pertinent to the interchange and application of rules.

The contributions in this volume include the abstracts of the three invited keynote presentations, five introductory/survey track papers by the track chairs introducing the topics of the RuleML 2009 tracks and a selection of 12 full papers and 17 short papers chosen from a pool of 56 submissions, from 14 countries. The accepted papers address a wide range of rule topics, including traditional topics, such as complex event processing using rules (a RuleML stronghold with several past invited speakers form the area and many past papers on this topic), to rules for transformations and rule extraction, applications of rule systems to handle data and processes, investigations on how to deploy rules on the Web to use of rules to model uncertainty and norms.

The accepted papers were carefully selected after a rigorous peer-review process where each paper was evaluated by a panel of five members of the international Programme Committee. Each paper received a minimum of three reviews. We thank the referees for their effort and very valuable contribution; without them it would not be possible to maintain and improve the high scientific standard the symposium has now reached. We thank the authors for submitting good papers, responding to the reviewers' comments, and abiding by our production schedule. We thank the keynote speakers for their interesting talks. And we thank the Business Rules Forum organizers for enabling this fruitful collocation of the 12th Business Rules Forum and RuleML 2009.

The real success of rule technology will be measured by the applications that use the technology rather than the technology itself. To place emphasis on the practical use of rule technologies, RuleML 2009 continued the tradition of hosting the International Rule Challenge. The challenge offered participants the opportunity to demonstrate their commercial and open source tools, use cases, benchmarks, and applications. It was the ideal forum for those wanting to understand how rules technology can produce benefits, both technically and commercially.

The RuleML 2009 Symposium was financially supported by industrial companies and research institutes and was technically supported by several professional societies. We thank our sponsors, whose financial support helped us to organize this event, and whose technical support enabled us to attract many high-quality submissions.

At the conference, the Programme Committee awarded prizes for the best paper, and for the best entry and runner up for the International Rule Challenge. The results are available at www.ruleml.org.

August 2009

Guido Governatori
John Hall
Adrian Paschke

Conference Organization

General Chair

Adrian Paschke   Freie Universität Berlin, Germany

Programme Chairs

Guido Governatori   NICTA, Australia
John Hall   Model Systems, UK

Liaison Chair

Hai Zhuge   Chinese Academy of Sciences, China

Publicity Chair

William Langley   NRC-IRAP, Canada

Track Chairs

Rule Transformation and Extraction
Erik Putrycz   Apption Software, Canada
Mark Linehan   IBM, USA

Rules and Uncertainty
Matthias Nickles   Univiversity of Bath, UK
Davide Sottara   University Bologna, Italy

Rules and Norms
Thomas Gordon   Fraunhofer FOKUS, Germany
Antonino Rotolo   CIRSFID, University of Bologna, Italy

Rule-Based Game AI
Benjamin Craig   National Research Council, Canada
Weichang Du   University of New Brunswick, Canada

Rule-Based Event Processing and Reaction Rules

Alex Kozlenkov   Betfair Ltd., UK
Adrian Paschke   Freie Universität Berlin, Germany

Rules and Cross-Industry Standards

Tracy Bost   Valocity, USA
Robert Golan   DBMind, USA

RuleML Challenge

Yuh-Jong Hu   National Chengchi University, Taiwan
Ching-Long Yeh   Tatung University, Taiwan
Wolfgang Laun   Thales Rail Signalling Solutions GesmbH,
     Austria

Programme Committee

Asaf Adi
Hassan Ait-Kaci
Grigoris Antoniou
Sidney Bailin
Matteo Baldoni
Cristina Baroglio
Claudio Bartolini
Nick Bassiliades
Bernhard Bauer
Mikael Berndtsson
Jean Bezivin
Pedro Bizarro
Jonathan Bnayahu
Harold Boley
Peter Bollen
Adrian Bowles
Jordi Cabot
Carlos Castro
Donald Chapin
Federico Chesani
Horatiu Cirstea
Claudia d'Amato
Mike Dean
Jens Dietrich
Juergen Dix
Daniel Dougherty
Schahram Dustdar

Andreas Eberhart
Jenny Eriksson Lundström
Opher Etzion
Todd Everett
Maribel Fernandez
Dragan Gasevic
Adrian Giurca
Neal Hannon
Marek Hatala
Ioannis Hatzilygeroudis
Stijn Heymans
Minsu Jang
Claude Kirchner
Yiannis Kompatsiaris
Manolis Koubarakis
Rick Labs
Holger Lausen
Heiko Ludwig
Thomas Lukasiewicz
Ian Mackie
Mirko Malekovic
Christopher Matheus
Craig McKenzie
Jing Mei
Zoran Milosevic
Anamaria Moreira
Leora Morgenstern

Jörg Müller
Chieko Nakabasami
Ilkka Niemelä
Jeff Pollock
Alun Preece
Maher Rahmouni
Dave Reynolds
Graham Rong
Markus Schacher
Marco Seiriö
Rachael Sokolowski
Jorge Sousa Pinto
Silvie Spreeuwenberg
Giorgos Stamou

Giorgos Stoilos
Nenad Stojanovic
Umberto Straccia
Heiner Stuckenschmidt
Terrance Swift
Vagan Terziyan
Jan Vanthienen
Paul Vincent
George Vouros
Kewen Wang
Segev Wasserkrug
Nikolaus Wulff
Ching Long Yeh

External Reviewers

Anirban Basu
Tristan Behrens
Carlos Chesñevar
Stamatia Dasiopoulou
Pablo Fillottrani
Efstratios Kontopoulos
Emilian Pascalau
Ruth Raventós

RuleML 2009 Sponsors & Partners

Silver Sponsors

Bronze Sponsors

Partner Organisations

BRFORUM.BE
Belgian Business Rules Forum

Media Partners

Table of Contents

General Rule Topics

Rule Transformation and Extraction

Session 6

Session 7

Session 8

Bringing Order to Chaos: RIF as the New Standard for Rule Interchange

Sandro Hawke

World Wide Web Consortium (W3C) at MIT
`sandro@w3.org`

Abstract. As the W3C Rule Interchange Format (RIF) nears completion, we consider what it offers users and how it may change the design of systems and change the industry. More than just a standard XML format for rules, RIF is integrated with the W3C Semantic Web technology stack, offering a vision for combining some of the best features of the Web with the best features of rule systems. RIF is designed to directly handle rule bases which use only standard features, but it can be extended. Some example extensions and possible areas for future standards work will be discussed.

G. Governatori, J. Hall, and A. Paschke (Eds.): RuleML 2009, LNCS 5858, p. 1, 2009.
© Springer-Verlag Berlin Heidelberg 2009

Why Rules Matter in Complex Event Processing... and Vice Versa

Paul Vincent

TIBCO
pvincent@tibco.com

Abstract. Many commercial and research CEP solutions are moving beyond simple stream query languages to more complete definitions of "process" and thence to "decisions" and "actions". And as capabilities increase in event processing capabilities, there is an increasing realization that the humble "rule" is as relevant to the event cloud as it is to specific services. Less obvious is how much event processing has to offer the process and rule execution and management technologies. Does event processing change the way we should manage businesses, processes and services, together with their embedded (and hopefully managed) rulesets?

G. Governatori, J. Hall, and A. Paschke (Eds.): RuleML 2009, LNCS 5858, p. 2, 2009.
© Springer-Verlag Berlin Heidelberg 2009

Terminology: The Semantic Foundation for an Organizations Executable Rules

Donald Chapin

Business Semantics Ltd
Donald.Chapin@BusinessSemantics.com

Abstract. For rules to be applied consistently they have to be expressed consistently — they need common terminology. It is also essential that the semantics and organizations business definitions, policies and rules is carried forward into the IT systems that support the organization.

This talk introduces the general principles of terminology and the role played by terminology in Semantics of Business Vocabulary and Business Rules (SBVR), the specification published by the Object Management Group (OMG) in 2008.

The major topics are:

- Introduction to the discipline of terminology (sister discipline of lexicography) and the fundamentals of ISO terminology standards. The terminology capabilities that SBVR adds to ISO terminology standards.
- The significance of the industry trend to use terminology to improve communication and the quality of document/content authoring.
- How SBVR supports the people who operate the organization in specifying the meaning of business policies and rules in the terminology they use every day.
- An overview of the transformation from an organization's business policies and rules —as expressed in its managed terminology — to executable data and rules, while maintaining the organizations semantics.

G. Governatori, J. Hall, and A. Paschke (Eds.): RuleML 2009, LNCS 5858, p. 3, 2009.
© Springer-Verlag Berlin Heidelberg 2009

Challenges for Rule Systems on the Web

Yuh-Jong Hu[1], Ching-Long Yeh[2], and Wolfgang Laun[3]

[1] Emerging Network Technology (ENT) Lab.
Department of Computer Science
National Chengchi University, Taipei, Taiwan
hu@cs.nccu.edu.tw
[2] Department of Computer Science Engineering
Tatung University, Taipei, Taiwan
chingyeh@cse.ttu.edu.tw
[3] Thales Rail Signalling Solutions
GmbH, Austria
wolfgang.laun@gmail.com

Abstract. The RuleML Challenge started in 2007 with the objective of inspiring the issues of implementation for management, integration, interoperation and interchange of rules in an open distributed environment, such as the Web. Rules are usually classified as three types: deductive rules, normative rules, and reactive rules. The reactive rules are further classified as ECA rules and production rules. The study of combination rule and ontology is traced back to an earlier active rule system for relational and object-oriented (OO) databases. Recently, this issue has become one of the most important research problems in the Semantic Web. Once we consider a computer executable policy as a declarative set of rules and ontologies that guides the behavior of entities within a system, we have a flexible way to implement real world policies without rewriting the computer code, as we did before. Fortunately, we have de facto rule markup languages, such as RuleML or RIF to achieve the portability and interchange of rules for different rule systems. Otherwise, executing real-life rule-based applications on the Web is almost impossible. Several commercial or open source rule engines are available for the rule-based applications. However, we still need a standard rule language and benchmark for not only to compare the rule systems but also to measure the progress in the field. Finally, a number of real-life rule-based use cases will be investigated to demonstrate the applicability of current rule systems on the Web.

1 Introduction

The RuleML Challenge competitions started in 2007[1], so the RuleML-2009 Challenge will be the third year for the rule system competition. We offer participants the chance to demonstrate their commercial and open source tools, use cases, and applications for rule related technologies. For the past two RuleML Challenge

---

[1] RuleML-2007 Challenge, http://2007.ruleml.org/index-Dateien/Page787.htm

G. Governatori, J. Hall, and A. Paschke (Eds.): RuleML 2009, LNCS 5858, pp. 4–16, 2009.

competitions, only a minimum set of requirements was given for evaluating the submitted demo systems. The criteria were that declarative rules should have to play a central role in the application, and that the demo systems should preferably be embedded into a Web-based or distributed environment, etc. The Challenge winners were selected and 1st and 2nd places were awarded with prestigious prizes.

The RuleML-2009 Challenge follows similar processes and the evaluation criteria are the same as in the previous two events. But we consider inviting more participants to submit their rule related systems in this year. In the RuleML-2009 Challenge, we organize events as two tracks, one is by invitation, to demonstrate a commercial or open source environment for its rule systems, and the other is open to general public for a real system competition. In addition to the demo systems with reports submitted to the RuleML Challenge website[2], it is also possible to submit demo papers describing research and technical details, and the selected papers will be published in additional special Challenge proceedings, such as CEURS. A final selection of revised papers from the Challenge proceedings will be resubmitted to a special issue of a journal for publishing. In this RuleML Challenge survey paper, we point out the possible research and implementation challenges for rule systems on the Web that are related to the Challenge competition events in the forthcoming years.

1.1   Challenges for Rule Systems

Rules as human understandable policies are everywhere in our daily life to impose human behaviors. For example, before you take a flight, you need to read airline check-in and boarding time rules in the policy statement of your booking itinerary receipt. If you violate any rule you might miss your plane. Related situations in this scenario of using rules are early-bird conference registration, special discount hotel reservation, payment and refund policies, etc. These rules as policies are represented as human understandable natural language. However, we still need to transform these natural language policies into computer programming rules for computer system understanding and automatic execution. Sometimes, not all of the rules imposed on a human are necessarily and possibly represented as software programs to accomplish automatic execution in our computer systems. Usually, these rules restrict only human behavior, without direct connection with any software system. For example, we have law for privacy protection and digital rights management but not all of privacy rights and digital rights for human are required to be represented and evaluated in computer systems.

There are several challenges while implementing rule systems on the Web. Rules should be allowed to cope with the data model, such as RDB/OO-DB, or a knowledge base, such as an ontology, to permit query and modification services on the data models. Policies imposed on human behavior are declared in some policy language by the combination of rules and an ontology (or database), and these policies can be automatically interpreted and executed by a computer.

---

[2] http://ruleml-challenge.cs.nccu.edu.tw

There should be a standard language and framework for rule systems to enable rule interchange services on the Web. A certain number of use cases are easily represented and executed by rule and ontology reasoning engines with rule interchange and ontology merging standards to ensure rule interoperability and ontology compatibility.

In the early computer development stage, imperative programming languages such as C and Java were used to represent rules and execute them on a computer system. But these rules are inflexible and not easy to maintain when they are distributed on the Web and require interchange and integration between rule systems. Moreover, imperative programming languages are not appropriate to express concepts of human policies as computerized rules. Recently, people use declarative programming to specify the rules and execute them automatically, where XML is used as a standard syntax representation for interchange of declarative rules, such as RuleML [1], RIF [2], etc.

Even though an XML-based standard rule language and framework provides rule interchange service, pure XML cannot specify a well-defined semantics for rules. So people in the standard rule community constructed a logic foundation behind rule languages and their framework, to preserve the integrity of syntax and semantics of rules interchange for various rule systems. Similarly, OMG SBVR intends to define the vocabulary and rules for documenting semantics of business vocabulary, facts, and rules, as well as an XMI schema for interchange of business vocabularies and rules among organizations and between software tools[3].

In this paper, we first introduce the classification of rules, then, in section 2, we address the issue of rules, and databases and ontologies . In section 3, the current status of a declarative policy as the combination of ontology and rules will be introduced. In addition, Semantic Web Service (SWS) processes also require a declarative policy to express and execute Web Service rules to control information sharing and service execution. In section 4, we examine current different rule management systems and engines. In section 5, we investigate different rule interchange languages. In section 6, we look into the use cases that are possibly represented and executed by the rule systems. Finally, we conclude this study in section 7.

2   Rule and Data Model

2.1   The Classification of Rules

Rules are classified as three types: deductive rules (or derivation rules), normative rules (or integrity rules), and reactive rules (or active rules). One can use deductive rules and facts to trigger a forward or backward reasoning engine to derive implicit facts. Normative rules pose constraints on the data or on the business logic to ensure their consistency in the database or knowledge

---

[3] http://www.omg.org/spec/SBVR/1.0/

base. Without reactive rules, we cannot update a database or knowledge base by using deductive rules only.

Reactive rules are further subdivided into event-condition-action (ECA) rules and production rules. ECA rules are rules of the form *ON Event IF Condition DO Action*, where *Action* should be executed if the *Event* occurs, provided that the *Condition* holds. Production rules are rules of the form *IF Condition DO Action*, where *Condition* queries the working memory containing the data on which the rules operate. *Action* should be executed whenever a change to the underlying database makes *Condition* true [2].

In reactive rules, we verify the satisfaction of conditions and also execute the action whenever message arrival or timer event triggers the rule. Declarative rules extend their executive power by the combination of rule semantics and imperative programming in the action part.

2.2   Rules and Databases

As early as 1980, Ullman pointed out the principles of the integration of database and knowledge base systems [3] [4]. The foundation of database is relation algebra with SQL as a declarative database query language. However, first order logic (FOL) was also proposed as a way to represent "knowledge" and as a language for expressing operations on relations. The roots of relational theory is logic, and so we cannot deny that the foundation of relational DBMS is based on logic [5]. The simplest data model of FOL is "Datalog", which was coined to suggest a version of Prolog suitable for database systems where it does not allow function symbols in Datalog's predicate arguments. In the IDEA methodology, deductive rules and reactive rules were built on top of the object-oriented (OO) database as a way to express operations on the OO data model [6].

2.3   Rules and Ontologies

Concepts of the Semantic Web have been proposed by Tim Berners-Lee et al. since 2001 [7]. Graph-based RDF(S), including RDF and RDF-schema were the first standardized ontology languages to represent an ontology's schema and instances. Then, standardized ontology languages based on Description Logic (DL) [8], i.e., OWL-DL (later OWL 2), enhanced RDF(S) that plays the major role of knowledge representation for the Semantic Web. However, the logic program (LP) rule language was also introduced because of the limited expressive power of a DL-based ontology language in some situations, such as property chaining, and the manipulation of events, states, and actions.

Initially, the "rule" layer was laid on top of the "ontology" layer in the Semantic Web layered architecture but it has undergone several revisions reflecting the evolution of layers[4]. The most recent layered architecture of rule and ontology layers is one where they sit side by side to reflect their equal status but with some basic assumption differences between ontology and rule, such as the open world

---

[4] http://www.w3.org/2007/03/layerCake.svg

assumption (OWA) vs. the closed world assumption (CWA), or the non-unique name assumption (non-UNA) vs. the unique name assumption (UNA) [9].

It will be a challenge to resolve these basic assumption differences when we combine rule and ontology to execute rule systems on the Web.

Rules and RDF(S). Inspired from F-Logic, TRIPLE[5] was one of the earliest rule languages using Horn rules to access the RDF datasets. Another rule language called Notation3 (N3) uses a CWA forward reasoning engine to access the ontologies generated from RDF(S)[6]. SPARQL is another W3C standardized query language for querying RDF datasets. SPARQL queries are represented as Datalog rules so SPARQL's **CONSTRUCT** queries are viewed as deductive rules, which create new RDF triples from the RDF datasets.

Rules and OWL. In addition to the Semantic Web Rule Language (SWRL) [10], Rule Interchange Format (RIF) is an emerging rule interchange language from W3C RIF WG [2]. It intends to provide core and extend languages with a common exchange syntax for all of the classification rule languages, i.e., deductive, normative, and reactive rules. The requirements of integrating different types of rules with possible data (and meta data) accessing representation, i.e., RDB, XML, RDF, and OWL, drive the development of a RIF core interchange format, the *RIF Core,* and its extensions, *RIF dialects.* Another recent development is to combine RIF and OWL 2 in RIF, RDF, and OWL that specifies the interactions between RIF, RDF and OWL for their compatibilities[7].

2.4   Combination of Rule and Ontology

A one-way knowledge flow exists from an ontology module to a rule module for knowledge acquisition, where an ontology module's instances are imported as basic facts and filtered with conditions in the rules. This passive knowledge query only uses deductive rules. If a rule engine derives implicit new facts not in an ontology module and furthermore updates new facts back to an ontology module, then it provides another reverse knowledge flow from a rule module to an ontology module. In this two-way knowledge flow process, normative and reactive rules are also required to check the knowledge consistency and trigger the message passing for updating the ontology's knowledge base.

The idea of combining rules and ontologies is to fulfill a goal of two-way knowledge flow. The combination is classified as two types: tightly coupled integration and loosely coupled integration [11]. In the tightly coupled integration model, all of the terms in the rule's body and head are specified in the ontology schema, but in the loosely coupled integration model we do not have this requirement. So, some rules have their own defined terms in the rules' body or head. This loosely coupled integration model enhances the expressive power of ontology and rule as compared to the tightly coupled one.

---

[5] http://triple.semanticweb.org/

[6] http://www.w3.org/2000/10/swap/doc/cwm

[7] http://www.w3.org/2005/rules/wiki/OWLRL

Description Logic Program (DLP) [12] and SWRL are two well-known tightly coupled integration models. In general, both DL and LP are subsets of FOL in knowledge representation but each has its own part that cannot be expressed in the other part. DLP only takes intersection of DL and LP so knowledge representation in this model is limited. In SWRL, the major knowledge representation is OWL-DL with additional Datalog rules from LP to enhance the lack of property chaining in OWL-DL. In SWRL, DL-safe is the condition where variables occurring in each rule's head are also required to occur in its body to ensure the decidable reasoning of the rule engine. The availability of SWRL rule and ontology integration development in the popular Protégé environment[8] makes the SWRL model the most attractive one for people to use.

In the loosely coupled integration, DL-log [13], AL-log [14], and DL+Log [15] are three well-known models. In these models, rules are extended to Horn rules. Besides, not all of the terms in rules are required from ontology so rule module in these models provides more powerful knowledge representation and rule reasoning than the ones in SWRL. However, none of loosely coupled integration models provide standardized XML markup languages and a development environment, as SWRL does in Protégé. Obviously, this will be a challenge to represent and execute rule systems for loosely coupled integration on the Web. Moreover, the reactive rules [16] have not been seriously considered in all of the ontology and rule integration models. This will be the biggest impediment to implement rule and ontology systems for distributed applications on the Web.

3   Policy as Ontology and/or Rule

Since computers understand the data semantics in the Semantic Web, people are much more satisfied with the search results when a semantic search engine is fully developed. Policy-aware Web extends Semantic Web that provides computerized policies, such as privacy protection or digital rights management policies for computers to understand and execute automatically [17]. However, pure rule and/or ontology languages are not explicit enough to represent policies that regulate human behavior in the real world. We need a well-defined policy language that describes the concepts of rights, obligation, conditions, resources, etc. between resource owner and user to represent and execute access control policies of resources on the Web.

Following [18], policies are considered as knowledge bases, allowing deontic classes, properties, and access control rules. This has the advantage that many operations are automated, thereby reducing ad hoc program coding to a minimum and enabling automated documentation. Regulations imposed on human behavior and activity are simulated by computerized policies that are specified by using policy languages, such as Rei or KAoS [19]. The semantics of these policy languages is only DL-based, and needs to be further extended by using LP-based semantics of rule languages, such as RuleML, RIF or Protune [20]. Recently, AIR (AMORD In RDF) is a policy language that considers using both

---

[8] http://protege.stanford.edu/

RDF ontology language and N3 rule language for the privacy protection policy execution[9].

3.1   Policy for Semantic Web Services

The idea of Web Services in the SOA of distributed software systems has become a tremendous success. Semantic Web Services (SWS) employ Semantic Web technology in the Web Services area: service functionality, service inputs and outputs, preconditions and effects, etc.; all are expressed and executed in knowledge representation languages, i.e., ontology and rule languages [21]. A policy can be considered in the SWS because of using similar ontology and rule languages' semantics on the Policy-aware Web. Thus policies are represented and executed as Web Service rules for the compliance of human regulations to control information sharing and service execution.

One of the challenges to implement rule systems on the Policy-aware Web is how to design and implement rules as computerized policy by the integration of rule and ontology. This computerized policy imitates human regulation for controlling information sharing and service execution for a composite web service on the Web. And the ultimate goal is the satisfaction of legal regulation compliance from the execution of a computerized policy. This idea is similar to the Legal Knowledge Interchange Format (LKIF) proposed in the past EU FP6 project [22].

4   Rule Management Systems and Engines

Before looking into the details of rule management systems, we need to decide about a rule management systems implementation platform. If we choose a rule system that is also embedded in the Semantic Web development environment, then we have several advantages. First, it provides sufficient facilities to implement subsystems for rules and the data model. Second, both the ontology and rule languages used in the Semantic Web are complementary to each other so we can leverage on the declarative knowledge representation. Third, we have a standard query language or a rule language to support the access of underlying knowledge bases for ontology or rule bases. Finally, if applications are embedded in Java or some other popular imperative programming language, we have language typing, control flow, and interaction mechanism available for the implementation of application system on the Web.

4.1   Rule Systems in the Semantic Web Framework

The SemWebCentral[10] is one of the well-known websites providing Open Source development tools for the Semantic Web. The Semantic Web system development framework can be subdivided into three subsystem modules: an application module, a controller module and a view module. The application module

---

[9] http://dig.csail.mit.edu/TAMI/2007/amord/air-specs.html
[10] http://www.semwebcentral.org/

contains reasoning functions, including task and inference, domain schema and knowledge base. The controller module handles interactions with the user and functions in the application model. The view module provides output for the user. The Semantic Web system development framework usually includes two development parts, one is for ontology and the other is for rule. For example, Protégé has been successfully developed for ontology and rule, such as Jena[11] and Jess[12]. The Jena rule engine was integrated in the Semantic Web system development framework Protégé for having rule-based inference with the access to knowledge base in the ontologies of RDF and OWL[13]. In addition, the system for development of ontology and rule combination, such as SWRL is also available in the Protégé with SWRLTab[14].

4.2 Standalone Rule Systems

A number of standalone rule systems have been investigated by the RIF Working Group[15]. A rule system is defined as a piece of software that implements or supports a rule language in some way (e.g., a rule engine or a rule editor). Among the RIF list, some rule systems are developed for commercial usage but others are for open source purposes. Based on the classification of rule types, some rule systems are developed for a deductive rule engine but others are implemented for a reactive rule engine.

Commercial Rule Systems. IBM ILOG Business Rule Management Systems (BRMS)[16] provides a complete BRMS for analysts, architects and developers, featuring tools of rule authoring and rule management besides its rule engine. In fact, ILOG JRules is one of the best-known production rule systems. JBoss Drools[17] Enterprise BRMS is a well-known open source rule system which provides perfect integration with the service-oriented architecture (SOA) Web service solutions. On the other hand, existing rule systems, such as Prova[18] and ruleCore[19] are also available for ECA rules inference. For more details about reactive rules on the Web please refer to [16].

Some commercial rule systems are developed from a matured prototype of the Semantic Web middleware, such as OntoBroker[20]. Oracle Business Rules integrates with the Business Process Execution Language (BPEL) and tries to enrich decision making for processes in the SOA[21]. In general, commercial rule

[11] http://jena.sourceforge.net/
[12] http://www.jessrules.com/
[13] http://protege.stanford.edu/plugins/owl/jena-integration.html
[14] http://protege.cim3.net/cgi-bin/wiki.pl?SWRLTab
[15] http://www.w3.org/2005/rules/wg/wiki/List_of_Rule_Systems
[16] http://ilog.com/products/businessrules/index.cfm
[17] http://www.jboss.com
[18] http://www.prova.ws
[19] http://www.rulecore.com
[20] http://www.ontoprise.de
[21] http://www.oracle.com/technology/products/ias/business_rules/index.html

systems use proprietary rule languages for the development of the rule bases so we need a standard rule interchange language, such as RIF to obtain rule interoperability among these rule systems.

Academic Rule Systems. The academic ECA rule system XChange, with its integration of the Web query language Xcerpt, provides the access of data sources to obtain information on the dynamic Web. Other academic rule systems are deductive reasoning rule engines, such as jDREW and its object-oriented extension OO jDREW[22]. An Object-Oriented Knowledge Base Language FLORA-2 provides frame-based logic reasoning engine with the knowledge base development environment[23].

Logic programming systems are also used to develop rule-based applications. For example, Logic Programming Associates Prolog provides a complete rule development environment with a graphical interface for rule editing[24]. Thea is a Prolog library for generating and manipulating OWL content on the Semantic Web. The Thea OWL parser uses SWI-Prolog's Semantic Web library for parsing RDF/XML serialization of OWL documents into RDF triples, and then it builds a representation of the OWL ontology[25].

One of the challenges for implementing rule systems on the Web is to be aware of the current rule management systems, including commercial and academic ones, and, furthermore, an understanding of their system features and which rule type reasoning they can support. Moreover, we need to investigate the possible application domains they intent to accomplish through the underlying rule interchange standard.

4.3   Performance Benchmark for Rule Systems

It is not easy to propose an acceptable measurement benchmark to evaluate the performance of current rule systems because rule systems vary considerably with respect to rule syntax and features. In [23], a set of benchmarks were proposed for analyzing and comparing the performance of numerous rule systems. In this OpenRuleBench, they include five rule technologies to compare with: Prolog-based, deductive databases, production rules, triple engines, and general knowledge bases. Jena and OntoBroker we mentioned before were also two of the selective rule systems in their comparison list. We envision that the benchmark performance evaluation output will be just one of the criteria for people to decide for which rule system they are going to adopt in their application development.

5   Rule Interchange Languages

In early expert systems, a specific language, such as Prolog or LISP was used to encode expert domain knowledge into rules and facts, for execution in a standalone

---

[22] http://www.jdrew.org/oojdrew/

[23] http://flora.sourceforge.net/

[24] http://www.lpa.co.uk/

[25] http://www.semanticweb.gr/TheaOWLLib/

system. However, when rules and facts are created in different rule systems and distributed on the Web, we need a rule standard exchange language for the interchange of heterogeneous rule formats. Otherwise, we cannot implement an application, such as composite (semantic) web services that might require rules created and distributed in the different rule systems [21]. Therefore, a common rule format facilitates decision making on the network environment with multiple rule formats. For example, the therapeutic guideline recommendation rules for diabetes type 2 are constructed with the combination of clinical and therapeutic criteria as the condition part and therapeutic options as the actions. When users or organizations switch rules from one rule product to another, they can employ the rule interchange technologies without re-developing their rules.

Proposed rule interchange languages include RuleML [1], REWERSE Rule Markup Language (R2ML) [24], and W3C RIF[26], where R2ML attempts at integrating aspects of RuleML, SWRL, and Object Constraint Language (OCL). The most recent W3C RIF[27] was proposed to achieve the objective of rules interchange and interoperability for major rule systems. These rule interchange languages provide XML schemata to guarantee the comparability of rule syntax and semantics from source to target rule systems and vice versa. The other important rule language is Semantics of Business Vocabulary and Business Rules (SBVR), submitted by Business Rule Group (BRG) to OMG on the standardization of semantics for business vocabulary and rules[28].

One of the challenges to apply rule systems on the Web is to finalize a rule interchange language to provide a rule interchange framework and format of rules for current major rule systems. When agents proceed towards a two way rule interchange, a rule interchange language with the framework ensures the compatibility of rules' syntax and semantics between rule systems. The related challenge is the requirement to have a software development system and a runtime environment for people to build, design, and implement standardized rule interchange formats to automatically extract and transform rules from different rule systems on the Web.

6   Use Cases with Rules

Rules are used to express computational or business logic in the information systems which do not have explicit control flow, so rules are more suitable for execution in the dynamic situations for business collaborations. Along with the rapid development of the Web, multiparty collaborations for carrying out business services in this environment are more significant than ever before. For example, when a credit card transaction is requested from a merchant, a customer needs a payment authorization from the merchant and the card issuer (the bank) to accomplish a successful transaction service. In this case, both merchant and bank have their own policies as rules to conduct their authorization processes.

---

[26] http://www.w3.org/TR/rif-bld/
[27] http://www.w3.org/2005/rules/wiki/RIF_Working_Group
[28] http://www.businessrulesgroup.org/sbvr.shtml

If both parties are required to combine their policies, we hope they can transform the rules into a formal common rule format, such as RIF. For example, rules from the bank are directly imported by the merchant and processed with his local rule engine to derive an authorization decision. In addition, this situation can be extended to other relevant web services for conducting composite web services. Another use case is a seller, posting his price discount and refund policies for execution as rules on his website, to attract potential customers for making a purchase decision from his selling goods. Moreover, a vendor advertises his lead time policies in formal rules to attract customers and also as a part of contract negotiation in the supply chain management.

Use cases such as the ones we have shown above are categorized by the W3C RIF Working Group as a type of policy-based transaction authorization policy for access control with the interchange of human-oriented business rules. Several other interesting use cases focusing on different application domains are also available on this website[29]. Another interesting use case study was proposed by the Business Rule Group (BRG) to use SBVR for illustrating business rule concepts of EU-Rent, EU-Fly, and EU-Stay. They are available on the BRG website[30]. The challenge here is whether we have enough use cases that can be accomplished by current rule systems on the Web to convince people to adopt and use this technology.

7   Conclusion

In this study, we outlined the objectives of RuleML-Challenge competitions started in 2007. Alos, we have elaborated the possible research and implementation challenges for rule systems on the Web that are closely related to the Challenge competition events in the forthcoming years.

The first challenge is to perfectly implement rule systems with the data model, either from a relational or object-oriented database or from a DL-based knowledge base. The second challenge is to enable computerized policies, created in a policy language that is compliant with human legal regulations. In addition to the legalized policy implementation with policies created from the policy language, computerized policy can be shown as a combination of ontology and/or rule languages for the purpose of information sharing and web service execution. The third challenge is full awareness of current available commercial and open source rule management systems and, moreover, finding out the pros and cons of each rule system by a standard evaluation benchmark to verify its scalability and performance. The fourth challenge is to achieve rule interoperability using available rule interchange languages for rules created and distributed on the Web. The fifth challenge is to demonstrate sufficient use cases implemented from rule systems, while interchanging their rules through one of rule interchange language.

---

[29] http://www.w3.org/2005/rules/wg/wiki/Use_Cases
[30] http://www.businessrulesgroup.org/egsbrg.shtml

Acknowledgements

This research was partially supported by the NSC Taiwan under Grant No. NSC 95-2221-E004-001-MY3 and NSC 98-2918-E-004-003.

References

1. Boley, H.: The ruleML family of web rule languages. In: Alferes, J.J., Bailey, J., May, W., Schwertel, U. (eds.) PPSWR 2006. LNCS, vol. 4187, pp. 1–17. Springer, Heidelberg (2006)
2. Boley, H., Kifer, M., Pătrânjan, P.-L., Polleres, A.: Rule interchange on the web. In: Antoniou, G., Aßmann, U., Baroglio, C., Decker, S., Henze, N., Patranjan, P.-L., Tolksdorf, R. (eds.) Reasoning Web. LNCS, vol. 4636, pp. 269–309. Springer, Heidelberg (2007)
3. Ullman, D.J.: Principles of Database and Knowledge-Base Systems Volume I. Computer Science Press, Rockville (1988)
4. Ullman, D.J.: Principles of Database and Knowledge-Base Systems Volume II. Computer Science Press, Rockville (1989)
5. Date, C.J.: Logic and Databases: The Roots of Relational Theory. Trafford Publishing (2007)
6. Ceri, S., Fraternali, P.: Designing Database Applications with Objects and Rules: The IDEA Methodology. Addison-Wesley, Reading (1997)
7. Berners-Lee, T., et al.: The semantic web. Scientific American (2001)
8. Brachman, J.R., Levesque, H.J.: Knowledge Representation and Reasoning. Morgan Kaufmann, San Francisco (2004)
9. Patel-Schneider, F.P., Horrocks, I.: A comparison of two modelling paradigms in the semantic web. Journal of Web Semantics, 240–250 (2007)
10. Horrocks, I., et al.: SWRL: A semantic web rule language combining OWL and RuleML (2004)
11. Maluszynski, J.: Hybrid integration of rules and DL-based ontologies. In: Maluszynski, J. (ed.) Combining Rules and Ontologies. A survey. EU FP6 Network of Excellence (NoE), pp. 55–72. REWERSE (2005)
12. Grosof, N.B., et al.: Description logic programs: Combining logic programs with description logic. In: World Wide Web 2003, Budapest, Hungary, pp. 48–65 (2003)
13. Motik, B., Sattler, U., Studer, R.: Query answering for OWL-DL with rules. In: McIlraith, S.A., Plexousakis, D., van Harmelen, F. (eds.) ISWC 2004. LNCS, vol. 3298, pp. 549–563. Springer, Heidelberg (2004)
14. Donini, M.F., et al.: AL-log: Integrating datalog and description logics. Journal of Intelligent Information Systems 10, 227–252 (1998)
15. Rosati, R.: DL+log: Tight integration of description logics and disjunctive datalog. In: Proc. of the 10th International Conference on Principles of Knowledge Representation and Reasoning, KR (2006)
16. Berstel, B., Bonnard, P., Bry, F., Eckert, M., Pătrânjan, P.-L.: Reactive rules on the web. In: Antoniou, G., Aßmann, U., Baroglio, C., Decker, S., Henze, N., Patranjan, P.-L., Tolksdorf, R. (eds.) Reasoning Web. LNCS, vol. 4636, pp. 183–239. Springer, Heidelberg (2007)
17. Weitzner, D.J., et al.: Creating a policy-aware web: Discretionary, rule-based access for the world wide web. In: Ferrari, E., Thuraisingham, B. (eds.) Web and Information Security, pp. 1–31. Idea Group Inc., USA (2006)

18. Bonatti, P., Olmedilla, D.: Policy language specification, enforcement, and integration. project deliverable D2, working group I2. Technical report, REWERSE (2005)
19. Tonti, G., Bradshaw, J.M., Jeffers, R., Montanari, R., Suri, N., Uszok, A.: Semantic web languages for policy representation and reasoning: A comparison of KAoS, Rei, and Ponder. In: Fensel, D., Sycara, K., Mylopoulos, J. (eds.) ISWC 2003. LNCS, vol. 2870, pp. 419–437. Springer, Heidelberg (2003)
20. Antonious, G., et al.: Rule-based policy specification. In: Security in Decentralized Data Management, Springer, Heidelberg (2007)
21. Studer, R., Grimm, S., Abecker, A.: Semantic Web Services: Concepts, Technologies and Applications. Springer, Heidelberg (1990)
22. Gordon, F.T.: The legal knowledge interchange format (LKIF). Estrella deliverable d4.1, Fraunhofer FOKUS Germany (2008)
23. Liang, S., Fodor, P., Wan, H., Kifer, M.: Openrulebench: an analysis of the performance of rule engines. In: Word Wide Web 2009, pp. 601–610 (2009)
24. Wagner, G., Damásio, C.V., Antoniou, G.: Towards a general web rule language. International Journal Web Engineering and Technology 2 (2005)

A Modest Proposal to Enable RIF Dialects with Limited Forward Compatibility

Ch. de Sainte Marie

IBM, 9 rue de Verdun, 94250 Gentilly, France
csma@fr.ibm.com

Abstract. We introduce the notion of limited forward compatibility, and we argue for its usefulness. We describe, and argue for, a low cost, non-disruptive, extensible implementation, using XSLT to specify individual transforms, a new XML format to associate them with individual RIF constructs, and the RIF import mechanism to convey the fallback information from RIF producer to RIF consumer. We argue, also, for consumer-side fallbacks, as opposed to producers-side approaches such as mandatory least dialect serialization.

Keywords: rule interchange, W3C RIF, extensibility, forward compatibility, fallback transforms, XSLT.

1 Introduction

The rule interchange format (RIF) working group of W3C has published recently the specifications of the first three RIF dialects: RIF Core [1], RIF basic logic dialect [2] (RIF-BLD) and RIF production rule dialect [5] (RIF-PRD). The rule interchange format aims at providing a common, standard XML serialization for many different rule languages, "allowing rules written for one application to be published, shared, and re-used in other applications and other rule engines" [7], in a semantics preserving way.

The working group has also published a list of requirements for a rule interchange format [4], that provide useful insight in the design and architecture principles that the group adopted.

The "Rule language coverage" requirement explains and clarifies the notion of RIF dialects: *"Because of the great diversity of rule languages, no one interchange language is likely to be able to bridge between all. Instead, RIF provides dialects which are each targeted at a cluster of similar rule languages. RIF must allow intra-dialect interoperation, i.e. interoperability between semantically similar rule languages (via interchange of RIF rules) within one dialect, and it should support inter-dialect inter-operation, i.e. interoperation between dialects with maximum overlap"* [4]; whereas the "Semantic precision" requirement explains the specific role of the RIF Core dialect [1]: *"RIF core must have a clear and precise syntax and semantics. Each standard RIF dialect must have a clear and precise syntax and semantics that extends RIF core"* [4].

Any conformant implementation of any RIF dialect must, therefore, be able to process any RIF Core document: RIF dialects are backward compatible by design. But what about forward compatibility?

G. Governatori, J. Hall, and A. Paschke (Eds.): RuleML 2009, LNCS 5858, pp. 17–28, 2009.
© Springer-Verlag Berlin Heidelberg 2009

RIF Core should also provide, by design, a bottom-line bridge between any two RIF dialects: any document that is conformant to a RIF dialect A, and that could be represented equivalently in a conformant RIF Core document can be processed as a conformant document by any implementation of any other RIF dialect B. But, as we will see, this is not the case, even considering only the two published extensions of RIF Core, that is RIF-BLD and RIF-PRD.

In this paper, we consider, more specifically, the extensibility requirement, "Extensible Format": *"It must be possible to create new RIF dialects which extend existing dialects (thus providing backward compatibility) and are handled gracefully by systems which support existing dialects (thus providing forward compatibility)"* [4].

A specification is said to be forward compatible if a conformant implementation will process data produced by any future version according to the specification of that future version. Accordingly, *a RIF dialect would be considered forward compatible if a conformant implementation could process instances of any future or unknown extension according to the specification of that extension.* By contrast,, *backwards compatibility* is the ability to process data produced by any past version according to the specification of that past version.

Forward compatibility for RIF dialects requires, obviously, that the way a dialect is expected to handle extensions be specified along with the extensions themselves. Since, presumably, the semantics of most of the constructs that extend a RIF dialect will not be axiomatizable using that dialect, the only practical way to achieve forward compatibility seems to be for the extending dialect to specify how each new construct must be translated into the extended dialect.

For the same reason, for most extending constructs, fully semantics-preserving fallback translation will not be possible, and a general mechanism to enable RIF with useful, even if lossy, semantic fallbacks would require features, such as some kind of system to describe and qualify the impact of the fallback on the semantics.

These are out of the scope of this paper. More modestly, the ideas presented here aim at enabling W3C rule interchange format (RIF) with a lightweight, easy to implement, limited syntactic forward compatibility mechanism.

In the remainder of this paper, after describing a motivating example (section 2), we define what we call "limited forward compatibility", first, and describe our lightweight, low-cost proposal for enabling RIF dialects with such limited forward compatibility (section 3). In section 4, we discuss the benefits of our proposal, especially with respect to other possible approaches. In section 5, we examine implementation issues, before concluding (section 6).

2 Motivating Example

Consider the production rule that states that, for each value of an integer variable x between 1 and 5, $P(x)$ must be asserted in the facts base. In RIF-PRD/XML, the rule can be represented as follows[1]:

---

[1] The RIF/XML representation of the reference to the externally defined predicate *pred:list-contains(List(1 2 3 4 5) x)* is the same whether in a RIF-PRD or a RIF-Core document. The same holds for the RIF/XML representation of the predicate *P(x)*. They have no impact on the example and they are replaced, here, with a more concise non-XML representation.

```
<Forall>
  <declare> <Var>x</Var> </declare>
  <formula>
    <Implies>
      <if> pred:list-contains(List(1 2 3 4 5) x) </if>
      <then>
        <Do>
          <actions ordered='yes'>
            <Assert> <target> P(x) </target> </Assert>
          </actions>
        </Do>
      </then>
    </Implies>
  </formula>
</Forall>
```

Consider, now, the Datalog rule that states that, forall x, if x is an integer between 1 and 5, then $P(x)$ is true. In RIF-Core/XML, that rule can be represented as follows:

```
<Forall>
  <declare> <Var>x</Var> </declare>
  <formula>
    <Implies>
      <if> pred:list-contains(List(1 2 3 4 5) x) </if>
      <then> P(x) </then>
    </Implies>
  </formula>
</Forall>
```

The two rules are *effectively equivalent*, in the sense that, given any set of facts F, the final state of the facts base reached when executing the production rule under the RIF-PRD operational semantics with F as the initial state of the facts base will be identical to the minimal model that satisfies the conjunction of F and the Datalog rule, under the RIF-Core model theoretic semantics.

A conformant RIF-PRD implementation must accept the RIF-Core rule, since RIF-PRD extends RIF-Core.

But a conformant RIF-Core implementation will reject the RIF-PRD representation, because the Do and the Assert constructs are undefined ni RIF-Core. And a conformant RIF-BLD implementation will also reject the RIF-PRD rule, for the same reasons, although, in its RIF-Core form, it is perfectly valid and equivalent under the RIF-BLD semantics, as well.

3   Limited Forward Compatibility

Of course, the conclusion of the production rule under RIF-PRD semantics has a RIF/XML representation that uses only RIF Core constructs and that results in the representation of a rule that is effectively equivalent under RIF-Core semantics only because the only action is the assertion of a predicate, and the RIF/XML representation of the production rule uses, otherwise, RIF-Core constructs only (e.g., no negation in the condition, etc).

The example shows that, even if the complete semantics of an extending construct cannot be expressed in the extended dialect, there might be contexts where an equivalent expression for the extending construct exists in the extended dialect. Here, the complete semantics of the RIF-PRD Do construct cannot be expressed in the RIF-Core dialect that RIF-PRD extends, but, when the only action is the assertion of a predicate, and the RIF/XML representation of the production rule uses, otherwise, only RIF-Core constructs, the conclusion of the production rule under RIF-PRD semantics has a RIF/XML representation that uses only RIF Core constructs and that results in the representation of a rule that is effectively equivalent under RIF-Core semantics.

We call: *limited forward compatibility*, the possibility, for an implementation of a RIF dialect, to process any extension of that dialect, in every RIF document where the extension is conservative; that is, in every document that can be rewritten to remove all the extension-specific constructs, without changing the intended effective semantics of the RIF document being processed.

Definition (limited forward compatibility). *Let D be a RIF dialect that extends another RIF dialect, E. D provides* **limited forward compatibility** *to E, if D provides a fallback translation for all the specific cases where the semantics of a construct d that belongs to D but not to E, happens to be expressible in E.*

RIF dialects could be enabled with limited forward compatibility by associating a fallback property to every single RIF construct, say, where the allowed values of the property would be:

- reject, meaning that there is no fallback to the construct, and that an implementation that does not know the element must reject the document. The default value for the fallback property, when it is unknown, is reject;
- ignore, meaning that an implementation that does not know the element can ignore it and its content;
- zoom-in, meaning that the proper fall-back for the element is contextual to its content, and that it will be indicated by a sub-element. The element itself can be ignored by an implementation that does not know it, but not its sub-elements. If the element has no sub-element, then the fallback is ignore;
- the specification of a transform, such that the fallback is the new RIF document that results from applying the transform to the original RIF document.

Examples

- The fallback for RIF-PRD inflationary negation construct, INeg, can be reject: there is no case or context where it can be rewritten using only RIF-Core constructs;
- The fallback for the RIF-PRD Priority element is always ignore. Indeed, either the conflict resolution strategy can be ignored, and so does the priority associated with rules or group of rules; or the conflict resolution strategy cannot be expressed in RIF-Core, and the document will be rejected anyway; or there is a way to rewrite the RIF-PRD document as a Core document with the same semantics, but without a conflict resolution strategy, and the priorities will be removed during the rewrite;

- The value of the fallback property associated with the RIF PRD `behavior` element is `zoom-in`: either a conflict resolution is specified, and the appropriate fallback depends on the conflict resolution, or the `behavior` element contains only a `Priority` sub-element, which can be ignored, and the `behavior` element itself can be ignored; or it is empty, and it can be ignored;
- The value of the fallback property for the RIF-PRD `ConflictResolution` element has to be the specification of a fallback transform, since the appropriate fallback depends on the value of the text content of the element: if the conflict resolution that is specified can be ignored, e.g. `rif:forwardChaining`, the whole element will be removed; otherwise, the document must be rejected[2];
- The value of the fallback property for the RIF-PRD `Do` construct is the specification of a transform that replaces the `<Do>`...`</Do>` pair of tags by `<And>`... `</And>`. It will be `ignore` for the RIF-PRD elements `actions` and `Assert`, and another transform for the `target` element, that replaces the `<target>`... `</target>` pair with `<formula>`...`</formula>`. All the other elements that can be contained in a `Do` construct and that are specific to RIF-PRD would have `reject` as their fallback property.

Given the RIF-PRD fragment on the left and the specification of the transforms, an appropriate processor will produce the RIF-Core fragment on the right:

```
<Do>                          --transform-->      <And>
  <actions ordered="yes">     ---ignore--->
    <Assert>                   ---ignore--->
      <target>                --transform-->       <formula>
        Assertion 1           (belongs to RIF-Core)    Assertion 1
      </target>               --transform-->       </formula>
    </Assert>                  ---ignore--->
    ...                           ...    ...
  </actions>                   ---ignore--->
</Do>                         --transform-->       </And>
```

- Other examples where the value of the fallback property is the specification of a transform are the RIF-PRD `pattern` construct and the RIF-BLD `Name` element.

In none of the cases mentioned above where the fallback property is the specification of a transform, does the transform need to be conditional. A RIF-PRD `Do`, for instance, can be replaced by an `And` tag whether or not the rule that contains it can be represented in RIF-Core: of course, this may produce an invalid RIF document, if the element contains sub-elements that cannot be ignored or transformed into RIF-Core

---

[2] From the point of view of this limited forward compatibility proposal, it would be a more extensible design to use the tag of a, possibly empty, element to indicate the intended conflict resolution strategy, rather than a URI: a specific fallback could thus be associated to each conflict resolution strategy identifier, and a dialect that extends the set of accepted conflict resolution strategies could define the fallbacks only for the extension, instead of having to redefine the fallback for the `ConflictResolution` element. The pseudo-schema definition for the `behavior` element, according to this design, would be: `<behavior>` *xs:any? Priority?* `<behaviour>`, where `xs:any` must by an XML element that identifies a conflict resolution strategy (such as `<rif:ForwardChaining/>`).

compatible constructs. But the containing RIF document will be rejected anyway, in that case, because the fallback of these sub-elements can only be `reject`.

Notice that the `zoom-in` and `ignore` values are not absolutely necessary, as the corresponding behaviour could be implemented with a simple fallback transform. But using them may avoid unnecessary calls to a processor, when a RIF document contains no fallback transform otherwise.

4 Discussion

4.1 Is Limited Forward Compatibility Useful?

A first question is whether the proposed mechanism satisfies, formally, the RIF "Extensible format" requirement that we mentioned in the introduction, since it enables only syntactic fallbacks for semantically equivalent forms. We mentioned, in the introduction, that semantically lossy fallback could be envisioned, but that they were out of the scope of this paper. On the other hand, the kind of forward compatibility that the designers of RIF had in mind does not require them, as is further clarified in the "Default behavior" requirement: *"RIF must specify at the appropriate level of detail the default behavior that is expected from a RIF compliant application that does not have the capability to process all or part of the rules described in a RIF document, or it must provide a way to specify such default behavior"* [4].

Where a semantically preserving syntactic transform cannot be provided, the default behaviour is to gracefully reject the document. And the proposed mechanism, therefore, satisfies the requirements.

But, beyond the formal satisfaction of a requirement, the more important questions are, of course, whether the very concept of limited forward compatibility is useful at all.

Indeed, a default behaviour is already specified for the published RIF dialects (RIF-Core [1], RIF-BLD [2] and RIF-PRD [5]) to handle extensions, since the conformance clauses relative to these dialects specify that *"a conformant RIF-Core (resp. RIF-BLD, RIF-PRD) consumer must reject all inputs that do not match the syntax of RIF-Core (resp. RIF-BLD, RIF-PRD)."* However, our limited forward compatibility proposal extends, unarguably, the interoperability between RIF dialects, and, therefore, their usefulness.

4.2 Consumer-Side Fallback vs. Producer-Side Fallbacks

When discussing RIF implementations and applications[3], it is often useful to distinguish between:

---

[3] In this paper, we call: *RIF implementation*, a processor that translates between a custom rule language and a RIF dialect. This view is based on the, so-called, "translation paradigm" that is one of RIF's design principles, as reflected in the "Translators" requirement: *"For every standard RIF dialect it must be possible to implement translators between rule languages covered by that dialect and RIF without changing the rule language"* [4]. We call: *"RIF application"*, an application that includes a RIF implementation, and that can, therefore, receive and use RIF input.

- the *producer implementation*, that is, the implementation that produces a RIF document, e.g. by serializing rules written in a custom rule language in RIF/XML; and
- the *consumer implementation*, that is, the implementation that retrieves and uses a RIF document, e.g. by de-serializing it in the same or another custom rule language.

Our proposal assumes that fallbacks are processed on the consumer side. It could be argued that the fallback should be handled by the producer implementation, instead. That approach, that we will call: *mandatory least dialect serialization*, would require that, whenever a construct, d, in a dialect D can be expressed using only constructs from a dialect D' that D extends and that does not include d, conformant producer implementations of D must produce the D' RIF/XML. The argument in favor of such a requirement is that the burden of additional implementation complexity should be on the party that requires the added expressiveness.

There are, however, a number of strong arguments in favor of customer-side fallbacks, as opposed to mandatory least dialect serialization.

Mandatory least dialect serialization would require that the conformance clauses of the published RIF dialects be modified, and that the fallback transforms be specified normatively as part of the specification of the dialects. In other words, that would require changes in published specifications, which is not practical. We will see in the next section, where we discuss implementation issues, that consumer-side fallbacks can be implemented without impact on the specification of RIF dialects.

More significantly, mandatory least dialect serialization does not scale. RIF-PRD and RIF-BLD implementers may have only a few fallbacks to take into account to provide limited forward compatibility to Core, but the complexity of implementation is likely to grow wildly in future extensions with more "industrial strength" expressive power, as such exceptions will have to be taken into account in a cumulative way.

Consumer-side fallbacks, on the other hand, is scalable because a dialect D that extends a dialect E has only to specify the fallbacks that provide limited forward compatibility of E with respect to D, and implementers of D have only to care about the syntax and semantics of D; whereas mandatory least dialect serialization requires that implmenters take care of the limited forward compatibility with respect to D of E, and of all the dialects that are extended by E as well).

Of course, there is a cost: the consumer implementations outside the family of rule languages catered to by a dialect D (that is, consumer implementation that interoperate with D producer implementation through a dialect that is extended by D) bears the burden that a RIF document may have to go through cascading transforms before being consumed, thus impacting the performance of the interchange negatively. But, presumably, a RIF dialect will be used, mostly, for rule interchange amongst users of rule languages in the same family (that is, the family catered to by that dialect), and exceptionally for rule interchange with dialects in a different family.

On the other hand, consumer-side fallbacks are user-friendlier than producer-side ones, like, e.g., mandatory least dialect serialization. Producer-side fallbacks will lose the idioms, in a dialect, that have no impact on the formal semantics, but that make it easier to implement translators to and from rule languages in the target family (because they share the same structure and idiosyncrasy). Loosing the idiomatic structure and organisation of a rule may obscure the intent or even meaning of the rule, once the RIF document is translated back into the consumer's rule language, thus adversely affecting communication as well as round-tripping. More importantly, removing such idioms may adversely affect the affect the performance of the rules, even if they do not change the end result. For instance, RIF-PRD nested `Forall` and binding `pattern` formulas preserve the order and conditions by which variables are bound in the original rule, and producer-side least dialect serialization would flatten the nested structure and merge the binding formulas indiscriminately in the condition, and lose information that makes a difference when executing the rules.

Producer-side fallback makes idiomatic syntax extensions pointless, and, thus, useless. Consumer-side fallback makes idiomatic syntax extensions useful: a widely deployed consumer-side fallback mechanism would increase the appeal of developing idiomatic extensions of RIF dialects.

We believe that the possibility to use an idiomatic extension inside one's community, without renouncing inter-operability with the rest of the world, increases the appeal of RIF as a whole[4]; and, as a likely consequence, its adoption. Widespread use of a consumer-side fallback mechanism, such as the one we propose in this paper, would, therefore, increase the overall utility and value of RIF[5].

Finally, in addition to being useful and to offering more and better benefits, consumer-side fallbacks are a better design for limited forward compatibility than producer-side mandatory least dialect serialization. The main reason is that mandatory least dialect serialization violates the design principle that the same construct should not be used with different semantics in different dialects. Indeed, replacing a RIF-PRD `Do` construct in a rule's conclusion by a RIF-Core `And` construct, does not change the semantics of RIF-PRD, and the `And` construct, for a RIF-PRD consumer application, is only a `Do` in disguise, with the semantics of the `Do` construct: for instance, the individual assertions will be ordered. That order is irrelevant and the rule will produce the same result with any ordering. But this shows that mandatory least dialect serialization effectively requires that the `And` construct be used with different semantics in RIF-Core and RIF-PRD.

On the contrary, consumer-side fallback from RIF-PRD to RIF-Core will either produce valid RIF-Core, for consumption by a RIF-Core implementation, where the `And` construct in the conclusion is interpreted according to its semantics in RIF-Core; or it will produce an invalid document (from a RIF-Core viewpoint) that will be rejected[6].

---

[4] Notice that, in this respect, idiomatic extensions include idiomatic XML serialization: the normative RIF/XML syntax is rather verbose, and the possibility to use a compact idiomatic XML serialization is likely to appeal to some users communities.

[5] Notice that such a consumer-side fallback mechanism could also be used to cater to RIF-compatible legacy or proprietary XML serialization format, thus easing the transition to RIF and making the adoption of RIF more appealing to communities that have already their own XML rule interchange format (provided that it is compatible with RIF).

[6] From this point of view, backward compatibility should also be interpreted in terms of consumer-side fallbacks, even if actual implementations will, presumably, not translate explicitly a RIF document that contains extensions into one that does not, before consuming it: a RIF-PRD implementation, for instance, will, presumably, not translate explicitly the `And` construct in the conclusion of a rule in a RIF-Core document into a `Do`, before consuming it as a RIF-PRD document.

5 Implementation Issues

5.1 Specification and Interchange of Fallback Transforms: RIF or XSLT?

Sandro Hawke suggested [3] that the fallback transforms, being essentially rewriting rules, could be interchanged in RIF, using the fact the RIF documents can be seen a sets of RIF frames, or object-attribute-value triples, thanks to the RIF/XML syntax being essentially in the alternating normal form [6]. Class elements (starting with an upper case letter) are the objects) and role elements (starting with a lower case letter) represent the attributes.

The idea is that consumer implementations would run the rules; e.g the fallback transforms from RIF-PRD to RIF-Core would be written in RIF-Core, so that a RIF-Core consumer implementation could implement the transforms.

However appealing the idea, we see two major problems with that approach. A first problem is that it works only if the consumer application is a rule engine, which is not necessarily the case: it might be, e.g. a rule viewer or a rule editor. On the other hand, there are quite a few free XSLT processors available, and running a RIF document through an XSLT processor as a pre-processing step would be a minimal change to existing implementations.

In terms of performance, we have no reason to believe that the cost of running a RIF document through an XSLT processor should be higher than the cost of running the RIF document that contains the transforms through the consumer RIF implementation, plus the cost of running the RIF document to be consumed through a rule-based application running the, thus translated, rewriting rules.

A more fundamental problem with using RIF as the interchange medium for fallback rewriting rules, is that the rules must be, always, written in a RIF dialect implemented by the target consumer application. That means that all fallback rewriting rules for each RIF dialect must be made available:

- either in each of the RIF dialects to which they provide a relevant fallback;
- or in RIF Core (since any RIF implementation is a RIF-Core implementation).

The second alternative does not seem feasible: rewriting rules do not seem to be expressible in RIF-Core (and, generally, in monotonic dialects). The first alternative is unlikely to scale well, and, has the same feasibility problem as the second alternative with respect to the monotonic dialects.

If the fallbacks are done in XSLT, fallback information is required, for each RIF dialect, only with respect to the RIF dialects that it extends immediately[7]. XSLT stylesheets can, then, be cascaded to reach RIF dialects that are farther away from the dialect of the producer implementation.

Thus, we believe that XSLT is a better choice than RIF as an interchange medium for the fallback transforms.

Notice that the following proposition holds for every extension to RIF-Core in RIF-PRD and in RIF-BLD: *for every single element in an extension, if there is a*

---

[7] Or the "closest" RIF dialects, if we extend the approach to limited "sidewards" compatibility, to provide transforms between two dialects where one is not an extension of the other.

fallback transform, then there is an XSLT stylesheet that implements that transform and that can be always executed irrespective of whether there is a valid fallback for all the contained and containing constructs. While we do not prove it here, we conjecture that it holds generally (that is, beyond RIF-Core, RIF-PRD and RIF-BLD). The implication is that, in many cases, the required XSLT transforms will be very simple[8].

5.2   Conveying the Fallback Information from RIF Producers to RIF Consumers

Orthogonal to the question of the interchange format for the fallback information, there is the question of how the fallback information can be conveyed from the producer of a RIF document to a RIF consumer that needs that information to exploit the document.

A first possibility would be to include the fallback information directly in the RIF/XML, e.g. by adding an optional `fallback` attribute, to represent the fallback property as described in section 0, above. The allowable values of the attribute are `reject` (default), `ignore`, `zoom-in`, or an URL that dereferences to an XSLT stylesheet when the value of the property is the specification of a fallback transform.

The drawbacks of that approach are that:

- it requires extensive modifications in the specification of published RIF dialects and in existing consumer implementations;
- It bloats noticeably the already verbose RIF/XML syntax.

Another possibility, suggested by Sandro Hawke [3], would be to define a file format for fallbacks and to associate an URL to the RIF namespace where RIF consumer implementations can retrieve the fallback information associated with each RIF construct in any RIF dialect.

A major benefit of that approach is that it would provide a central repository where all the XML construct in the RIF namespace would be identified. Information regarding the first dialect to specify the construct could be included, making it a useful place for designer of new RIF dialect, to make sure that they reuse existing constructs where appropriate, and do not reuse the name of existing constructs with different syntax or semantics. E.g., using the prefix `xtan:`[9] for the namespace in which the format of the repository is defined:

```
<xtan:element name="rif:And">
    <xtan:definedBy dialect="RIF-Core"/>
</xtan:element>
```

In addition, fallback information could be targeted, e.g. for the case of RIF dialect that extends two incompatible RIF dialects, or for the case where a RIF dialect offers, for a specific construct fallback transforms towards dialects in different branches, not only to dialects that it extends:

---

[8] An example of a complete XSLT stylesheet that specifies the RIF-PRD Do to RIF-Core And transform is provided in [8].

[9] XTAN is the name S. Hawke proposes for XML fallback transformation. Notice that the proposal is not limited to use for RIF.

```
<xtan:element name="rif:Do">
  <xtan:definedBy dialect="RIF-PRD"/>
  <xtan:fallback target="RIF-Core">
      http://www.w3.org/.../fallbacks/prd2core.xsl
  </xtan:fallback>
</xtan:element>
```

Another benefit of such a fallback file format is it can take into account additional cases such as a dialect that extends an element defined by another dialect by adding an optional attribute (to extend the semantics), e.g.

```
<xtan:element name="rif:Foo">
  <xtan:definedBy dialect="D">
  ...
  <xtan:attribute name="bar">
      <xtan:added-by dialect="E"/> <!-- where E extends D,
                                        directly or indirectly -->
      <xtan:fallback target="...">...</fallback>
  </xtan:attribute>
<xtan:element>
```

The impact of that approach on existing implementations is minimal: nothing changes in the RIF/XML, so that producer implementations are not impacted at all. Unenabled consumer implementations would simply reject RIF documents that are not valid in any of the RIF dialects that they can handle. On the other hand, when preprocessing a RIF document and finding an element it does not know, an enabled consumer implementation would retrieve, from the repository, the fallback targeted to one of the dialects it can handle, if any (where not finding one means `reject`), apply the relevant transform, and so on.

The one drawback that we see with the central repository approach is that it requires... a central repository. Even assuming that W3C or a similar organisation would own and maintain such a repository, the approach works only for standards extensions: it does not work for proprietary, custom or ad hoc extensions.

This restriction can be relaxed using the import mechanism in RIF. We propose to define a `fallback` profile for the `Import` construct, meaning that the URL in the `location` is a file containing fallback information potentially useful to consumers of the importing RIF document. All the fallback information could be conveyed this way, or only the non-standard one, if a repository of standard constructs is, eventually, associated with the RIF namespace.

6   Conclusion

Based on earlier effort by the RIF working group, we have introduced the notion of limited forward compatibility, and we have argued for its usefulness. We have also described and argued for a low cost implementation, using XSLT to specify individual transforms, a new XML format to associate them with individual RIF constructs, and the RIF import mechanism to convey the fallback information from RIF producer to RIF consumer. We have argued for consumer-side fallbacks, as opposed to producers-side approaches such as mandatory least dialect serialization.

We hope that we have argued convincingly. But limited forward compatibility, however useful it might be, works only with extensions that are conservative in some well-identified contexts: more complete forms of forward comaptibility would be, arguably, much more useful.

A final benefit of the approach proposed in this paper is that it extends easily to more complete forms of forward and sideward compatibility. Some fallback transforms can be readily implemented based on our proposal, e.g. crispification of fuzzy or uncertain rules [9] by ignoring the constructs that bear the information about uncertainty[10]. Of course, new constructs would have to be added in the fallback XML format, for such semantic fallbacks, e.g. to provide indication of the severity of the impact on the semantics, as proposed by Sandro Hawke [3].

Already useful by itself, limited forward compatibility, implemented following the approach that we propose, would be an enabler for the future development and deployment of more complete forms of forward compatibility.

References

1. Boley, H., et al. (eds.): RIF Core dialect, W3C last call draft (July 2009), http://www.w3.org/TR/rif-core/
2. Boley, H., Kifer, M. (eds.): RIF basic logic dialect, W3C last call draft (July 2009), http://www.w3.org/TR/rif-core/
3. Hawke, S.: Unpublished draft, http://www.w3.org/2005/rules/wg/wiki/Arch/Extensibility2, http://www.balisage.net/Proceedings/html/2008/Hawke01/Balisage2008-Hawke01.html
4. Paschke, A., et al. (eds.): RIF use cases and requirements, W3C public working draft. Latest version, http://www.w3.org/TR/rif-ucr/
5. de Sainte Marie, C., Hallmark, G., Paschke, A. (eds.): RIF production rule dialect, W3C last call draft (July 2009), http://www.w3.org/TR/rif-prd
6. Thompson, H.: Normal Form Conventions for XML Representations of Structured Data (October 2001), http://www.ltg.ed.ac.uk/~ht/normalForms.html
7. W3C Rule interchange format working group charter, http://www.w3.org/2005/rules/wg/charter.html
8. http://www.w3.org/2005/rules/wiki/FallbackDo2And.xsl
9. Zhao, J., Boley, H.: Uncertainty Treatment in the Rule Interchange Format: From Encoding to Extension. In: 4th Intl. Workshop on Uncertainty Reasoning for the Semantic Web (URSW) (October 2008), http://c4i.gmu.edu/ursw/2008/papers/URSW2008_F9_ZhaoBoley.pdf

---

[10] Considering such extensions makes it even clearer why any transform to enable forward or sideward compatibility, limited or not, must be applied at the consumer, not the producer side.

RIF RuleML Rosetta Ring: Round-Tripping the Dlex Subset of Datalog RuleML and RIF-Core

Harold Boley

Institute for Information Technology
National Research Council of Canada
Fredericton, NB, E3B 9W4, Canada

Abstract. The RIF RuleML overlap area is of broad interest for Web rule interchange. Its kernel, Dlex, is defined syntactically and semantically as the common sublanguage of Datalog RuleML and RIF-Core restricted to positional arguments and non-conjunctive rule conclusions, and allowing equality plus externals in rule premises (only). Semantics-preserving mappings are then defined between the Dlex subset of the RIF Presentation Syntax and RIF/XML, RIF/XML and RuleML/XML, as well as RuleML/XML and the Prolog-like RuleML/POSL. These mappings are the basis for RIF RuleML feature comparison and translation. The slightly augmented mappings can be composed into a ('Rosetta') ring for round-tripping between all pairs of Dlex representations.

1 Introduction

The RuleML family of languages [Bol07] has recently been enhanced by RIF dialects starting with the Basic Logic Dialect [BK09a]. As RuleML moves to version 1.0 and RIF to Recommendation, the overlap area between RIF and RuleML is of broad interest for Web rule interchange. In this paper, we focus on its kernel, Dlex, defined as the common sublanguage of Datalog RuleML [Bol06] and RIF-Core [BHK+09] restricted to positional arguments and non-conjunctive rule conclusions, and allowing equality plus externals – built-ins as defined in RIF-DTB [PBK09] – in rule premises (only).

To advance RIF RuleML development and interoperation, we define semantics-preserving mappings within this overlap, in particular within Dlex. These include central mappings bridging RIF/XML and RuleML/XML, mappings between the RIF Presentation Syntax and RIF/XML, as well as mappings between RuleML/XML and the Prolog-like RuleML/POSL. The slightly augmented mappings can be composed into a ('Rosetta') ring for round-tripping between all pairs of Dlex representations.

This study on the Dlex kernel and its mappings includes a RIF RuleML feature comparison, acts as a specification of concrete translators between RIF

G. Governatori, J. Hall, and A. Paschke (Eds.): RuleML 2009, LNCS 5858, pp. 29–42, 2009.
© Springer-Verlag Berlin Heidelberg 2009

and RuleML, and helps in language convergence; definitions of an 'even' predicate exemplify.[1] Moreover, it prepares identifying the largest common subset of Datalog RuleML and RIF-Core, also to include "slotted" (RuleML) or "named" (RIF) arguments as well as objects/frames.

First, we formalize the syntax and semantics of Dlex by specializing relevant definitions of RIF-BLD [BK09a], since the specialization of RIF-BLD to RIF-Core is currently only done 'by exclusion' [BHK$^+$09]. This formalization can be seen both as a new RIF dialect, RIF-Dlex, and a new RuleML sublanguage, Dlex RuleML.We focus on Dlex since (1) it contains Datalog [CGT89], which is close to relational databases with recursive views, (2) with equality plus externals added, it is relevant to various commercial rule systems, (3) it is a kernel of RIF and RuleML, also helping to understand and converge their larger subsets. Because RIF-Core (and even RIF-BLD) does not support negation, its Dlex sublanguage does not either (cf. section 6). Following RuleML's SWSL subfamily[2] and RIF's overall design, Dlex uses Hilog/Relfun-like higher-order syntactic sugar with individual (`Ind`) and predicate (`Rel`) symbols merged into a single set of constant (`Const`) symbols. The Dlex formalization provides a foundation for RIF RuleML mapping as needed for our round-tripping.

Second, instead of defining pairwise Dlex mappings or mappings to and fro a distinguished Dlex representative, we define a neutral 'mapping ring', enabling round-tripping between all of these languages using compositions that chain one to four mappings. The mappings that have XML as their sources can be refined to XSLT translators. The other ones are handled by text parsers and generators. We focus on a mapping χ_{dlex} from RIF Presentation Syntax to RIF/XML [BK09a], a mapping ξ_{dlex} from RIF/XML to Fully Striped RuleML/XML, and a mapping π_{dlex} from Stripe-Skipped RuleML/XML to RuleML/POSL. A further mapping σ_{dlex} from Fully Striped to Stripe-Skipped RuleML/XML essentially omits all (lower-case) role tags.[3] For a ***closed ring***, a final (text-to-text) mapping ω_{dlex} from RuleML/POSL to RIF Presentation Syntax can omit/change separators and insert RIF `Documents`, `Groups`, etc. For an ***open ring***, the 4-mapping composition $\pi_{dlex} \circ \sigma_{dlex} \circ \xi_{dlex} \circ \chi_{dlex}$ can instead be complemented by the inverse composition $\chi_{dlex}^{-1} \circ \xi_{dlex}^{-1} \circ \sigma_{dlex}^{-1} \circ \pi_{dlex}^{-1}$.

The rest of the paper is organized as follows. Dlex will be defined next. The syntax, in section 2; the semantics, in section 3. The mappings of the RIF Presentation Syntax to RIF/XML (χ_{dlex}) to RuleML/XML (ξ_{dlex}) to RuleML/POSL (π_{dlex}) will then be specified for Dlex via two chains of tables, and their inverses discussed (for semantics preservation and open-ring round-tripping). The chain in section 4 defines those mappings for the Condition Language (conjunctive queries, as in rule premises); the other, in section 5, for the Rule Language (entire rules and rulebases); we will omit annotations and their mappings. Conclusions will be given in section 6.

---

2 Syntax of Dlex

We begin with defining the syntax of Dlex, starting from the alphabet, proceeding to terms, and then to formulas.

2.1 Alphabet of Dlex

Definition 1 (Alphabet). *The **alphabet** of the Dlex presentation syntax consists of*

- *a countably infinite set of **constant symbols** Const*
- *a countably infinite set of **variable symbols** Var (disjoint from Const)*
- *connective symbols And and : -*
- *the quantifier Forall*
- *the symbols = and External*
- *the symbols Group and Document*
- *the auxiliary symbols (,), <, >, and ^^.*

The set comprising the connective symbols, Forall, = and External, etc., is disjoint from Const and Var. Variables are written as Unicode strings preceded with the symbol "?".

Constants are written as "literal"^^symspace, where literal is a sequence of Unicode characters and symspace is an identifier for a symbol space.   □

2.2 Terms

Dlex defines several kinds of terms: *constants* and *variables*, *positional* terms, plus *equality* and *external* terms.

In the next definition, the phrase *base term* refers to simple or positional terms, or to terms of the form External(t), where t is a positional term.

Definition 2 (Term)

1. *Constants and variables. If $t \in$ Const or $t \in$ Var then t is a **simple term**.*
2. *Positional terms. If $t \in$ Const and t_1, ..., t_n, $n \geq 0$, are base terms then $t(t_1$ ... $t_n)$ is a **positional term**.*
   Positional terms correspond to the usual terms and atomic formulas of classical first-order logic.
3. *Equality terms. t = s is an **equality term**, if t and s are base terms.*
4. *Externally defined terms. If t is a positional term then External(t) is an **externally defined term**.*
   External terms are used for representing built-in functions and predicates. □

2.3 Formulas

Dlex distinguishes certain subsets of the set Const of symbols, including subsets of *predicate symbols* and *function symbols*.

Any term of the form p(...), where p is a predicate symbol, is also an *atomic formula*. Equality terms are also atomic formulas. An externally defined term of the form External(φ), where φ is an atomic formula, is also an atomic formula, called an *externally defined* atomic formula. Note that simple terms (constants and variables) are *not* formulas.

More general formulas are constructed out of atomic formulas with the help of logical connectives.

Definition 3 (Formula)
   *A **formula** can have several different forms and is defined as follows:*

1. *Atomic: If φ is an atomic formula then it is also a formula.*
2. *Condition formula: A **condition formula** is either an atomic formula or a formula that has the form of a conjunction: If φ_1, ..., φ_n, $n \geq 0$, are condition formulas then so is And(φ_1 ... φ_n), called a conjunctive formula.*
3. *Rule implication: φ :- ψ is a formula, called rule implication, if:*
   - *φ is an atomic formula that is not an equality term,*
   - *ψ is a condition formula, and*
   - *the atomic formula in φ is not an externally defined term (i.e., a term of the form External(...)) and does not have such a term as an argument.*
4. *Universal rule: If φ is a rule implication and $?V_1$, ..., $?V_n$, $n>0$, are distinct variables then Forall $?V_1$ ... $?V_n(\varphi)$ is a universal rule formula. It is required that all the free variables in φ occur among the variables $?V_1$ ... $?V_n$ in the quantification part. An occurrence of a variable $?v$ is free in φ if it is not inside a substring of the form Q $?v$ (ψ) of φ, where Q is a quantifier (Forall) and ψ is a formula. Universal rules will also be referred to as **Dlex rules**.*
5. *Universal fact: If φ is an atomic formula that is not an equality term and $?V_1$, ..., $?V_n$, $n>0$, are distinct variables then Forall $?V_1$ ... $?V_n(\varphi)$ is a universal fact formula, provided that all the free variables in φ occur among the variables $?V_1$ ... $?V_n$.*
   Universal facts are treated as rules without premises.
6. *Group: If φ_1, ..., φ_n are Dlex rules, universal facts, variable-free rule implications, variable-free atomic formulas, or group formulas then Group(φ_1 ... φ_n) is a group formula.*
   Group formulas are used to represent sets of rules and facts. Note that some of the φ_i's can be group formulas themselves, which means that groups can be nested.
7. *Document: An expression of the form Document(directive$_1$ ... directive$_n$ Γ) is a Dlex document formula (or simply a document formula), if*
   - *Γ is an optional group formula; it is called the group formula associated with the document.*
   - *directive$_1$, ..., directive$_n$ is an optional sequence of directives. We will assume this sequence [BK09a] to be empty for simplicity.*   □

The Dlex *Condition Language* comprises formulas according to definition 3, items 1 and 2; its *Rule Language* is the superset defined via formulas 4-7.

3   Semantics of Dlex

Given the syntax, we now define the semantics of Dlex via semantic structures and the interpretation of formulas, finishing with logical entailment.

3.1   Semantic Structures

We will use TV to denote $\{t,f\}$, the set of truth values used in the semantics. The key concept in a model-theoretic semantics of a logic language is the notion of *semantic structures*, which is defined below.

Definition 4 (Semantic structure). *A **semantic structure**, $\mathcal{I}$, is a tuple of the form $<TV, D, D_{ind}, D_{func}, I_C, I_V, I_F, I_=, I_{external}, I_{truth}>$. Here D is a non-empty set of elements called the **domain** of $\mathcal{I}$, and D_{ind}, D_{func} are nonempty subsets of D. D_{ind} is used to interpret the elements of* `Const` *that occur as individuals and D_{func} is used to interpret the constants that occur as function symbols. As before,* `Const` *denotes the set of all constant symbols and* `Var` *the set of all variable symbols.*

The remaining components of $\mathcal{I}$ are total mappings defined as follows:

1. I_C *maps* `Const` *to D.*
   This mapping interprets constant symbols. In addition, it is required that:
   - *If a constant, $c \in$ `Const`, is an individual then it must be that $I_C(c) \in D_{ind}$.*
   - *If $c \in$ `Const`, is a function symbol then it must be that $I_C(c) \in D_{func}$.*
2. I_V *maps* `Var` *to D_{ind}.*
   This mapping interprets variable symbols.
3. I_F *maps D to functions $D^*_{ind} \to D$ (here D^*_{ind} is a set of all finite sequences over the domain D_{ind}).*
   This mapping interprets positional terms. In addition:
   - *If $d \in D_{func}$ then $I_F(d)$ must be a function $D^*_{ind} \to D_{ind}$.*
   - *This implies that when a function symbol is applied to arguments that are individual objects then the result is also an individual object.*
4. $I_=$ *is a mapping of the form $D_{ind} \times D_{ind} \to D$.*
   It gives meaning to the equality operator.
5. I_{truth} *is a mapping of the form $D \to TV$.*
   It is used to define truth valuation for formulas.
6. $I_{external}$ *is a mapping that is used to give meaning to* `External` *terms. It maps symbols in* `Const` *designated as external to fixed functions of appropriate arity. Typically, external terms are invocations of built-in functions or calls to external definitions, and their fixed interpretations are determined by the specification of those built-ins and external definitions.*

We also define the following generic mapping from terms to $\boldsymbol{D}$, which we denote by $\boldsymbol{I}$ (this is the same symbol as the one used to denote semantic structures, but note the different font).

- $\boldsymbol{I}(k) = \boldsymbol{I}_C(k)$, *if k is a symbol in* `Const`
- $\boldsymbol{I}(?v) = \boldsymbol{I}_V(?v)$, *if $?v$ is a variable in* `Var`
- $\boldsymbol{I}(f(t_1 \ \ldots \ t_n)) = \boldsymbol{I}_F(\boldsymbol{I}(f))(\boldsymbol{I}(t_1),...,\boldsymbol{I}(t_n))$
- $\boldsymbol{I}(x{=}y) = \boldsymbol{I}_=(\boldsymbol{I}(x), \boldsymbol{I}(y))$
- $\boldsymbol{I}(\texttt{External}(p(s_1 \ \ldots \ s_n))) = \boldsymbol{I}_{external}(p)(\boldsymbol{I}(s_1), \ ..., \ \boldsymbol{I}(s_n))$.   □

3.2   Interpretation of Formulas

It can now be established how semantic structures determine the truth value of Dlex formulas, excluding group and document formulas [BK09a] for simplicity. Here we define a mapping, $TVal_\mathcal{I}$, from the set of all formulas to $\boldsymbol{TV}$.

Definition 5 (Truth valuation). *Truth valuation for well-formed formulas in Dlex is determined using the following function, denoted $TVal_\mathcal{I}$:*

1. *Positional atomic formulas:* $TVal_\mathcal{I}(r(t_1 \ \ldots \ t_n)) = \boldsymbol{I}_{truth}(\boldsymbol{I}(r(t_1 \ \ldots \ t_n)))$
2. *Equality:* $TVal_\mathcal{I}(\boldsymbol{x}{=}\boldsymbol{y}) = \boldsymbol{I}_{truth}(\boldsymbol{I}(\boldsymbol{x}{=}\boldsymbol{y}))$.
   - *To ensure that equality has precisely the expected properties, it is required that $TVal_\mathcal{I}(\boldsymbol{x}{=}\boldsymbol{y}) = \boldsymbol{t}$ if and only if $\boldsymbol{I}(x) = \boldsymbol{I}(y)$.*
3. *Externally defined atomic formula:* $TVal_\mathcal{I}(\texttt{External}(t)){=}\boldsymbol{I}_{truth}(\boldsymbol{I}_{external}(t))$.
4. *Conjunction:* $TVal_\mathcal{I}(\texttt{And}(c_1 \ \ldots \ c_n)) = \boldsymbol{t}$ *if and only if $TVal_\mathcal{I}(c_1) = \ldots = TVal_\mathcal{I}(c_n) = \boldsymbol{t}$. Otherwise, $TVal_\mathcal{I}(\texttt{And}(c_1 \ \ldots \ c_n)) = \boldsymbol{f}$.*
   The empty conjunction is treated as a tautology, so $TVal_\mathcal{I}(\texttt{And}()) = \boldsymbol{t}$.
5. *Quantification:*
   - $TVal_\mathcal{I}(\texttt{Forall} \ ?v_1 \ \ldots \ ?v_n \ (\varphi)) = \boldsymbol{t}$ *if and only if for every $\mathcal{I}^*$, described below, $TVal_{\mathcal{I}*}(\varphi) = \boldsymbol{t}$.*
   Here $\mathcal{I}^*$ is a semantic structure of the form $<\boldsymbol{TV}, \boldsymbol{D}, \boldsymbol{D}_{ind}, \boldsymbol{D}_{func}, \boldsymbol{I}_C, \boldsymbol{I}^*_V,$ $\boldsymbol{I}_F, \boldsymbol{I}_=, \boldsymbol{I}_{external}, \boldsymbol{I}_{truth}>$, which is exactly like $\boldsymbol{I}$, except that the mapping $\boldsymbol{I}^*_V$, is used instead of $\boldsymbol{I}_V$. $\boldsymbol{I}^*_V$ is defined to coincide with $\boldsymbol{I}_V$ on all variables except, possibly, on $?v_1,...,?v_n$.
6. *Rule implication:*
   - $TVal_\mathcal{I}(conclusion \ :\text{-} \ condition) = \boldsymbol{t}$, *if either $TVal_\mathcal{I}(conclusion){=}\boldsymbol{t}$ or $TVal_\mathcal{I}(condition){=}\boldsymbol{f}$.*
   - $TVal_\mathcal{I}(conclusion \ :\text{-} \ condition) = \boldsymbol{f}$ *otherwise.*   □

3.3   Logical Entailment

We now define what it means for a set of Dlex rules (in a group or a document formula) to entail another Dlex formula. In Dlex we are mostly interested in entailment of condition formulas, which can be viewed as queries to Dlex groups or documents. In the definitions, the symbol |= will generically stand for **models** or for **entails**.

Definition 6 (Models). *A semantic structure $\mathcal{I}$ is a **model** of a formula, φ, written as $\mathcal{I}$ |= φ, iff $TVal_{\mathcal{I}}(\varphi) = \boldsymbol{t}$.* □

Definition 7 (Logical entailment). *Let φ and ψ be formulas. We say that φ **entails** ψ, written as φ |= ψ, if and only if for every semantic structure $\mathcal{I}$ for which both $TVal_{\mathcal{I}}(\varphi)$ and $TVal_{\mathcal{I}}(\psi)$ are defined, $\mathcal{I}$ |= φ implies $\mathcal{I}$ |= ψ.* □

4   Mapping of the Dlex Condition Language

We proceed to formalizing the chain of semantics-preserving Dlex mappings χ_{dlex}, ξ_{dlex}, and π_{dlex} for the Condition Languages of Datalog RuleML and RIF-Core.

A Dlex mapping μ is **semantics-preserving** if for any pair φ, ψ of *Dlex* formulas for which φ |=$_K$ ψ is defined, φ |=$_K$ ψ if and only if $\mu(\varphi)$ |=$_L$ $\mu(\psi)$. Here |=$_K$ and |=$_L$ denote logical entailment in, respectively, the source language K and target language L of μ.

In each table below defining a mapping μ, each row specifies a translation $\mu(\text{column}_1) = \text{column}_2$. The **bold-italic** symbols represent metavariables. (A sequence of terms containing metavariables with subscripts is indicated by an ellipsis.) column_2 often contains applications of the mapping μ to these metavariables. When an expression $\mu(\boldsymbol{metavar})$ occurs in column_2, it should be understood as a recursive application of μ to $\boldsymbol{metavar}$. The result of such an application is substituted for the expression $\mu(\boldsymbol{metavar})$. The μ inverse, μ^{-1}, is obtained by reading the table right-to-left while replacing right-hand-side recursive μ applications with corresponding left-hand-side recursive μ^{-1} applications.

4.1   Conditions from RIF Presentation Syntax to RIF/XML

The text-to-XML mapping χ_{dlex} from the RIF Presentation Syntax of Dlex to RIF/XML specializes a mapping χ_{core} for RIF-Core, which itself is a specialization of χ_{bld} from [BK09a]. We will thus start the mapping chain with the subset of the RIF Presentation Syntax that was defined as the Dlex Presentation Syntax in section 2 and used to specify the Dlex semantics in section 3.

Since the RIF Presentation Syntax is context sensitive, the mapping χ_{dlex} must differentiate between terms that occur in the position of individuals and terms that occur as atomic formulas. To this end, the terms that occur in the context of atomic formulas are denoted by expressions of the form $\boldsymbol{pred}(...)$ and those that occur as individuals are denoted by expressions of the form $\boldsymbol{func}(...)$. In RIF-Core, however, $\boldsymbol{func}(...)$ is only needed for built-in calls.

RIF Presentation Syntax	RIF/XML

And (
    $conjunct_1$
    . . .
    $conjunct_n$
)
	`<And>`
	    `<formula>`$\chi_{dlex}(conjunct_1)$`</formula>`
	        . . .
	    `<formula>`$\chi_{dlex}(conjunct_n)$`</formula>`
	`</And>`

External (
    $atomexpr$
)
	`<External>`
	    `<content>`$\chi_{dlex}(atomexpr)$`</content>`
	`</External>`

$pred$ (
    $argument_1$
    . . .
    $argument_n$
)
	`<Atom>`
	    `<op>`$\chi_{dlex}(pred)$`</op>`
	    `<args ordered="yes">`
	        $\chi_{dlex}(argument_1)$
	            . . .
	        $\chi_{dlex}(argument_n)$
	    `</args>`
	`</Atom>`

$func$ (
    $argument_1$
    . . .
    $argument_n$
)
	`<Expr>`
	    `<op>`$\chi_{dlex}(func)$`</op>`
	    `<args ordered="yes">`
	        $\chi_{dlex}(argument_1)$
	            . . .
	        $\chi_{dlex}(argument_n)$
	    `</args>`
	`</Expr>`

$left = right$	`<Equal>`
	    `<left>`$\chi_{dlex}(left)$`</left>`
	    `<right>`$\chi_{dlex}(right)$`</right>`
	`</Equal>`

$"unicodestring"\char`^\char`^symspace$	`<Const` `type="`$symspace$`">`$unicodestring$`</Const>`
$?unicodestring$	`<Var>`$unicodestring$`</Var>`

The table shows that the functional-style terms of the RIF Presentation Syntax are mapped to RIF/XML elements by inserting role tags such as `formula` in `And` terms. The `Atom` and `Expr` elements serialize n-ary *$pred$*(...) and *$func$*(...) applications, respectively, using two role tags and moving the n-ary branching underneath the second role tag, `args`. Since the RIF/XML serialization provides explicit markup, the unique inverse χ_{dlex}^{-1} is easily obtained. The round-tripping composition $\chi_{dlex}^{-1} \circ \chi_{dlex}$ is the identity, so χ_{dlex} is semantics-preserving.

4.2   Conditions from RIF/XML to RuleML/XML

The XML-to-XML mapping ξ_{dlex} from RIF/XML to RuleML/XML is central to our chain of mappings, as it allows to compare, and bridge between, the kernels of RIF and RuleML. To reduce the mapping 'distance', Fully Striped RuleML/XML is used as the target (i.e., second column) of ξ_{dlex}. Stripe-skipping via a mapping σ_{dlex} (cf. section 1) can then be employed to obtain the more compact Stripe-Skipped RuleML/XML, chaining to the source (i.e., first column) of the mapping π_{dlex} in section 4.3.

RIF/XML	RuleML/XML (Fully Striped)

`<And>`
  `<formula>`*conjunct*$_1$`</formula>`
  . . .
  `<formula>`*conjunct*$_n$`</formula>`
`</And>`
	`<And>`
	  `<formula>`ξ_{dlex}(*conjunct*$_1$)`</formula>`
	  . . .
	  `<formula>`ξ_{dlex}(*conjunct*$_n$)`</formula>`
	`</And>`

`<External>`
  `<content>`*atomexpr*`</content>`
`</External>`
	ξ_{dlex}(*atomexpr*)

`<Atom>`
  `<op>`ξ_{dlex}(*pred*)`</op>`
  `<args ordered="yes">`
    ξ_{dlex}(*argument*$_1$)
    . . .
    ξ_{dlex}(*argument*$_n$)
  `</args>`
`</Atom>`
	`<Atom>`
	  `<op>`ξ_{dlex}(*pred*)`</op>`
	  `<arg index="1">`
	    ξ_{dlex}(*argument*$_1$)
	  `</arg>`
	  . . .
	  `<arg index="n">`
	    ξ_{dlex}(*argument*$_n$)
	  `</arg>`
	`</Atom>`

`<Expr>`
  `<op>`ξ_{dlex}(*func*)`</op>`
  `<args ordered="yes">`
    ξ_{dlex}(*argument*$_1$)
    . . .
    ξ_{dlex}(*argument*$_n$)
  `</args>`
`</Expr>`
	`<Expr>`
	  `<op>`ξ_{dlex}(*func*)`</op>`
	  `<arg index="1">`
	    ξ_{dlex}(*argument*$_1$)
	  `</arg>`
	  . . .
	  `<arg index="n">`
	    ξ_{dlex}(*argument*$_n$)
	  `</arg>`
	`</Expr>`

`<Equal>`
  `<left>`*left*`</left>`
  `<right>`*right*`</right>`
`</Equal>`
	`<Equal>`
	  `<left>`ξ_{dlex}(*left*)`</left>`
	  `<right>`ξ_{dlex}(*right*)`</right>`
	`</Equal>`

`<Const`
  `type="`*st*`">`*unicodestring*`</Const>`
	`<Const`
	  `type="`*st*`">`*unicodestring*`</Const>`

`<Var>`*unicodestring*`</Var>`	`<Var>`*unicodestring*`</Var>`

As can be seen in the table, the commonalities between RIF and RuleML prevail, but the following differences should be noted:

While RIF employs `External` wrappers around every built-in call, RuleML makes the distinction between user-defined and built-in calls via an optional attribute on the operators (not shown here). A RIF advantage here may be explicitness. The RuleML advantage is, as in Lisp, that there is no need to (structurally) change the syntax at all call occurrences when a user-defined operator becomes 'compiled' into a built-in. Note that the RIF built-ins themselves, currently in version RIF-DTB 1.0 [PBK09], are adopted by RuleML: Unlike the earlier SWRL built-ins, which write n-ary functions as $(1+n)$-ary relations [HPSB$^+$04], functional RIF built-ins remain functions. While RIF uses an `args` role tag around the sequence of all arguments to predicates and functions, RuleML uses `arg` role tags around each argument separately, which can be omitted thanks to stripe-skipping. We propose that both the RIF and RuleML methods should be permitted in RIF RuleML. While RIF `Constants` use symbol spaces as values of their mandatory `type` attribute, RuleML `Individuals` and neutralized `Constants` use optional types, which can be symbol spaces.

Taking care of these differences, the unique inverse $\xi_{\texttt{dlex}}^{-1}$ can be obtained, making $\xi_{\texttt{dlex}}$ semantics-preserving.

4.3 Conditions from RuleML/XML to RuleML/POSL

The XML-to-text mapping $\pi_{\texttt{dlex}}$ from Stripe-Skipped RuleML/XML to RuleML/ POSL generates Prolog-like conditions (where, however, variables are indicated by a "?"-prefix rather than by capitalization).

RuleML/XML (Stripe-Skipped)	RuleML/POSL

`<And>`
  $conjunct_1$
  . . .
  $conjunct_{n-1}$
  $conjunct_n$
`</And>`
	$\pi_{\texttt{dlex}}(conjunct_1),$
	. . .
	$\pi_{\texttt{dlex}}(conjunct_{n-1}),$
	$\pi_{\texttt{dlex}}(conjunct_n)$

`<Atom>`
  $pred$
  $argument_1$
  . . .
  $argument_{n-1}$
  $argument_n$
`</Atom>`
	$\pi_{\texttt{dlex}}(pred)$ (
	    $\pi_{\texttt{dlex}}(argument_1),$
	    . . .
	    $\pi_{\texttt{dlex}}(argument_{n-1}),$
	    $\pi_{\texttt{dlex}}(argument_n)$
	  )

`<Expr>`
  $func$
  $argument_1$
  . . .
  $argument_{n-1}$
  $argument_n$
`</Expr>`
	$\pi_{\texttt{dlex}}(func)$ (
	    $\pi_{\texttt{dlex}}(argument_1),$
	    . . .
	    $\pi_{\texttt{dlex}}(argument_{n-1}),$
	    $\pi_{\texttt{dlex}}(argument_n)$
	  )

`<Equal>`
  $left$
  $right$
`</Equal>`
	equal (
	  $\pi_{\texttt{dlex}}(left),$
	  $\pi_{\texttt{dlex}}(right)$)

`<Const` `type="`st`">`$unicodestring$`</Const>`	$unicodestring{:}st$
`<Var>`$unicodestring$`</Var>`	$?unicodestring$

Note that the `type` attribute of RIF RuleML XML serializations becomes a ":"-infix in POSL, which – closing the mapping ring with mapping $\omega_{\texttt{dlex}}$ – in RIF Presentation Syntax is written as a "^^"-infix. As a term-type separator, the POSL use of ":" is modeled on ":" in sorted logics, while the Presentation Syntax use of "^^" is modeled on N3/Turtle (where ":" is needed for namespaces).

The generator $\pi_{\texttt{dlex}}$ is semantics-preserving since it can be uniquely inverted to the parser $\pi_{\texttt{dlex}}^{-1}$, although parsing is more difficult, as always. Both $\pi_{\texttt{dlex}}$

and the inverse text-to-XML mapping $\pi_{\mathtt{dlex}}^{-1}$ are supported by online translators under a single GUI (http://www.ruleml.org/posl/converter.jnlp).

5   Mapping of the Dlex Rule Language

We now extend the chain of Dlex Condition Language mappings $\chi_{\mathtt{dlex}}$, $\xi_{\mathtt{dlex}}$, and $\pi_{\mathtt{dlex}}$ of section 4 to a chain of semantics-preserving Dlex Rule Language mappings for Datalog RuleML and RIF-Core.

5.1   Rules from RIF Presentation Syntax to RIF/XML

The mapping $\chi_{\mathtt{dlex}}$ of section 4.1 from the RIF Presentation Syntax to RIF/XML is extended here for rules.

RIF Presentation Syntax	RIF/XML

`Document(`
    group
        `)`
	`<Document>`
	    `<payload>`$\chi_{\mathtt{dlex}}$(*group*)`</payload>`
	`</Document>`

`Group(`
    clause₁
    . . .
    clauseₙ
        `)`
	`<Group>`
	    `<sentence>`$\chi_{\mathtt{dlex}}$(*clause₁*)`</sentence>`
	    . . .
	    `<sentence>`$\chi_{\mathtt{dlex}}$(*clauseₙ*)`</sentence>`
	`</Group>`

`Forall`
    variable₁
    . . .
    variableₙ `(`
            rule
        `)`
	`<Forall>`
	    `<declare>`$\chi_{\mathtt{dlex}}$(*variable₁*)`</declare>`
	    . . .
	    `<declare>`$\chi_{\mathtt{dlex}}$(*variableₙ*)`</declare>`
	    `<formula>`$\chi_{\mathtt{dlex}}$(*rule*)`</formula>`
	`</Forall>`

conclusion `:- ` *condition*	`<Implies>`
	    `<if>`$\chi_{\mathtt{dlex}}$(*condition*)`</if>`
	    `<then>`$\chi_{\mathtt{dlex}}$(*conclusion*)`</then>`
	`</Implies>`

The table rows map through a RIF `Document`, via a `Group` and `Forall`s, down to the level of " `:-` "-rules. The extended RIF/XML serialization again provides explicit markup, so that the unique inverse $\chi_{\mathtt{dlex}}^{-1}$ and the semantics-preservation property of $\chi_{\mathtt{dlex}}$ are easily obtained.

5.2   Rules from RIF/XML to RuleML/XML

The central mapping $\xi_{\mathtt{dlex}}$ of section 4.2 from RIF/XML to RuleML/XML is similarly extended for rules.[4]

---

[4] Regarding the last table row, the RuleML 0.91 upgrade to 0.95 adopts the RIF role tags `<if>` ... `<then>`, also long contemplated in RuleML, instead of `<body>` ... `<head>`.

RIF/XML	RuleML/XML (Fully Striped)

`<Document>`
  `<payload>`*group*`</payload>`
`</Document>`
	`<RuleML>`
	  `<performative index="1">`
	    `<Assert>`
	      `<formula>`
	        ξ_{dlex}(*group*)
	      `</formula>`
	    `</Assert>`
	  `</performative>`
	`</RuleML>`

`<Group>`
  `<sentence>`*clause$_1$*`</sentence>`
    . . .
  `<sentence>`*clause$_n$*`</sentence>`
`</Group>`
	`<Rulebase>`
	  `<formula>`ξ_{dlex}(*clause$_1$*)`</formula>`
	    . . .
	  `<formula>`ξ_{dlex}(*clause$_n$*)`</formula>`
	`</Rulebase>`

`<Forall>`
  `<declare>`*variable$_1$*`</declare>`
    . . .
  `<declare>`*variable$_n$*`</declare>`
  `<formula>`*rule*`</formula>`
`</Forall>`
	`<Forall>`
	  `<declare>`ξ_{dlex}(*variable$_1$*)`</declare>`
	    . . .
	  `<declare>`ξ_{dlex}(*variable$_n$*)`</declare>`
	  `<formula>`ξ_{dlex}(*rule*)`</formula>`
	`</Forall>`

`<Implies>`
  `<if>`*condition*`</if>`
  `<then>`*conclusion*`</then>`
`</Implies>`
	`<Implies>`
	  `<if>`ξ_{dlex}(*condition*)`</if>`
	  `<then>`ξ_{dlex}(*conclusion*)`</then>`
	`</Implies>`

Again, the commonalities between RIF and RuleML prevail, but there are the following differences: The first row maps the RIF root element `Document` to the RuleML root element `RuleML`, which permits (ordered) transactions of performatives including `Assert`s. The second row maps a RIF `Group` to a RuleML `Rulebase`. While a general RIF `Document` can also contain directives for a base, prefixes, and imports, a general `RuleML` root can also contain `Query` and `Retract` transactions. But for our special case the unique inverse ξ_{dlex}^{-1} can be obtained, keeping ξ_{dlex} semantics-preserving.

5.3   Rules from RuleML/XML to RuleML/POSL

The mapping π_{dlex} of section 4.3 and the online translators from Stripe-Skipped RuleML/XML to RuleML/POSL are likewise extended for rules.

RuleML/XML (Stripe-Skipped)	RuleML/POSL

`<RuleML>`
  `<Assert>`
    group
  `</Assert>`
`</RuleML>`
	π_{dlex}(***group***)

continued on next page

continued from previous page	

`<Rulebase>`
  `clause`$_1$
  `. . .`
  `clause`$_n$
`</Rulebase>`
	$\pi_{\mathrm{dlex}}(clause_1)$
	`. . .`
	$\pi_{\mathrm{dlex}}(clause_n)$

`<Forall>`
  `<declare>`$variable_1$`</declare>`
  `. . .`
  `<declare>`$variable_n$`</declare>`
  `<formula>`$rule$`</formula>`
`</Forall>`
	$\pi_{\mathrm{dlex}}(rule)$

`<Implies>`
  `condition`
  `conclusion`
`</Implies>`
	$\pi_{\mathrm{dlex}}(conclusion)$ `:-` $\pi_{\mathrm{dlex}}(condition).$

Note that the `RuleML`, `Assert`, and `Rulebase` levels are understood in POSL. Similarly, RuleML/XML (optionally) and RuleML/POSL (always) make the same universal closure assumption as Prolog, so the row for `Forall` is rarely needed. Finally, RuleML/POSL ":-"-rules are terminated by Prolog-style periods, unlike those in RIF Presentation Syntax (cf. section 5.1). The generator π_{dlex} has a parser inverse π_{dlex}^{-1} preserving the semantics (http://www.ruleml.org/posl/converter.jnlp).

6   Conclusion

The Dlex kernel of the RIF RuleML overlap studied here revealed commonalities and differences between the two languages, enabling round-tripping and convergence. This positional Dlex can now be expanded for the remaining syntactic features of Datalog RuleML / RIF-Core, including slotted/named arguments, objects/frames, and IRIs as attributes/elements. Since our Dlex mappings already include (interpreted) function symbols for external calls, the study should also be extended to the overlap of Hornlog RuleML and RIF-BLD using (uninterpreted) function symbols for Herbrand terms.

In the RuleML family, Datalog RuleML is complemented by Naf Datalog RuleML, a sublanguage with the often needed Negation as failure, which is also carried through to Naf Hornlog RuleML. Corresponding RIF Non-Monotonic Dialects are called for as well, specializing default negation in the RIF Framework for Logic Dialects [BK09b]. Since Naf constructs are wide-spread in practice, most rule engines have implemented them (including Scoped Naf in cwm [BLCK+08]), so their interoperation is already a focus of RIF RuleML efforts (cf. use case: http://ruleml.org/WellnessRules).

On the next level, the development of the RIF Production Rule Dialect towards a RIF Reaction Rule Dialect aligned with Reaction RuleML will provide further opportunities for joint work between W3C, OMG, and RuleML, as needed by the industry.

Acknowledgements

Many thanks go to Michael Kifer and all colleagues in the W3C RIF Working Group and in the RuleML Technical Groups for continued collaboration on Web rule interchange. NSERC is thanked for its support through a Discovery Grant.

References

[BHK⁺09]   Boley, H., Hallmark, G., Kifer, M., Paschke, A., Polleres, A., Reynolds, D.: RIF Core Dialect. W3C Working Draft (July 2009), http://www.w3.org/2005/rules/wiki/Core

[BK09a]   Boley, H., Kifer, M.: RIF Basic Logic Dialect. W3C Working Draft (July 2009), http://www.w3.org/2005/rules/wiki/BLD

[BK09b]   Boley, H., Kifer, M.: RIF Framework for Logic Dialects. W3C Working Draft (July 2009), http://www.w3.org/2005/rules/wiki/FLD

[BLCK⁺08]   Berners-Lee, T., Connolly, D., Kagal, L., Scharf, Y., Hendler, J.: N3Logic: A Logical Framework For the World Wide Web. Theory and Practice of Logic Programming (TPLP) 8(3) (May 2008)

[Bol06]   Boley, H.: The RuleML Family of Web Rule Languages. In: Alferes, J.J., Bailey, J., May, W., Schwertel, U. (eds.) PPSWR 2006. LNCS, vol. 4187, pp. 1–17. Springer, Heidelberg (2006)

[Bol07]   Boley, H.: Are Your Rules Online? Four Web Rule Essentials. In: Paschke, A., Biletskiy, Y. (eds.) RuleML 2007. LNCS, vol. 4824, pp. 7–24. Springer, Heidelberg (2007)

[CGT89]   Ceri, S., Gottlob, G., Tanca, L.: What You Always Wanted to Know About Datalog (And Never Dared to Ask). IEEE Trans. on Knowledge and Data Eng. 1(1) (March 1989)

[HPSB⁺04]   Horrocks, I., Patel-Schneider, P.F., Boley, H., Tabet, S., Grosof, B., Dean, M.: SWRL: A Semantic Web Rule Language Combining OWL and RuleML. W3C Member Submission (May 2004), http://www.w3.org/Submission/SWRL/

[PBK09]   Polleres, A., Boley, H., Kifer, M.: RIF Datatypes and Built-ins 1.0. W3C Working Draft (July 2009), http://www.w3.org/2005/rules/wiki/DTB

WellnessRules: A Web 3.0 Case Study in RuleML-Based Prolog-N3 Profile Interoperation

Harold Boley, Taylor Michael Osmun, and Benjamin Larry Craig

Institute for Information Technology
National Research Council of Canada
Fredericton, NB, E3B 9W4, Canada

Abstract. An interoperation study, WellnessRules, is described, where rules about wellness opportunities are created by participants in rule languages such as Prolog and N3, and translated within a wellness community using RuleML/XML. The wellness rules are centered around participants, as profiles, encoding knowledge about their activities conditional on the season, the time-of-day, the weather, etc. This distributed knowledge base extends FOAF profiles with a vocabulary and rules about wellness group networking. The communication between participants is organized through Rule Responder, permitting wellness-profile translation and distributed querying across engines. WellnessRules interoperates between rules and queries in the relational (Datalog) paradigm of the pure-Prolog subset of POSL and in the frame (F-logic) paradigm of N3. An evaluation of Rule Responder instantiated for WellnessRules revealed acceptable Web response times.

1   Introduction

Web 2.0 combined with Semantic Web techniques is currently leading to Web 3.0 techniques. As part of NRC-IIT's Health & Wellness and Learning & Training efforts, we are exploring Wellness 3.0, employing Web 3.0 rules plus ontologies to plan wellness-oriented activities and nutrition.

We focus here on WellnessRules[1], a system supporting the management of wellness practices within a community based on rules plus ontologies. The idea is the following. As in Friend of a Friend (FOAF)[2], people can choose a (community-unique) nickname and create semantic profiles about themselves, specifically their wellness practices, for their own planning and to network with other people supported by a system that 'understands' those profiles. As in FindXpRT [LBBM06], such FOAF-like fact-only profiles are extended with rules to capture conditional person-centered knowledge such as each person's wellness activity depending on the season, the time-of-day, the weather, etc. People can use rules of various refinement levels and rule languages ranging from pure Prolog to N3, which will be interoperated through RuleML/XML [Bol07]. Like our (RuleML-20xy)

---

[1] http://ruleml.org/WellnessRules/
[2] http://www.foaf-project.org/

G. Governatori, J. Hall, and A. Paschke (Eds.): RuleML 2009, LNCS 5858, pp. 43–52, 2009.

SymposiumPlanner [CB08] (and unlike FindXpRT), WellnessRules is based on Rule Responder [PBKC07, CB08], which is itself based on the Mule Enterprise Service Bus (ESB).

We will discuss an example where John (p0001) advertises Prolog-style rules on his wellness community profile, including a refinement of the following: p0001 may do outdoor running if it is summer and not raining. Hence, Peter and Paul can find p0001 via Prolog or N3 queries to Rule Responder expressing their own preferences, so that an initial group might be formed. Interoperating with translators, WellnessRules thus frees participants from using any single rule language. In particular, it bridges between Prolog as the main Logic Programming rule paradigm and N3 as the main Semantic Web rule paradigm.

The distributed nature of Rule Responder profiles, each queried by its own (copy of an) engine, permits scalable knowledge representation and processing. Since participants of a wellness community are supposed to meet in overlapping groups for real-world events such as skating, this kind of community (unlike a virtual-only community) has a maximal effective size (which we estimate to be less than 1000 participants). Beyond that size, it can be split into two or more subcommunities based on preferred wellness practices, personal compatibility, geographic proximity, etc. Rule Responder support thus needs to extend only to that maximal size, but can be cloned as subcommunities emerge.

The rest of the paper is organized as follows. Section 2 discusses the hybrid global knowledge bases of WellnessRules. Section 3 explains its local knowledge bases distributed via Rule Responder. Section 4 focusses on the interoperation between Prolog and N3. Section 5 explains and evaluates Rule Responder querying of WellnessRules knowledge. Section 6 concludes the paper.

2   Hybrid Global Knowledge Bases in WellnessRules

WellnessRules employs a hybrid combination [Bol07] of ontologies and rules. While the entire ontology and a portion of the rulebase is globally shared by all participants (agents), the other portion of the rulebase is locally distributed over the participants (agents).

As its (light-weight) ontology component, WellnessRules employs subClassOf taxonomies. We reuse parts of the Nuadu ontology collection [SLKL07], mainly the Activity and Nutrition ontologies. WellnessRules currently employs an Activity taxonomy using Nuadu classes Running, Walking, WaterSports subsuming SwimmingCalm, WinterSports subsuming IceSkating, and Sports subsuming a WellnessRules class Baseball, as well as WellnessRules classes Hiking, and Yoga. The corresponding RuleML-/N3-readable RDFS subClassOf statements are shared at http://ruleml.org/WellnessRules/WR-Taxonomy.rdf.

As its rule component, WellnessRules employs Naf Datalog POSL and N3 with scoped Naf. We restrict the use of Naf Datalog POSL to atoms with positional arguments, leaving F-logic-like frames with property-value slots to N3, thus demonstrating the range of our approach through complementary rule styles. For that reason, the POSL syntax corresponds to pure-Prolog syntax except that POSL variables are prefixed by a question mark while Prolog variables are upper-cased.

This Datalog POSL sublanguage uses (positional) n-ary relations (or, predicates) as its central modeling paradigm. N3 instead uses (unordered) sets of binary relations (or, properties) centered around object identifiers (OIDs, in the role of 'subjects' in N3).

For example, this is a global 4-ary `meetup` fact:

```
meetup(m0001,walk,out,conniesStation).
```

Similarly, this is its slotted counterpart:

```
:meetup_1
    rdf:type        :Meetup;
    :mapID          :m0001;
    :activity       :run;
    :inOut          :out;
    :location       :conniesStation.
```

Both express that one `meetup` for `walk` activities of the supported wellness community is `conniesStation` as found on map `m0001`.

An example of a global POSL rule defines a `participation` as follows:

```
participation(?ProfileID,?Activity,?Ambience,?MinRSVP,?MaxRSVP) :-
    groupSize(?ProfileID,?Activity,?Ambience,?Min,?Max),
    greaterThanOrEqual(?MinRSVP,?Min),
    lessThanOrEqual(?MaxRSVP,?Max).
```

As in FindXpRT, the first argument of a WellnessRules conclusion predicate always is the person the rule is about. Similar to Prolog, the rule succeeds for its five positional arguments if the acceptable `groupSize` of the participant with `?ProfileID`, for an `?Activity` in an `?Ambience`, is between `?Min` and `?Max`, and the emerging group has size between $?MinRSVP \geq ?Min$ and $?MaxRSVP \leq ?Max$, where `greaterThanOrEqual` and `lessThanOrEqual` are SWRL built-ins as implemented in OO jDREW 0.961.

The corresponding global N3 rule represents this in frame form as follows:

```
{
?rsvpQuery
    rdf:type        :RSVPQuery;
    :profileID      :p0001;
    :minRSVP        ?MinRSVP;
    :maxRSVP        ?MaxRSVP.

?groupSize
    rdf:type        :GroupSize;
    :profileID      ?ProfileID;
    :activity       ?Activity;
    :inOut          ?Ambience;
    :min            ?Min;
    :max            ?Max.
```

```
?MinRSVP math:notLessThan ?Min.

?MaxRSVP math:notGreaterThan ?Max.
}
=>
{
  _:participation
      rdf:type      :Participation;
      :profileID    :p0001;
      :activity     ?Activity;
      :inOut        ?Ambience;
      :min          ?MinRSVP;
      :max          ?MaxRSVP.
}.
```

Here, the first premise passes the input arguments ?MinRSVP and ?MaxRSVP into the rule (cf. its use in section 5). The remaining premises correspond to those in the POSL version, where math:notLessThan and math:notGreaterThan are N3 built-ins as implemented in Euler.

The global OA rulebase is being maintained in both languages at http://ruleml.org/WellnessRules/WR-Global.posl and *.n3.

3   Locally Distributed Knowledge Bases in WellnessRules

Each PA has its own local rules, which were selected from profiles created by participants of the NRC-IIT Fredericton wellness community.

This is an example of a local POSL rule from the PA rulebase of a participant p0001, defining the main predicate myActivity for running:

```
myActivity(p0001,?:Running,out,?MinRSVP,?MaxRSVP,?StartTime,?EndTime,
                                          ?Place,?Duration,?Level) :-
    calendar(p0001,?Calendar),
    event(?Calendar,?:Running,possible,?StartTime,?EndTime),
    participation(p0001,run,out,?MinRSVP,?MaxRSVP),
    season(?StartTime,summer),
    forecast(?StartTime,sky,?Weather),
    notEqual(?Weather,raining),
    map(p0001,?Map),
    meetup(?Map,run,out,?Place),
    level(p0001,run,out,?Place,?Duration,?Level),
    fitness(p0001,?StartTime,?ExpectedFitness),
    greaterThanOrEqual(?ExpectedFitness,?Level),
    goodDuration(?Duration,?StartTime,?EndTime).
```

The rule conclusion starts with the person's profile ID, p0001, followed by the kind of activity, run, and its ambience, outdoors, followed by variables for the group limits ?MinRSVP and ?MaxRSVP, the earliest ?StartTime and ?EndTime, its actual ?Duration and its ?Level. The rule premises query p0001's ?Calendar, an event of a possible (or tentative) ?:Running (the anonymous variable "?"

has type Running), the participation (see above), an appropriate season and forecast, p0001's ?Map, a meetup ?Place, the required level less than the expected fitness, as well as a goodDuration.

The corresponding local N3 rule is given abridged below (complete, online at http://ruleml.org/WellnessRules/PA/p0001.n3):

```
{
...

?forecast
    rdf:type    :Forecast;
    :startTime  ?StartTime;
    :aspect     :sky;
    :value      ?Weather.

?Weather log:notEqualTo :raining.

...
}
=>
{
_:myActivity
    rdf:type      :MyActivity;
    :profileID    :p0001;
    :activity     :Running;
    :inOut        :out;
    :minRSVP      ?MinRSVP;
    :maxRSVP      ?MaxRSVP;
    :startTime    ?StartTime;
    :endTime      ?EndTime;
    :location     ?Place;
    :duration     ?Duration;
    :fitnessLevel ?FitnessLevel.
}.
```

The online version of the above POSL rule employs the premise naf(forecast(?StartTime,sky,raining)) instead of separate forecast and notEqual premises. For the N3 online version, the above log:notEqualTo built-in call is more convenient. An irreducible Naf used in POSL's online version adds the following premises in the myActivity rule (after the current event premise):

```
yesterday(?StartTime,?StartTimeYtrday,?EndTime,?EndTimeYtrday),
naf(event(?Calendar,?:Running,past,?StartTimeYtrday,?EndTimeYtrday))
```

They make sure that p0001's calendar does not contain a running event on the day before. The counterpart in N3 could use log:notIncludes, which in Euler, as in our online version, is replaced with e:findall, checking that the result is the empty list, '()'.

The resulting PA rulebases, which require Datalog with Naf and N3 with '()'-e:findall, are being maintained in both of these languages at http://

ruleml.org/WellnessRules/PA, e.g. those for p0001 at http://ruleml.org/
WellnessRules/PA/p0001.posl and *.n3.

4   Cross-Paradigm Rulebase Alignment and Translation

The WellnessRules case study includes a testbed for the interoperation (i.e.,
alignment and translation) of rulebases in the main two rule paradigms:
Prolog-style (positional) relations and N3-style (slotted) frames. In our interoperation methodology, we make iterative use of alignment and translation: An
initial alignment permits the translation of parts of a hybrid knowledge base.
This then leads to more precise alignments, which in turn leads to better translations, etc. Using this methodology for WellnessRules, we are maintaining a
relational as well as a frame version of the rules, both accessing the same, independently maintained, RDFS ontology.

For rulebase translation, we first use a pair of online translators (http://
ruleml.org/posl/converter.jnlp) between the human-oriented syntax POSL
and its XML serialization in OO RuleML. Based on the RDF-RIF combinations
in [dB09], similar translators are being developed between N3, RIF, and RuleML.

The interoperation between WellnessRules PAs that use different rule
paradigms is then enabled by RuleML, which has sublanguages for both the
relational and the frame paradigms, so that the cross-paradigm translations can
use the common XML syntax of RuleML.

The alignment of sample relations and frames in sections 2 and 3 suggests
translations between both paradigms. We consider here translations that are
'static' or 'at compile-time' in that they take an entire rulebase as input and
return its entire transformed version. We can thus make a 'closed-arguments'
assumption of fixed signatures for relations and frames. In particular, the arity
of relations cannot change at run-time and no slots can be dynamically added
or removed from a frame. The translations work in both directions:

Objectify (Prolog to N3): Mapping from an n-ary relation rel to a frame,
this constructs a new frame with a generated fresh OID rel_j, where $j > 0$ is
the first integer making rel_j a unique name, and with the argument positions
p_1, p_2, ..., p_n as slot (or property) names.

Positionalize (N3 to Prolog): Mapping from a frame to a relation, this constructs a new relationship with the first argument taking the frame OID and
the remaining arguments taking the slot values of the sorted slot names from all
frames of OID's class (null values for properties not used in the current frame).

Formally, positionalizing is specified as follows, using POSL's frame notation
with slot arrows (->) and an OID separated from its slots by a hat (^):

1. Unite all slot names from all frames whose OID is an instance of a class cl
   into a finite set SN_{cl} of n_{cl} elements.
2. Introduce (SN_{cl},<) as a total order '<' over SN_{cl}, where '<' usually is the
   lexicographic order. Assume without loss of generality that the elements of
   SN_{cl} are $prop_1 < ... < prop_{n_{cl}}$.

3. For each frame frel $= \mathrm{cl}(\mathrm{oid}\hat{\ }\mathrm{prop}_{k_1}\text{->}\mathrm{TERM}_{k_1};...;\mathrm{prop}_{k_m}\text{->}\mathrm{TERM}_{k_m})$ assume without loss of generality that $\mathrm{prop}_{k_1} < ... < \mathrm{prop}_{k_m}$. Replace frel by a relation frel' $= \mathrm{cl}(\mathrm{oid},\mathrm{TERMCOMP}_1,...,\mathrm{TERMCOMP}_{n_{cl}})$, where for $1 \leq i \leq n_{cl}$ and $1 \leq j \leq m$ we have $\mathrm{TERMCOMP}_i = \mathrm{TERM}_{k_j}$ if $i = k_j$ and $\mathrm{TERMCOMP}_i = \perp$ otherwise ('$\perp$' is the null value formalized as the bottom element of the taxonomy, e.g. owl:NOTHING, which is equal only to itself, not to any other sort, constant, or variable).

Step 3 can be thought of as 'replenishing' the lexicographically sorted slots of a frame frel with slots $\mathrm{prop}_x\text{->}\perp$ for all slot names prop_x 'missing' for their class cl, and then making cl the relation name, inserting the oid, and omitting all slot names (keeping only their slot values).

An XSLT implementation of such a translator is available online (`http://ruleml.org/ooruleml-xslt/oo2prml.html`).

For the translation of a rule, the above relation-frame translation is applied to the relation (frame) in the conclusion and to all the relations (frames) in the premises. For a rulebase the translation then applies to all of its rules.

With the above-discussed human-oriented syntax translators, rulebases containing rules like the `myActivity` rule in section 3 can thus be translated via Prolog, POSL, RuleML (relations, frames), and N3. These translators permit rule and query interoperation, via RuleML/XML, for the Rule Responder infrastructure of WellnessRules communities.

5   Distributed Rule Responder Querying of WellnessRules

WellnessRules instantiates the Rule Responder multi-agent architecture as follows: Rule Responder's virtual organization is instantiated to a wellness community. An organizational agent (OA) becomes an assistant for an entire wellness community. Each personal agent (PA) becomes an assistant for one participant. Fig. 1 describes the OA/PA metamodel of WellnessRules for the activity and nutrition profiles of participants. Newcomers and participants can assume the role of an external agent (EA), (indirectly) querying participants' profiles.

Rule Responder uses the following sequence of steps: An EA asks queries to an OA. The OA maps and delegates each query to the PA(s) most knowledgeable about it. Each PA poses the query to its local rulebase plus ontology, sending the derived answer(s) back to the OA. The OA integrates relevant answers and gives the overall answer(s) to the EA, by default not revealing the coordinates of the answering PA(s).

In this way, the OA acts as a mediator that protects the privacy of profiles of participants in a wellness community. Participants within the same community can of course later decide to reveal their real name and open up their wellness profiles for (direct) querying by selected other participants.

The above Rule Responder steps have been instantiated earlier, including to the SymposiumPlanner system [CB08].

On the basis of Rule Responder the knowledge bases of sections 2 and 3 can be queried, using the translators of section 4 for interoperation. The implemented

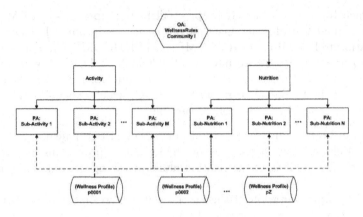

Fig. 1. WellnessRules OA/PA metamodel

Rule Responder for WellnessRules is available for online use at `http://ruleml.org/WellnessRules/RuleResponder/`.

For example, this is a POSL query regarding p0001's wellness profile, for execution by a top-down engine such as OO jDREW TD:

```
myActivity(p0001,?:Running,out,1:Integer,20:Integer,"2009-06-10T10:15:00",
                "2009-06-10T11:15:00",?Place,?Duration,?Level)
```

It uses the rule from sections 3 to check whether p0001 will possibly be ?:`Running`, outdoors, in a group of 1:`Integer` to 20:`Integer`, between start time `"2009-06-10T10:15:00"` and end time `"2009-06-10T11:15:00"`. Using further rules and facts from p0001's profile (`http://ruleml.org/WellnessRules/PA/p0001.posl`), it produces multiple solutions, binding the meetup ?`Place`, the ?`Duration`, and the required fitness ?`Level`.

The corresponding N3 query for execution by a bottom-up engine such as EulerSharp EYE uses a temporary fact to pass the input arguments:

```
:rsvpQuery
    rdf:type        :RSVPQuery;
    :profileID      :p0001;
    :minRSVP        1;
    :maxRSVP        20.
```

The N3 query itself then is as follows:

```
@prefix : <wellness_profiles#>.
@prefix rdf: <http://www.w3.org/1999/02/22-rdf-syntax-ns#>.
_:myActivity
    rdf:type        :MyActivity;
    :profileID      :p0001;
    :activity       :Running;
    :inOut          :out;
    :minRSVP        ?MinRSVP;
```

```
:maxRSVP     ?MaxRSVP;
:startTime   "2009-06-10T10:15:00";
:endTime     "2009-06-10T11:15:00";
:location    ?Place;
:duration    ?Duration;
:fitnessLevel   ?FitnessLevel.
```

After declaring two prefixes, it builds an existential '(_)' node, _:myActivity, using slots for the fixed parameters and the fact-provided ?MinRSVP and ?MaxRSVP bindings to create slots with the ?Place, ?Duration, and ?Level bindings.

An evaluation of the response times of the Mule infrastructure and the Rule Responder engines (OO jDREW, Euler, and Prova) instantiated for Wellness-Rules has been conducted using the previously discussed scenario. We found that this Rule Responder instantiation operates with acceptable Web response times.

Specifically, the execution times for the above myActivity query in Euler (N3), OO jDREW (POSL), and Rule Responder on average were 157ms, 1483ms, and 5053ms, respectively, measured as the Java system time, running in Java JRE6, Windows XP Professional SP3, on an Intel Core 2 Duo 2.80GHz processor.

For this and similar WellnessRules queries, the major contribution to the overall execution time has come from the ESB (Mule), which is not the focus of this work. Rule Responder operates using a star-like connection architecture, where the OA dispatches network traffic to and fro the most appropriate PA. A separate study in distributed querying has worked on minimizing this communication overhead.[3]

The above query could be specialized to produce exactly one solution, e.g. by changing the parameter outdoors to indoors. It would fail for ?MaxRSVP greater than 20. Using such queries, WellnessRules participants can check out profiles of other participants to see if they can join an activity group.

6   Conclusion

The WellnessRules case study demonstrates FOAF-extending Web 3.0 profile interoperation between a pure Prolog subset (Datalog with Naf) and N3 through RuleML/XML. With all of its source documents available, it has become a major use case for exploring various aspects, including scalability, of (distributed) knowledge on the Web (3.0), starting with derivation rules and light-weight ontologies. While WellnessRules so far has probed the OO jDREW, Euler, and Prova engines, its open Rule Responder architecture will make it easy to bring in new engines. A GUI can generate rule profiles, e.g. extending FOAF-a-Matic.[4]

WellnessRules currently emphasizes rulebase translation and querying. These constitute basic services that we intent to extend by superimposed update services, e.g. for changing calendar entries for activities from status possible to planned; this will require production rules. The next extension will be relevant

---

[3] http://ruleml.org/papers/EvalArchiRule.pdf
[4] http://www.ldodds.com/foaf/foaf-a-matic

for **performing** wellness events, which will call for event-condition-action rules. Both of these extended rule types are covered by Reaction RuleML [PKB07].

This case study will also provide challenges for RIF [BK09]: WellnessRules' current derivation rules, for RIF-BLD[5] implementations; its planned production rules, for RIF-PRD[6] updates & implementations; and its envisioned reaction rules, for a possible RIF Reaction Rule Dialect (RIF-RRD).

Acknowledgements

We thank the wellness community at NRC-IIT Fredericton for their advice & enthusiasm. Thanks also go to Jos de Roo for his help with the Euler engine. NSERC is thanked for its support through a Discovery Grant for Harold Boley.

References

[BK09]   Boley, H., Kifer, M.: RIF Basic Logic Dialect. W3C Working Draft (July 2009), http://www.w3.org/2005/rules/wiki/BLD

[Bol07]   Boley, H.: Are Your Rules Online? Four Web Rule Essentials. In: Paschke, A., Biletskiy, Y. (eds.) RuleML 2007. LNCS, vol. 4824, pp. 7–24. Springer, Heidelberg (2007)

[CB08]   Craig, B.L., Boley, H.: Personal Agents in the Rule Responder Architecture. In: Bassiliades, N., Governatori, G., Paschke, A. (eds.) RuleML 2008. LNCS, vol. 5321, pp. 150–165. Springer, Heidelberg (2008)

[dB09]   de Bruijn, J.: RIF RDF and OWL Compatibility. W3C Working Draft (July 2009), http://www.w3.org/2005/rules/wiki/SWC

[LBBM06] Li, J., Boley, H., Bhavsar, V.C., Mei, J.: Expert Finding for eCollaboration Using FOAF with RuleML Rules. In: Montreal Conference of eTechnologies 2006, pp. 53–65 (2006)

[PBKC07] Paschke, A., Boley, H., Kozlenkov, A., Craig, B.: Rule Responder: RuleML-Based Agents for Distributed Collaboration on the Pragmatic Web. In: 2nd ACM Pragmatic Web Conference 2007. ACM, New York (2007)

[PKB07]   Paschke, A., Kozlenkov, A., Boley, H.: A Homogenous Reaction Rule Language for Complex Event Processing. In: Proc. 2nd International Workshop on Event Drive Architecture and Event Processing Systems (EDA-PS 2007), Vienna, Austria (September 2007)

[SLKL07] Sachinopoulou, A., Leppänen, J., Kaijanranta, H., Lähteenmäki, J.: Ontology-Based Approach for Managing Personal Health and Wellness Information. In: Engineering in Medicine and Biology Society (EMBS 2007). 29th Annual Int'l Conference of the IEEE (August 2007)

---

[5] http://www.w3.org/2005/rules/wiki/BLD
[6] http://www.w3.org/2005/rules/wiki/PRD

Rule-Based Event Processing and Reaction Rules

Adrian Paschke[1] and Alexander Kozlenkov[2]

[1] Freie Universitaet Berlin
Institut for Computer Science
AG Corporate Semantic Web
Koenigin-Luise-Str. 24/26, 14195 Berlin, Germany
Paschke@inf.fu-berlin.de
[2] Betfair Ltd. London
alex.kozlenkov@betfair.com

Abstract. Reaction rules and event processing technologies play a key role in making business and IT / Internet infrastructures more agile and active. While event processing is concerned with detecting events from large event clouds or streams in almost real-time, reaction rules are concerned with the invocation of actions in response to events and actionable situations. They state the conditions under which actions must be taken. In the last decades various reaction rule and event processing approaches have been developed, which for the most part have been advanced separately. In this paper we survey reaction rule approaches and rule-based event processing systems and languages.

1 Introduction

Reaction rules and event processing technologies have been investigated comprehensively over the last decades. Different rule-based approaches for reactive event processing have been developed, which have for the most part proceeded separately and have led to different formalisms and languages:

- Production rule systems have been investigated comprehensively in the realms of expert systems since the 1980s and successfully applied commercially. They typically implement a forward-chaining operational semantics for Condition-Action rules where changing conditions trigger update actions.
- Active databases in their attempt to combine techniques from expert systems and databases to support automatic triggering of global rules in response to events and to monitor state changes in database systems have intensively explored and developed the Event-Condition-Action (ECA) paradigm and event algebras to compute complex events and trigger reactions according to global ECA rules.
- In event/action logics, which have their origins in the area of knowledge representation (KR) and logic programming (LP), the focus is on the formalization of action/event axioms and on the inferences that can be made from the happened or planned events/actions. They define a declarative, often model-theoretic, semantics.
- Rule-based Event Processing Languages are aiming at a combination of Complex Event Processing (CEP) for real-time event detection and reaction rules for

G. Governatori, J. Hall, and A. Paschke (Eds.): RuleML 2009, LNCS 5858, pp. 53–66, 2009.

declarative representation and intelligent reaction. They often use event notification and messaging systems, such as Enterprise Service Bus (ESB), to facilitate the communication of events in a distributed environment. Typically, the interest here is in a context-dependent event sequence which follows e.g. a communication protocol or coordination workflow, rather than in single event occurrences which trigger immediate reactions.

- Rule markup languages, such as the quasi-standard RuleML, are the vehicle for using rules on the Web and in other distributed systems. They allow publishing, deploying, executing and communicating rules in a network. They may also play the role of a lingua franca for exchanging rules between different systems and tools.

In this paper we give a survey of these major lines of reaction rule approaches and event processing systems of the past decades.

2 Production Rule Systems and Update Rule Programs

Production rules have become very popular as a widely used technique to implement large expert systems in the 1980s for diverse domains such as troubleshooting in telecommunication networks or computer configuration systems. Classical production rule systems and most database implementations of production rules [21-23] typically have an operational or execution semantics defined for them. There are many forward-chaining implementations and many well-known forward-reasoning engines for production rules such as IBM ILOG's commercial jRules system, Fair Isaac/Blaze Advisor, CA Aion, Oracle Haley, ESI Logist or popular open source solutions such as OPS5, CLIPS or Jess which are based on the RETE algorithm. In a nutshell, this algorithm keeps the derivation structure in memory and propagates changes in the fact and rule base. This algorithm can be very effective, e.g. if you just want to find out what new facts are true or when you have a small set of initial facts and when there tend to be lots of different rules which allow you to draw the same conclusion. There are several optimized successor algorithms based on Rete such as TREAT [87] and LEAPS [88] being two examples and variations on Rete are implemented in many current production rule engines.

Although production rules might simulate derivation rules via asserting a conclusion as consequence of the proved condition, i.e. "if Condition then assert Conclusion", the classical and most commercial production rule languages are less expressive since they lack a clear declarative semantics, suffer from termination and confluence problems of their execution sequences and typically do not support expressive non-monotonic features such as classical or negation-as-finite failure or preferences, which makes it sometimes hard to express certain real life problems in a natural and simple way.

However, several extensions to this core production systems paradigm have been made which introduce e.g. negations (classical and negation-as-finite failure) [24] and provide declarative semantics for certain subclasses of production rules systems such as stratified production rules. It has been shown that such stratified production systems have a declarative semantics defined via their corresponding logic program (LP) into which they can be transformed [25] and that the well-founded, stable or preferred semantics for production rule systems coincide in the class of stratified production systems [24]. Stratification can be implemented on top of classical production rules in from of priority assignments between rules or by means of transformations into the corresponding classical ones. The strict definition of stratification for production rule systems has been further relaxed in [26] which defines an execution semantics for update rule programs based upon a monotonic fixpoint operator and a declarative semantics via transformation of the update program into a normal LP with stable model semantics.

Closely related are also logical update languages such as transaction logics and in particular serial Horn programs, where the serial Horn rule body is a sequential execution of actions in combination with standard Horn pre-/post conditions. [27] These serial rules can be processed top-down or bottom-up and hence are closely related to the production rules style of "condition → update action". Several approaches in the active database domain also draw on transformations of active rules into LP derivation rules, in order to exploit the formal declarative semantics of LPs to overcome confluence and termination problems of active rule execution sequences. [28-30]. The combination of deductive and active rules has been also investigated in different approaches mainly based on the simulation of active rules by means of deductive rules. [17, 18] Moreover, there are approaches which directly build reactive rules on top of LP derivation rules such as the Event-Condition-Action Logic Programming language (ECA-LP) which enables a homogeneous representation of ECA rules and derivation rules [31, 58,79-84].

Although production rules react to condition state changes and have no explicit formalization of events, as e.g. in ECA rules, recent extensions of production rule systems with object model and external fact updates, such as TIBCO's Business Events [74] and Drools [75], are extended to Complex Event Processing (CEP). Explicit event types and classes are defined in the production rule declarations. New event data instances are dynamically added to the fact base / working memory. The instances of event declarations are filtered and joined in the production rule conditions via typed pattern matching and if they pass the condition list the action part of the rule is triggered. In short, the condition part of production rules is used to define event filters and event processing is done via pattern matching.

3 Active Databases and ECA Rule Systems

Active databases [76] are an important research topic due to the fact that they find many applications in real world systems and many commercial databases systems have been extended to allow the user to express active rules whose execution can update the database and trigger the further execution of active rules leading to a cascading sequence of updates (often modeled in terms of execution programs). Several active database systems have been developed, e.g. ACCOOD [1], Chimera [2], ADL [3], COMPOSE [4], NAOS [5], HiPac [6]. These systems mainly treat event detection and processing purely procedural and often focus on specific aspects. In this spirit of procedural ECA formalisms are also systems such as AMIT [7], RuleCore [8] or JEDI [9]. Several papers discuss formal aspects of active databases on a more general level – see e.g. [10] for an overview. Several event algebras have been developed, e.g. Snoop [11], SAMOS [12], ODE [4], where one can create complex nested expressions, using operators like And, Or, Sequence, and others.

The object database ODE [4] implements event-detection mechanism using finite state automata.

Another early active database system is HiPAC [6]. It is an object-oriented database with transaction support. HiPAC can detect events only within a single transaction. Global event detectors are proposed which detect complex events across transaction boundaries and over longer intervals, but no further details are given.

SAMOS [12] combines active and object-oriented features in a single framework using colored Petri nets. Associated with primitive event types are a number of parameter-value pairs by which events of that kind are detected. SAMOS does not allow simultaneous atomic events.

Snoop [11] is an event specification language which defines different restriction policies that can be applied to the operators of the algebra. Complex events are strictly ordered and cannot occur simultaneously. The detection mechanism is based on trees corresponding to the event expressions, where primitive event occurrences are inserted at the leaves and propagated upwards in the trees as they cause more complex events to occur. Snoop has a detection-time-based semantics which only considers the time of event detection. This operational time point semantics poses problems with nested sequences as pointed out in [31, 58, 79, 80] and in [77]. Interval-based semantics for Snoop has been defined in SnoopIB [78].

Sentinel [85] is an active object-oriented database implementing complex event detection for the Snoop operators. Event detection is done in a directed acyclic graph which is constructed from event expressions. Complex events are represented as graph nodes which like to the sub-nodes of their sub-expressions.

An Even Calculus based Interval-based semantics implementing Snoop like event algebra operators was proposed in [31, 58, 79-82].

There has been a lot of research and development concerning knowledge updates in active rules (execution models) in the area of active databases and several techniques based on syntactic (e.g. triggering graphs [13] or activation graphs [14]) and semantics analysis (e.g. [15], [16]) of rules have been proposed to ensure termination of active rules (no cycles between rules) and confluence of update programs (always one unique minimal outcome).

The combination of deductive and active rules has been also investigated in different approaches manly based on the simulation of active rules by means of deductive rules [17-19]. In [86] deductive rules are utilised for creating implicit events. In [81-83] a homogeneous reaction rule language for ECA rules was proposed. This approach presents a declarative logical semantics for ECA rules and combines complex event and action calculus, formalisation of reaction rules in combination with other rule types such as derivation rules, integrity constraints, and transactional knowledge updates.

In contrast to event notification systems and distributed rule-based complex event processing systems, the ECA rules in active databases are typically defined with a global scope and react on internal events of the reactive system such as changes in the active database. However the ECA rule style and event algebra operators have been adopted in reaction rules and rule-based event processing languages (EPLs).

4 Temporal Knowledge Representation Event / Action / Transition Logic Systems

Another language dimension to events and actions which has for the most part proceeded separately has the origin in the area of knowledge representation (KR) and logic programming (LP) with close relations to formalisms of process and transition logics. Here the focus is on the development of axioms to formalize the notions of actions or events and causality, where events are characterized in terms of necessary and sufficient conditions for their occurrences and where events/actions have an effect on the actual knowledge states, i.e. they transit states into other states and initiate / terminate changeable properties called fluents. Instead of detecting the events as they occur as in the active database domain, the KR approach to events/actions focuses on the inferences that can be made from the fact that certain events are known to have occurred or are planned to happen in future. This has led to different views and terminologies on event/action definition and event processing in the temporal event/action logics domain. Reasoning about events, actions and change is a fundamental area of research in AI since the events/actions are pervasive aspects of the world in which agents operate enabling retrospective reasoning but also prespective planning. A huge number of formalisms for deductive but also abductive reasoning about events, actions and change have been developed. The common denominator to all this formalisms and systems is the notion of states a.k.a. fluents [34] which are changed or transit due to occurred or planned events/actions. Among them are the event calculus [35] and variants such as the interval-based Event Calculus [31, 58, 79-82], the situation calculus [36, 37], features and fluents [34], various (temporal) action languages [38-42], fluent calculi [43, 44] and versatile event logics [45]. Most of these formalisms have been developed in relative isolation and the relationships between them have only been partially studied, e.g. between situation calculus and event calculus or temporal action logics (TAL) which has its origins in the features and fluents framework and the event calculus.

Closely related and also based on the notion of (complex) events, actions and states with abstract models for state transitions and parallel execution processes are various process algebras like TCC [46], CSS [47] or CSP [48], (labelled) transition logics (LTL) and (action) computation tree logics (ACTL) [49, 50]. In [89] Calculus of Communicating Systems (CCS) was chosen to formally specify complex actions, and reason about their behavioural aspects. As a part of the Process Algebra family, CCS is suitable for the high-level description of interactions, communications, and synchronizations between a collection of concurrent processes, and hence an appropriate mechanism for reactive systems in general.

Related are also update languages [18, 31, 51-57] and transaction logics [27] which address updates of logic programs where the updates can be considered as actions which transit the initial program (knowledge state/base) to a new extended or reduced state hence leading to a sequence of evolved knowledge states. Many of these update languages also try to provide meaning to such dynamic logic programs (DLPs). Several Datalog extensions such as the LDL language of Naqvi and Krishnamurthy [52] which extends Datalog with update operators including an operational semantics for bulk updates or the family of update language of Abiteboul and Vinau [53] which has a Datalog-style have been proposed. A number of further works on adding a notion of state to Datalog programs where updates modelled as state transitions close to situation calculus has been taken place [18, 51].

The situation calculus [60] is a methodology for specifying the effects of elementary actions in first-order logic. Reiter further extended this calculus with an induction axiom specified in second-order logic for reasoning about action sequences. Applying, this Reiter has developed a logical theory of database evolution [61], where a database state is identified with a sequence of actions. This might be comparable to the Event Calculus, however, unlike the EC the situation calculus has no notion of time which in many event-driven systems is crucial, e.g. to define timestamps, deadline principles, temporal constraints, or time-based context definitions.

Transaction Logics [27] is a general logic of state changes that accounts for database updates and transactions supporting order of update operations, transaction abort and rollback, savepoints and dynamic constraints. It provides a logical language for programming transactions, for updating database views and for specifying active rules, also including means for procedural knowledge and object-oriented method execution with side effects. It is an extension of classical predicate calculus and comes with its own proof theory and model theory. The proof procedure executes logic programs and updates to databases and generates query answers, all as a result of proving theorems. Although it has a rich expressiveness in particular for combining elementary actions into complex transactional ones it primarily deals with database updates and transactions but not with general event processing functionalities in style of ECA rules such as situation detection, context testing in terms of state awareness or external inputs/outputs in terms of event/action notifications or external calls with side effects.

Multi-dimensional dynamic logic programming [54], LUPS [55], EPI [56], Kabul [57] Evolp [55], ECA-LP [32] are all update languages which enable intensional updates to rules and define a declarative and operational semantics of sequences of dynamic LPs. While LUPS and EPI support updates to derivation rules, Evolp and Kabul are concerned with updates of reaction rules.

However, to the best of our knowledge only ECA-LP [31, 58, 79-82] supports a tight homogenous combination of reaction rules and derivation rules enabling transactional updates to both rule types with post-conditional (ECAP) verification, validation, integrity tests (V&V&I) on the evolved knowledge state after the update action(s). Moreover, none of the cited update languages except of ECA-LP supports complex (update) actions and events as well as external events/actions. Moreover these update languages are only concerned on the declarative evolution of a knowledge base with updates, whereas the Event Calculus as used in ECA-RuleML/ECA-LP is developed to reason about the effects of actions and events on properties of the knowledge systems, which hold over an interval.

5   Rule-Based Event Processing Languages and Event Notification Systems

Event processing has many historical roots, as described in the previous sections. In the last years the overarching discipline (Complex) "Event Processing" has been introduce, which summarizes all technologies to achieve actionable, situational knowledge from events in real-time or almost real- time. Here Complex Event Processing

(CEP) and Event Stream Processing (ESP) are concerned with the handling of events. ESP addresses the extraction of events from an event stream with emphasis on efficiency for high throughput and low latency. Processing is done by analyzing the data of the events and selecting appropriate occurrences. CEP, on the other hand, is more focused on complex patterns of events in event clouds with partial temporal order. Detecting complex event patterns (a.k.a. complex event types) and situations (complex event + conditional context), i.e. detecting transitions in the universe (a so called event cloud [90]) that requires reaction either "reactive" or "proactive" is one of the basic ideas of CEP. A complex event pattern describes the detection condition of a complex event which is derived from low level events that occur (real-time view) or have been happened (retrospective view) in specific combinations. This process of event selection, aggregation, hierarching and event abstracting for generating higher level events is called "Complex Event Processing".

Stemming from the operational level of network management and IT infrastructure management (ITIM) event correlation systems / engines such as IBM Tivoli Event Console [91], HP OpenView Event Correlation Services [92], VERITAS NerveCenter [93], SMARTS InCharge [94], and the ongoing work on realizing the Common Event Infrastructure [95], are designed to primarily handle network events and events on the IT infrastructure level.

The event languages from active database systems (see previous section) support event algebra operators to describe complex event types that can be applied over the event clouds and event instance histories in real time. The active mechanisms in these systems serve to detect and react to updates in the database that may jeopardize the data integrity, or to execute some business-related application logic.

A closely related ESP area is sequence databases which extend the database query languages such as SQL with means to search for sequential patterns. Examples are the time series systems, such as the analytic functions in Oracle [96] or the time-series data blades in Informix [97]. Sequences are supported as user-defined types (UDF) and stored in tuple fields. Users are provided with a library of functions that can be invoked from SQL queries. However, the functions provided are rather general and do not support a more complex and dedicated class of queries. Another approach seeks to extend SQL to support sequenced data. Such extensions were first proposed for the PREDATOR system (SEQUIN [98]). SRQL [99] extended the relational algebra with sequence operators for sorted relations, and added constructs to SQL for querying sequences. Then, more powerful extensions for pattern recognition over sorted relations were proposed for SQL-TS [100,101]. It also proposed techniques for sequence queries optimizations. A recent approach extending SQL is the Continuous Query Language (CQL) [102] which is a declarative query language for data streams in Data Stream Management Systems. CQL is developed in the STREAM project at Stanford University. In summary, these works are aimed at finding more general solutions for sequence data. They do not take into account the specifics of the event data.

Various Event Processing Languages (EPLs) for CEP and ESP have been developed in the last years. There are SQL-like EPLs such as CQL [102], Aleri [103], Coral8 CCL [104], Streambase [105], Esper [106]; rule-based EPLs such as Reaction RuleML [108], Prova [107], Tibco Rules [74], Drools [75], XChangeEQ [72], Rule-Core [8], , or the AMIT SMRL [7]; agent oriented EPLs such as Agent Logic [110], EventZero [109], Spade [111] special, scripting languages such as Netcool Impact

[112], Apama [113]; and GUI-based approaches with 3GL (e.g. Java) code respectively EPL code generation. There is ample evidence that there is no single EPL which is best suited for all domains, but that different representation approaches are needed such as rule-based EPLs for describing higher-level conditional business event patterns and pattern-matching SQL-like EPLs for defining patterns of low-level aggregation views according to event types defined as nested queries.

In distributed environments such as the Web with independent agent / service nodes and open or closed domain boundaries event processing is often done using high-level event notification and communication mechanisms based on middleware such as an enterprise service bus (ESB). Systems either communicate events and event queries (e.g. on event streams) in terms of messages using a particular transport protocol/language such as JMS, HTTP, SOAP, JADE etc. according to a predefined or negotiated communication/coordination protocol [32] and or they might subscribe to publishing event notification servers which actively distribute events (push) to the subscribed and listening rule-based agents. Typically the interest here is in the particular and complex event sequence or event processing workflow which possibly follows a certain communication or workflow-like coordination protocol, rather than in single event occurrences which trigger immediate reactions as in the active database trigger or ECA rules. As a result reactive rules which build on (complex) events are typically local to a particular context, e.g. within a particular service/agent/node conversation. That is, the communicated events contribute to the detection of a complex event situation which triggers a local reaction within a context (a conversation waiting for at least three sequential answers or requests). Rule Responder [114] is a rule-based ESB middleware for distributed intelligent rule-based CEP and rule inference agents / services on the Web.

6 Reaction Rule Markup and Interchange Languages

Concerning mark-up languages for reactive rules several proposals for update languages exist, e.g. XPathLog [63], XUpdate [64], XML-RL [65] or an extension to XQuery proposed by Tatarinov et.al. [66]. This has been further extended with general event-driven and active functionalities for processing and reasoning about arbitrary events occurring in the Web and other distributed systems as well as triggering actions. Most of the proposals for an ECA or reactive web language are intended to operate on local XML or RDF databases and are basically simply trigger approaches, e.g. Active XML [67], Active Rules for XML [68], Active XQuery [69] or the ECA language for XML proposed by Bailey et.al [70] as well as RDFTL for RDF [71]. XChange [72] is a high level language for programming reactive behaviour and distributed applications on the Web. It allows propagation of changes on the Web (change) and event-based communication (reactivity) between Web-sites (exchange). It is a pattern-based language closely related to the web query language Xcerpt with an operational semantics for ECA rules of the form: "Event query → Web query → Action". Events are represented as XML instances that can be communicated and queried via XChange event messages. Composite events are supported, using event queries based on the Xcerpt query language [73] extended by operators similar to an event algebra applied in a tree-based approach for event detection.

Reaction RuleML [108, 116] is the quasi-standard for reaction rules. It is a general, practical, compact and user-friendly XML- serialized sublanguage of RuleML (http://ruleml.org) for the family of reaction rules. It incorporates various kinds of production, action, reaction, and KR temporal/event/action logic rules as well as (complex) event/action messages into the native RuleML syntax. **Rule Responder** [114, 117] is a middleware for distributed intelligent rule-based CEP and Pragmatic Web inference agents / services on the Web. It uses Reaction RuleML as standard rule interchange format.

IBM Situation Manager Rule Language (SMRL) [7] is a markup language for describing situations, which are semantic concepts in the customers' domain of discourse and syntactically equivalent to (complex) event patterns. The Situation Manager is part of the Amit (Active Middleware Technology) framework. Events in SMRL have a flat structure, and have a unique name and attributes that can be standard or user defined. The conceptual model defines an event type generalization hierarchy.

ruleCore Markup Language (rCML) is used for specification of events and ECA rules in ruleCore [8]. Like in SMRL, in ruleCore terminology a composite event is called a *situation* and the main focus ruleCore is situation detection. Two different event types are supported in rCML: basic events and composite events.

The W3C Rule Interchange Format (RIF) [115] is an effort, influenced by RuleML, to define a standard Rule Interchange Format for facilitating the exchange of rule sets among different systems and to facilitate the development of intelligent rule-based application for the Semantic Web. For these purposes, RIF Use Cases and Requirements (RIF-UCR) have been developed. The RIF architecture is conceived as a family of languages, called dialects. A RIF dialect is a rule-based language with an XML syntax and a well-defined semantics. So far, the RIF working group has defined the Basic Logic Dialect (RIF-BLD), which semantically corresponds to a Horn rule language with equality. RIF-BLD has a number of syntactic extensions with respect to 'regular' Horn rules, including F-logic-like frames, and a standard system of built-ins drawn from Datatypes and Built-Ins (RIF-DTB). The connection to other W3C Semantic Web languages is established via RDF and OWL Compatibility (RIF-SWC). Moreover, RIF-BLD is a general Web language in that it supports the use of IRIs (Internationalized Resource Identifiers) and XML Schema data types. The RIF Working Group has also defined the Framework for Logic Dialects (RIF-FLD), of which RIF-BLD was shown to be the first instantiation. RIF-FLD uses a uniform notion of terms for both expressions and atoms in a Hilog-like manner. The RIF Core dialect is in the intersection of RIF-BLD and the RIF Production Rule Dialect (RIF-PRD) aligned with OMG's PRR is a dialect addressing RIF production, which will be further supplemented by a RIF Reaction Rules Dialect (RIF-RRD).

7 Conclusion

In this paper we have surveyed several major approaches in event and action processing, namely ECA style reactive rules stemming from active databases, production rules, event/action logics from the KR logic domain, modern rule-based event processing languages combining CEP and rules technology, and proprietary as well as general standardized reaction rule markup languages for rule interchange and serialization.

References

1. Erikson, J.: CEDE: Composite Event Detector in An Active Database, University of Skövde (1993)
2. Meo, R., Psaila, G., Ceri, S.: Composite Events in Chimera. In: EDBT, Avingnon, France (1996)
3. Behrends, H.: A description of Event Based Activities in Database Related Information Systems, Report 3/1995, Univ. of Oldenburg (1995)
4. Gehani, N., Jagadish, H.V., Shmueli, O.: Event specification in an active object-oriented database. In: Int. Conf. on Management of Data, San Diego (1992)
5. Collet, C., Coupaye, T.: Composite Events in NAOS. In: Dexa, Zürich, Switzerland (1996)
6. Dayal, U., Buchmann, A., Chakravarty, S.: The HiPAC Project. In: Widom, J., Ceri, S. (eds.) Active Database Systems. Morgan Kaufmann, San Francisco (1996)
7. Adi, A., Opher, E.: Amit - the situation manager. VLDB Journal 13(2) (2004)
8. RuleCore (2006), http://www.rulecore.com
9. Cugola, G., Nitto, E.D., Fuggeta, A.: Exploiting an event-based infrastructure to develop complex distributed systems. In: Int. Conf. on Software Engineering (1998)
10. Paton, N., et al.: Formal Specification of Active Database Functionality: A Survey. In: Sellis, T.K. (ed.) RIDS 1995. LNCS, vol. 985, Springer, Heidelberg (1995)
11. Chakravarthy, S., et al.: Composite Events for Active Databases: Semantics Contexts and Detection. In: VLDB 1994 (1994)
12. Gatziu, S., Dittrich, K.: Event in an active object-oriented database system. In: Int. Conf. on Rules in Database Systems, Edinburgh (1993)
13. Aiken, A., Widom, J., Hellerstein, J.M.: Behaviour of database production rules: termination, confluence and observable determinism. In: Int. Conf. on Management of Data. ACM, New York (1994)
14. Baralis, E., Widom, J.: An algebraic approach to rule analysis by means of triggering and activation graphs. In: VLDB 1994 (1994)
15. Baley, J., et al.: Abstract interpretation of active rules and its use in termination analysis. In: Int. Conf. on Database Theory (1997)
16. Widom, J.: A denotational semantics for starbust production rule language. SIGMOD record 21(3), 4–9 (1992)
17. Lausen, G., Ludascher, B., May, W.: On Logical Foundations of Active Databases. Logics for Databases and Information Systems, 389–422 (1998)
18. Zaniolo, C.: A unified semantics for Active Databases. In: Int. Workshop on Rules in Database Systems, Edinburgh, U.K (1993)
19. Dietrich, J., et al.: Rule-Based Agents for the Semantic Web. Journal on Electronic Commerce Research Applications (2003)
20. Hayes-Roth, F.: Rule based systems. ACM Computing Surveys 28(9) (1985)
21. Declambre, L.M.L., Etheredge, J.N.: A self-controlling interpreter for the relational production language. In: ACM SIGMOD Int. Conf. on the Management of Data (1988)

22. Sellis, T., Lin, C.C., Raschid, L.: Coupling production systems and database systems. In: ACM SIGMOND Int. Conf. on the Management of Data (1993)
23. Widom, J., Finkelstein, S.J.: Set-oriented production rules in relational database systems. In: ACM SIGMOND Int. Conf. on the Management of Data (1990)
24. Dung, P.M., Mancaralle, P.: Production Systems with Negation as Failure. IEEE Transactions on Knowledge and Data Engineering 14(2) (2002)
25. Raschid, L.: A semantics for a class of stratified production system programs. Univ. of Maryland Institute for Advanced Computer Studies-UMIACS-TR-91-114.1: College Park, MD, USA (1992)
26. Raschid, L., Lobo, J.: Semantics for Update Rule Programs and Implementation in a Relational Database Management System. ACM Transactions on Database Systems 22(4), 526–571 (1996)
27. Bonner, A.J., Kifer, M.: Transaction logic programming (or a logic of declarative and procedural knowledge). University of Toronto (1995)
28. Baral, C., Lobo, J.: Formal characterization of active databases. In: Int. Workshop on Logic in Databases (1996)
29. Zaniolo, C.: Active Database Rules with Transaction-Conscious Stable-Model Semantics. In: Int. Conf. on Deductive and Object-Oriented Databases (1995)
30. Flesca, S., Greco, S.: Declarative semantics for active rules. In: Quirchmayr, G., Bench-Capon, T.J.M., Schweighofer, E. (eds.) DEXA 1998. LNCS, vol. 1460, p. 871. Springer, Heidelberg (1998)
31. Paschke, A.: ECA-LP: A Homogeneous Event-Condition-Action Logic Programming Language, Internet-based Information Systems, Technical University Munich (November 2005), http://ibis.in.tum.de/research/projects/rbsla
32. Paschke, A., Kiss, C., Al-Hunaty, S.: NPL: Negotiation Pattern Language - A Design Pattern Language for Decentralized Coordination and Negotiation Protocols. In: e-Negotiations. ICFAI University Press, New Deli (2006), http://ibis.in.tum.de/research/rbsla/docs/ICFAI_Chapter_NPL_final.pdf
33. Ogle, D., et al.: The Common Base Event, IBM (2003)
34. Sandewall, E.: Combining Logic and Differential Equations for Describing Real World Systems. In: KR 1989. Morgan Kaufmann, San Francisco (1989)
35. Kowalski, R.A., Sergot, M.J.: A logic-based calculus of events. New Generation Computing 4, 67–95 (1986)
36. Hayes, P., McCarthy, J.: Some philosophical problems from the standpoint of artificial intelligence. In: Meltzer, B., Michie, D. (eds.) Machine Intelligence 4. Edinburgh University Press, Edinburgh (1969)
37. Reiter, R.: Knowledge in Action: Logical Foundations for Specifying and Implementing Dynamic Systems. MIT Press, Camebridge (2001)
38. Gelfond, M., Lifschitz, V.: Representing action and change by logic programs. Journal of Logic Programming 17(2-4), 301–321 (1993)
39. Fikes, R.E., Nilsson, N.J.: STRIPS: A new approach to the application of theorem proving to problem solving. Artificial Intelligence, 1971(2), 189–208
40. Giunchiglia, E., Lifschitz, V.: An action language based on causal explanation: Preliminary report. In: Conf. on Innovative Applications of Artificial Intelligence. AAAI Press, Menlo Park (1998)
41. Giunchiglia, E., Lifschitz, V.: Action languages, temporal action logics and the situation calculus. Linköping Electronic Articles in Computer and Information Science 4(040) (1999)

42. Doherty, P., et al.: TAL: Temporal Action Logics language specification and tutorial. Linköping Electronic Articles in Computer and Information Science 3(015) (1998)
43. Hölldobler, S., Schneeberger, J.: A new deductive approach to planning. New Generation Computing 8(3), 225–244 (1990)
44. Thielscher, M.: From situation calculus to fluent calculus: State update axioms as a solution to the inferential frame problem. Artificial Intelligence 111, 277–299 (1999)
45. Bennett, B., Galton, A.P.: A unifying semantics for time and events. Artificial Intelligence 153(1-2), 13–48 (2004)
46. Saraswat, V., Jagadeesan, R., Gupta, V.: Timed default concurrency constraint programming. Journal of Symbolic Computation 22(5/6) (1996)
47. Milner, R.: Communication and Concurrency. Prentice-Hall, Englewood Cliffs (1989)
48. Hoare, C.A.R.: Communication and Concurrency. Prentice-Hall, Englewood Cliffs (1985)
49. Meolic, R., Kapus, T., Brezonic, Z.: Verification of concurent systems using ACTL. In: IASTED Int. Conf. on Applied Informatics (AI 2000), Anaheim, Calgary. IASTED/ACTA Press (2000)
50. Meolic, R., Kapus, T., Brezonic, Z.: An Action Computation Tree Logic With Unless Operator. In: Proc. of South-East European Workshop on Formal Methods (SEEFM 2003), Thessaloniki, Greece (2003)
51. Ludäscher, B., Hamann, U., Lausen, G.: A logical framework for active rules. In: Int. Conf. on Management of Data, Pune, India (1995)
52. Naqvi, S., Krishnamurthy, R.: Database updates in logic programming. In: ACM Symposium on Principles of Database Systems. ACM, New York (1988)
53. Abiteboul, S., Vianu, V.: Datalog extensions for database queries and updates. Journal of Computer and System Science 43, 62–124 (1991)
54. Leite, J.A., Alferes, J.J., Moniz Pereira, L.: On the use of multi-dimensional dynamic logic programming to represent societal agents' viewpoints. In: Brazdil, P.B., Jorge, A.M. (eds.) EPIA 2001. LNCS (LNAI), vol. 2258, p. 276. Springer, Heidelberg (2001)
55. Alferes, J.J., Brogi, A., Leite, J., Moniz Pereira, L.: Evolving logic programs. In: Flesca, S., Greco, S., Leone, N., Ianni, G. (eds.) JELIA 2002. LNCS (LNAI), vol. 2424, p. 50. Springer, Heidelberg (2002)
56. Eiter, T., et al.: A framework for declarative update specification in logic programs. In: IJCAI (2001)
57. Leite, J.A.: Evolving Knowledge Bases. Frontiers in Artificial Intelligence and Applications 81 (2003)
58. Paschke, A.: ECA-RuleML: An Approach combining ECA Rules with temporal interval-based KR Event/Action Logics and Transactional Update Logics, IBIS, Technische Universität München, Technical Report, 11/2005 (2005)
59. Ludascher, B.: Integration of Active and Deductive Database Rules, Phd thesis, in Institut für Informatik, Universität Freiburg, Germany (1998)
60. McCarthy, J., Hayes, P.: Some philosophical problems from the standpoint of artificial intelligence. Machine Intelligence 4, 463–502 (1969)
61. Reiter, R.: On specifying database updates. Journal of Logic Programming 25(1), 53–91 (1995)
62. Paschke, A., Bichler, M.: SLA Representation, Management and Enforcement - Combining Event Calculus, Deontic Logic, Horn Logic and Event Condition Action Rules. In: EEE 2005, Hong Kong, China (2005)
63. May, W.: XPath-Logic and XPathLog: A logic-programming style XML data manipulation language. Theory and Practice of Logic Programming 4(3) (2004)

64. Initiative, X.D.: XUpdate - XML Update Language (2000),
    http://www.xmldb.org/xupdate/
65. Liu, M., Lu, L., Wang, G.: A Declarative XML-RL Update Language. In: Song, I.-Y.,
    Liddle, S.W., Ling, T.-W., Scheuermann, P. (eds.) ER 2003. LNCS, vol. 2813, pp. 506–
    519. Springer, Heidelberg (2003)
66. Tatarinov, I., et al.: Updating XML. ACM SIGMOD, 133–154 (2001)
67. Abiteboul, S., et al.: Active XML: Peer-to-Peer Data and Web Services Integration. In:
    VLDB (2002)
68. Bonifati, A., Ceri, S., Paraboschi, S.: Pushing Reactive Services to XML Repositories
    Using Active Rules. In: WWW 2001 (2001)
69. Bonifati, A., et al.: Active XQuery. In: Int. Conf. on Data Engineering, ICDE (2002)
70. Bailey, J., Poulovassilis, A., Wood, P.T.: An Event-Condition-Action Language for
    XML. In: WWW 2002 (2002)
71. Papamarkos, G., Poulovassilism, A., Wood, P.T.: RDFTL: An Event-Condition-Action
    Rule Language for RDF. In: Hellenic Data Management Symposium (HDMS 2004)
72. Bry, F., Patranjan, P.L.: Reactivity on the Web: Paradigms and Applications of the Lan-
    guage XChange. In: ACM Symp. Applied Computing (2005)
73. Schaert, S.: A Rule-Based Query and Transformation Language for the Web, Phd thesis,
    in Institute for Informatics, University of Munich (2004)
74. TIBCO Business Events 3.0, http://www.tibco.de/software/
    complex-event-processing/businessevents/ (accessed August 2009)
75. jBoss Drools, http://jboss.org/drools/ (accessed August 2009)
76. Paton, N.W., D__az, O.: Active database systems. In: ACM Comput. Surv. ACM Press,
    New York (1989)
77. Galton, A., Augusto, J.C.: Two approaches to event definition. In: Hameurlain, A.,
    Cicchetti, R., Traunmüller, R. (eds.) DEXA 2002. LNCS, vol. 2453, pp. 547–556.
    Springer, Heidelberg (2002)
78. Adaikkalavan, R., Chakravarthy, S.: Snoopib: Interval-based event specification and de-
    tection for active databases. Data Knowl. Eng. 59(1), 139–165 (2006)
79. Paschke, A.: Eca-lp / eca-ruleml: A homogeneous event-condition-action logic pro-
    gramming language. CoRR, abs/cs/0609143 (2006)
80. Paschke, A.: Eca-ruleml: An approach combining eca rules with temporal interval-based
    kr event/action logics and transactional update logics. CoRR, abs/cs/0610167 (2006)
81. Paschke, A.: A homogenous reaction rule language for complex event processing. In:
    Proc. 2nd International Workshop on Event Drive Architecture and Event Processing
    Systems, EDA-PS (2007)
82. Paschke, A.: Eca-lp / eca-ruleml: A homogeneous event-condition-action logic pro-
    gramming language. In: RuleML 2006, Athens, Georgia, USA (2006)
83. Paschke, A.: Rule-Based Service Level Agreements - Knowledge Representation for
    Automated e-Contract, SLA and Policy Management. In: IDEA, Munich (2007)
84. Paschke, A., Bichler, M.: Knowledge representation concepts for automated sla man-
    agement. Decis. Support Syst. 46(1), 187–205 (2008)
85. Chakravarthy, S.: Sentinel: An object-oriented dbms with event-based rules. In:
    Peckham, J. (ed.) SIGMOD 1997: Proceedings of the 1997 ACM SIGMOD international
    conference on Management of data, pp. 572–575. ACM Press, New York (1997)
86. Bry, F., Eckert, M.: Towards formal foundations of event queries and rules. In: Second
    Int. Workshop on Event-Driven Architecture, Processing and Systems EDA-PS (2007)
87. Miranker, D.P.: TREAT: a new and e_cient match algorithm for AI production systems.
    Morgan Kaufmann Publishers Inc., San Francisco (1990)

88. Batory, D.: The leaps algorithms. Technical report, Austin, TX, USA (1994)
89. Behrends, E., Fritzen, O., May, W., Schenk, F.: Combining eca rules with process algebras for the semantic web. In: RuleML (2006)
90. Luckham, D.C.: The Power of Events: An Introduction to Complex Event Processing in Distributed Enterprise Systems. Addison-Wesley, Reading (2002)
91. IBM Tivoli Enterprise Console - documentation (2001)
92. Sheers, K.: HP OpenView event correlation services. Hewlett Packard Journal 47.5, 31–42 (1996)
93. VERITAS NerveCentertm VERITAS Software (1999), http://eval.veritas.com/Webfiles/docs/NCOverview.pdf
94. Yemini, S.A., Kliger, S., Mozes, E., Yemini, Y., Ohsie, D.: High speed and robust event correlation. IEEE Communications Magazine 34.5, 82–90 (1996)
95. Common Base Event Infrastructure, IBM (2003)
96. Oracle8 Time Series Data Cartridge, White Paper (February 1998)
97. IBM Informix Time Series Data Blade Module, User Guide, V.4.0 (2001)
98. Shadri, P., Livny, M., Ramakrishnan, R.: Sequence Query Processing. In: SIGMOD, pp. 430–441 (1994)
99. Ramakrishnan, R., et al.: SQRL: Sorted Relational Query Language. In: SSDBM 1998, pp. 84–95 (1998)
100. Sadri, R., Zaniolo, C., Zarkesh, A., Adibi, J.: Optimization of Sequence Queries in Database Systems (2001)
101. Sadri, R.: Optimization of Sequence Queries in Database Systems., PhD Thesis, UCLA (2001)
102. Motwani, R., Widom, J., Arasu, A., Babcock, B., Babu, S., Datar, M., Manku, G., Olston, C., Rosenstein, J., Varma, R.: Query Processing, Resource Management, and Approximation in a Data Stream Management System, Stanford (2002)
103. Aleri, http://www.aleri.com/ (accessed August 2009)
104. Coral 8, http://www.coral8.com/ (accessed August 2009)
105. Streambase, http://www.streambase.com/ (accessed August 2009)
106. Esper, http://www.espertech.com/ (accessed August 2009 -08-25)
107. Prova, http://prova.ws (accessed August 2009)
108. Reaction RuleML, http://reaction.ruleml.org (accessed August 2009)
109. EventZero, http://www.eventzero.com/ (accessed August 2009)
110. AgentLogic, http://www.agentlogic.com/ (accessed August 2009)
111. Hirzel, M., Andrade, H., Gedik, B., Kumar, V., Losa, G., Soulé, R., Wu, K.-L.: SPADE Language Specification, Published In: IBM Technical Report RC24760 in (2009)
112. IBM Netcool, http://www-01.ibm.com/software/tivoli/products/netcool-impact/ (accessed August 2009)
113. Apama, http://web.progress.com/apama/index.html (accessed 2009)
114. RuleResponde, http://responder.ruleml.org (accessed 2009)
115. W3C RIF, http://www.w3.org/2005/rules/wg (accessed 2009)
116. Paschke, A., Boley, H.: Rules Capturing Event and Reactivity. Handbook of Research on Emerging Rule-Based Languages and Technologies: Open Solutions and Approaches (March 2009)
117. Braubach, L., Pokhar, A., Paschke, A.: Rule-Based Concepts as Foundation for Higher-Level Agent Architectures. Handbook of Research on Emerging Rule-Based Languages and Technologies: Open Solutions and Approaches (March 2009)

Correlating Business Events for Event-Triggered Rules

Josef Schiefer, Hannes Obweger, and Martin Suntinger

UC4 Senactive Software GmbH, Prinz-Eugen-Strasse 72,
1040 Vienna, Austria
{josef.schiefer,hannes.obweger,martin.suntinger}@uc4.com

Abstract. Event processing rules may be prescribed in many different ways, including by finite state machines, graphical methods, ECA (event-condition-action) rules or reactive rules that are triggered by event patterns. In this paper, we present a model for defining event relationships for event processing rules. We propose a so-called correlation set allowing users to graphically model the event correlation aspects of a rule. We illustrate our approach with the event-based system SARI (Sense and Respond Infrastructure) which uses correlation sets as part of rule definitions for the discovery of event patterns. We have fully implemented the proposed approach and compare it with alternative correlation approaches.

Keywords: Rule Management, Event Correlation, Event Processing.

1 Introduction

Operational business systems, such as enterprise resource planning and process management systems are able to report state changes within a business environment as business events. The state changes might occur with the execution of business operations or the completion of customer requests. These events can be used as independent triggers for activities as well as for reports or mining purposes using systems with an event-driven architecture (EDA). Event-based systems integrate a wide range of components into a loosely-coupled distributed system with event producers which can be application components, post-commit triggers in a database, sensors, or system monitors, and event consumers such as application components, device controllers, databases, or workflow queues. Thus, event-based systems are seeing increasingly widespread use in applications ranging from time-critical systems, system management and control, to e-commerce.

The process of defining a relationship between events and implementing actions to deal with the related events is known as event correlation. Event correlation in the business domain is a technique for collecting and isolating related data from various (potentially lower level) events in order to condense many events each holding a fraction of information into a meaningful business incident. The correlated information can then be used for discovering business patterns, such as business opportunities or exceptional situations. In many cases, this "aggregated" information can be published again as an event to the messaging middleware. Hence, series of enrichment steps are possible triggering activities on various levels of abstraction.

G. Governatori, J. Hall, and A. Paschke (Eds.): RuleML 2009, LNCS 5858, pp. 67–81, 2009.

Many existing SQL-based approaches use joins for correlating events [1][3][5]. Thereby, events are processed as tuples which are joined when correlating them similar to the way it is done in relational databases. Although many people in the data management community are familiar with this approach, it has several drawbacks for event processing. Relational databases use foreign-key relationships and indices for creating optimal query execution plans. As such relationships are not available between event objects, it is difficult to build effective query execution plans for event streams. Within this paper, we show a correlation approach, which allows to model relationships between events for correlating them already during the event processing. Existing event-based systems do not allow the user to reuse correlations for defining rules in event-driven applications. With our approach, we try to fill the gap that existing event-based systems do not capture correlation information as separate meta-data and, thereby, loose many advantages for optimizing the rule processing. Event correlation is challenging due to the following reasons:

- Correlated business events can be processed by multiple rules for evaluating conditions, calculating metrics or discovering event patterns.
- Correlated events can occur at different points in time, requiring a temporary storage of correlated event data that is transparent for rules.
- Business events occur in various source systems and different formats. Nonetheless, they must be captured and correlated with minimal latency and minimal operational system impact.
- The event correlation should be independent from the execution systems or protocols.
- Late arrival of events due to network failures or downtimes of operational systems within heterogeneous and distributed software environments are common situations that have to be considered for the event correlation and for the rule processing.
- Only relevant event data for applications and users must be unified, transformed, and cleansed before they are correlated. In many situations, only a small set of selected event attributes are needed for the correlation.

The presented approach is generic and thus applicable in various application domains. Throughout the paper we discuss different application examples based on real-world business scenarios.

The remainder of this paper is organized as follows: Section 2 discusses related work. In Section 3, we discuss event-triggered rules and show how they are used in event-based system. Section 4 discusses correlation sets and correlation bands for modeling event relationships. How correlation sets and correlation bands are managed in runtime is discussed in Section 5. We furthermore present a real-world application from the logistics domain in Section 6. Finally, in Section 7, we conclude this paper and provide an outlook to future research.

2 Related Work and Contribution

The key characteristic of an event-based system is its capability of handling complex event situations, detecting patterns, aggregating events and making use of time windows for collecting event data over a period of time. Event correlation plays a crucial role for performing these tasks. In the following, we provide an overview of the correlation capabilities of event processing engines.

Esper [5] is an Open Source event stream processing solution for analyzing event streams. It has a lightweight processing engine and is currently available under GPL licence. Esper supports conditional triggers on event patterns and SQL queries for event streams. Event correlation can be performed by joining the attributes of events.

Borealis and Aurora [1] are further examples of stream processing engines for SQL-based queries over streaming data with efficient scheduling and QoS delivery mechanisms. Medusa [17] focuses on extending Aurora's stream processing engine to distribute the event processing. Borealis extends Aurora's event stream processing engines with dynamic query modification and revision capabilities and makes use of Medusa's distributed extensions. All of these stream engines use joins for correlating events.

RuleCore [12][14] is an event-driven rule processing engine supporting Event Condition Action (ECA) rules and providing a user interface for rule building and composite event definitions. In RuleCore, event correlation settings are embedded in the rule definition, which are used by the rule engine during the event processing. Chen et al. [4] show an approach for rule-based event correlation. In their approach, they correlate and adapt complex/structural XML events corresponding to an XML schema. The authors describe an approach for translating hierarchical structured events into an event model which uses name-value pairs for storing event attributes.

AMIT [2] is an event stream engine whose goal is to provide high-performance situation detection mechanisms. AMIT offers a sophisticated user interface for modelling business situations based on the following four types of entities: events, situations, lifespans and keys. In AMIT, lifespans allow the definition of time intervals wherein specific patterns of correlated events can be detected.

Detecting and handling exceptional events also plays a central role in network management [6]. Alarms indicate exceptional states or behaviors, for example, component failures, congestion, errors, or intrusion attempts. Often, a single problem will be manifested through a large number of alarms. These alarms must be correlated to pinpoint their causes so that problems can be addressed effectively. Many existing approaches for correlating events have been developed from network management. Event correlation tools help to condense many events, which individually hold little information, to a few meaningful composite events.

Rule-based analysis is a traditional approach to event correlation with rules in the "conclusion if condition" form which are used to match incoming events often via an inference engine. Based on the results of each test and the combination of events in the system, the rule-processing engine analyzes data until it reaches a final state [16].

Another group of approaches incorporate an explicit representation of the structure and function of the system being diagnosed, providing information about dependencies of components in the network [8] or about cause-effect relationships between network events. The fault discovery process explores the network model to verify correlations between events. NetFACT [7] uses an object-oriented model to describe the connectivity, dependency and containment relationships among network elements. Events are correlated based on these relationships. Nygate [11] models the cause-effect relationships among events with correlation tree skeletons that are used for the correlation.

In spite of intensive research in the past years for correlating events in the event and network management domain, there is, to our knowledge, no existing approach that allows to model relationships between business events for correlation purposes in order to support functions for the event processing. In many event-based systems, the relationship information of events is buried in query statements or programming code.

This paper is an attempt to show an approach which allows to externalize event relationship information in order to utilize this information for later reuse. We believe that information on how events are correlated is fundamental for processing and analyzing events and is, in current event-based systems, not appropriately managed. We propose a correlation model for defining relationships between events which can be used by event-based system for rule processing tasks which require correlated events. With our approach, event-based systems are able to capture correlation concerns with a separate model without increasing the complexity of the rule model. The captured correlation information is then used the processing of various rules.

3 Event-Triggered Rules

Event-triggered rules prescribe actions to be taken whenever an instance of a given event pattern is detected. While the typical users of event stream queries are developers, event-triggered rules try to describe event patterns on a more abstract level and link event-patterns with business actions. Event-triggered rules require event correlation for the detection of event patterns over a time period as well as for tracing the actions triggered by rules [13].

In an event-based system, information about business activities is encapsulated in events, which capture attributes about the context when the event occurred. Event attributes are items such as the agents, resources, and data associated with an event. For example, in an online betting scenario, a typical bet placement event could have the following attributes as context information: account ID, amount, market, league, sport event ID, odds, and bet type. For the rule processing, it is important to correlate temporally and semantically related events. Event attributes, i.e., data from the context of an event, can be used to define such relationships.

The rule engine uses the correlated events for discovering event patterns. When event patterns are matched, the rule engine instantly responds to the source systems. Fig. 1 illustrates the system architecture and essential data processing flows from the source system to the event-based system and vice-versa. The event-based system receives events on notable occurrences such as bet placements via unified interfaces to the betting platform and legacy systems. It can evaluate the received events and, if necessary, respond in real time by carrying out automated decisions [6]. The communication between the source systems and the event-based system is always asynchronous, avoiding a tight coupling between the systems.

Transactional data, such as a bet placement, is propagated and continuously integrated as an event stream. The event-based system correlates the events of the event stream and continuously calculates metrics and scores. Finally, the rule engine of event-based systems applies ECA rules on the correlated event sequences and calculated metrics and scores.

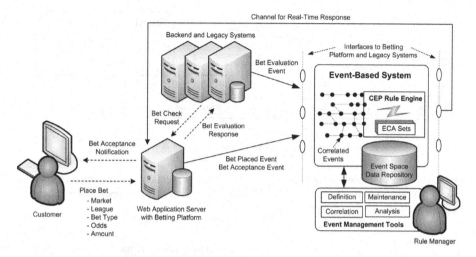

Fig. 1. Event-Driven Rule Engine for Monitoring a Betting Platform

In the following, we show a model for defining event relationships for event-triggered rules with so-called *correlation sets*. The rules use correlation sets for defining correlation concerns in separate model. During runtime the rule engine uses the rule and correlation model for the pattern detection.

4 Correlation Sets

An activity spanning a period of time is represented by the interval between two or more events. For example, a transport might have a TransportStart and TransportEnd event pair. Similarly, a shipment could be represented by the events ShipmentCreated, ShipmentShipped and multiple transport event-pairs. For purposes of maintaining information about business activities, events capture attributes about the context when the event occurred. Event attributes are items such as the agents, resources, and data associated with an event, the tangible result of an action (e.g., the result of a transport decision), or any other information that gives character to the specific occurrence of that type of event. Elements of an event context can be used to define a relationship with other events. We use *correlation sets* for modeling these event relationships between events of business activities.

In SARI, correlations between events are declaratively defined in a correlation model (correlation sets), which is used by the correlation engine during runtime. Correlation sets are able to define relationships based on matching methods for correlating attributes among event types. Correlation sets have a set of *correlation bands* which define a sequence of events which use the same matching approach for the event correlation. For instance, if we want to correlate order events with transportation events (of the same order), we might have one band defining the correlation between order events (e.g. OrderCreated, OrderShipped, OrderFulfilled) and a second band defining the correlation between transport events (e.g. TransportStart, TransportEnd).

SARI supports various types of correlation bands, such as elementary correlation, knowledge-based correlation, self-referencing correlation, language-specific correlation

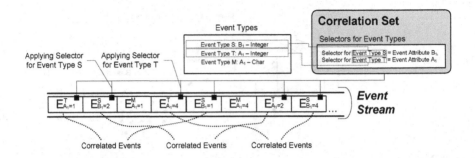

Fig. 2. Illustrates the application of a method of correlating events in a stream of events where E represents an event. Each event E has an identifier T, S, M etc. for the type of its event and at least one attribute A_1, B_1, etc. of this event.

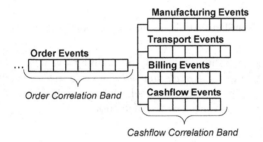

Fig. 3. Shows a correlation set with multiple correlation bands correlating events from different sources or domains. The idea of correlation bands is to capture semantically closely related events in a separate correlation arrangement. The correlation set is able to link multiple bands, thereby correlating the events of *all* its bands.

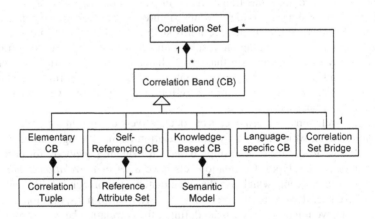

Fig. 4. Correlation Meta Model

and bridges for correlation sets. Fig. 4 shows the meta model for correlation sets. An elementary correlation band (CB) uses correlation tuples for defining relationships between event types based on matching event attributes. Knowledge-based correlation bands use a semantic model for defining the event relationships. Self-referencing correlation bands associate events based on reference information held by events (in other words, each event "knows" about its correlated events). Language-specific correlation bands use programming or query languages for defining event relationships. Correlation set bridges are a special kind of correlation bands which function as proxies for chaining multiple correlation sets. In other words, a correlation set bridge establishes a link between stand-alone correlation sets for correlating them. In the following sections, we define the various types of correlation bands in more detail.

4.1 Elementary Correlation Band

Elementary correlation bands define a direct relationship between events based upon matching event attributes. An event stream consists of events which conform to event types $T = \{t_1, t_2, \dots, t_n\}$. Each event type t_i in T defines a set of event attributes A_i. A relationship between events is defined by associating an attribute $a_j \in A_j$ of an event type t_j with attributes of one or more other events. Formally, an elementary correlation set is defined as follows:

Elementary Correlation Band $P = (T, \mathcal{A}, \mathcal{C})$ where:

T: a set $\{t_1, t_2, \dots, t_n\}$ of event types of an event stream.

$\mathcal{A}$:   a set $\{A_1, A_2, \dots, A_n\}$ of sets of attributes as defined by the event types in T. Each event type has its own set of attributes.

$\mathcal{C}$:   a set $\{C_1, C_2, \dots, C_m\}$ of correlation tuples $C_i = \left(a_{f_k}, a_{g_l}, \dots\right)$, $a_{i_j} \in A_i$, which define relationships between one or more event types from T by associating their attributes.

Fig. 5 shows a relationship between TransportStart and TransportEnd events. The relationship is defined by associated attributes (TransportId) of the two event types. A relationship may include one or more correlating attributes which are part of correlation tuples. In other words, a correlation tuple defines attributes from different event types which have to match in order to correlate. A correlation set may include one or more event types which define relationships with correlation tuples.

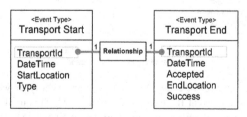

Correlation Tuple: ●——● TransportId

Fig. 5. Elementary Correlation Band Example

4.2 Self-referencing Correlation Band

When using self-referencing correlation bands, relationships between events are defined by the correlated events themselves. In self-referencing correlations, each event is able to create a link to other events by holding a unique identifier of each related event. For elementary correlation bands, we used attributes of the event context for modeling event relationships. Thereby, a user has to know which event attributes are appropriate for defining the relationships. Self-referencing correlation bands do not make this assumption and allow events to directly reference other events. Formally, a self-referencing correlation set is defined as follows:

Self-Referencing Correlation Band $S = (T, \mathcal{A}, a, R)$ where:

T: a set of event types of an event stream.

$\mathcal{A}$: a set $\{A_1, A_2, ..., A_n\}$ of sets of attributes as defined by the event types in T. Each event type has its own set of attributes.

a: an attribute that uniquely identifies an event. This attribute must be shared among all event types, i.e., $a \in A_i \ \forall \ i = 1, ..., n$.

R: a set of attributes $\{r_1, r_2, ..., r_m\}$, $r_i \in \bigcup_{j=1}^{n} A_i$, that is used to define relationships between events by having the value of another event's a-attribute.

In the following example, we extend our previous elementary correlation band of transport events with the corresponding shipment events ShipmentCreated and ShipmentShipped. For the correlation of the two shipment events, each ShipmentCreated event directly references a ShipmentShipped event using a unique event identifier.

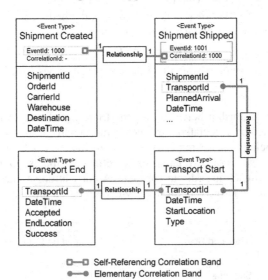

Fig. 6. shows a self-referencing correlation band. It includes event types with additional header attributes for a global unique event ID and a correlation ID which is used to directly reference other events. The advantage of self-referencing correlation sets is that relationships between events can be defined independently of the event context (which is given by the event attributes). Nevertheless, a consequence from wiring events directly is that the event producer has to track any generated events in order to be able to set the correct references for the event objects.

4.3 Knowledge-Based Correlation Band

A knowledge-based correlation set is a special case of an elementary correlation set. It defines an inferred relationship between events and associates a set of events by using semantic knowledge for correlating them. In other words, instead of directly associating a set of events by defining their relationships based solely upon the event attributes, knowledge-based correlation sets use algorithms which allow incorporating external knowledge for correlating events. These algorithms can use external databases, semantic networks or knowledge maps for inferring a relationship between events. Formally, a knowledge-based correlation set is defined as follows:

Knowledge-Based Correlation Band $K = (T, \mathcal{A}, k)$ where:

T: a set of event types of an event stream.

$\mathcal{A}$:   a set $\{A_1, A_2, ..., A_n\}$ of sets of attributes as defined by the event types in T. Each event type has its own set of attributes.

k: a knowledge base which uses sets of attributes from $\mathcal{A}$ to define relationships between events by inferring that two or more events are correlating.

The following example (see Fig. 7) shows two event types with an inferred relationship. We assume that we want to correlate the events for job openings and job applications which have a matching job description and objectives. Since the job description of a company and the job objectives of an application are usually captured as full-text, an algorithm has to decide based upon a semantic model whether a job description and the job objectives of an application are matching.

An approach for semantic event correlation was presented by Moser et al. [10]. They present three types of semantic correlation: 1) basic semantic correlation, 2) inherited semantic correlation, and 3) relation-based semantic correlation. For the first two kinds of semantic correlations, ontologies are used to find matching terms that share the same (inherited) meaning. Relation-based semantic correlations use modeled relations defined in ontologies for the correlation process.

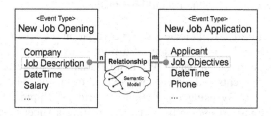

Fig. 7. Knowledge-Based Correlation Band

4.4 Language-Specific Correlation Band

Language-Based correlation bands use a language for defining the correlation. Many existing event-based systems use this type of correlation for querying event streams [1][5]. SQL-based query languages have become popular in recent years, which use join operations in order achieve event correlations. When using language-based correlation bands, only the correlation capabilities of the language are being used. In other words, language-based correlation bands use a language to specify relationships between events. The relationship information is embedded within the language and only the correlation or query engine is able to interpret the language to perform the correlation.

When using SQL-based query languages, the correlation band would use join operations for defining relationships between events. The language must be able to accept a set of input event streams and must generate an output stream with the correlated event data. The following example shows the correlation of customer and supplier events which occur within a certain time window [5].

```
SELECT  A.transactionId, B.customerId,
        A.supplierId,
        B.timestamp - A.timestamp,
FROM    TxnEventA.win:time(30 minutes) A,
        TxnEventB.win:time(30 minutes) B
WHERE   A.transactionId = B.transactionId
```

An advantage of a language-based approach is that the language can be used for manipulations, calculations, or filtering of event data which is required for the correlation process. In addition, many SQL-based languages gain further expressiveness by providing powerful time window or join operations.

4.5 Correlation Set Bridge

A correlation set bridge allows linking existing correlation sets, each having its own set of correlation bands. Thereby, correlation set bridges allow combining existing correlation sets in order to extend their correlation scope. Formally, a bridged correlation set is defined as follows:

Correlation Set Bridge $B = (S)$ where:
S: a set of correlation sets which should be linked.

The correlation set bridge will create a super-correlation set which includes all correlation bands of its child correlation sets. We want to extend the previous example with self-referencing correlation bands. We assume that the self-referencing correlation band of shipment events and all transport related events have been modeled in a separate correlation set. By using correlation set bridges, shipment events can be correlated with the transport events by virtually creating a super correlation set including all correlation bands. Conjunct event types (in our example the Shipment-Shipped event type) are used to create the bridge and expand the correlation scope by reusing existing correlation sets.

5 Correlating Events with Correlation Sets

In the following, we present an overview of the SARI correlation engine for correlating events with correlation sets in a distributed computing environment. Event correlation requires state information about the events being processed in order to determine which events are correlating. For each correlation band of a correlation set, SARI is creating a correlation session for collecting state information. If the correlation engine is able to link correlation bands, the corresponding sessions will be merged. In other words, correlation state information can evolve over time for different correlation bands which are combined as soon as events are correlated.

Another advantage of using correlation sessions is that they enable event correlation within a distributed and continuous event stream processing environment. Fig. 8 shows a distributed event-based system with a correlation service that manages correlation sessions for multiple nodes. Each node of the system performs event processing tasks which might require event correlations. When an event correlation is necessary, the node will access the correlation service in order to activate a session for each set of correlating events. The correlation service will synchronize the session access. The correlation session can be used by the node to store arbitrary data such as the correlated events or only some of their key information. For further details on managing the lifecycle of correlation sessions for distributed event-based systems refer to McGregor et. al [9].

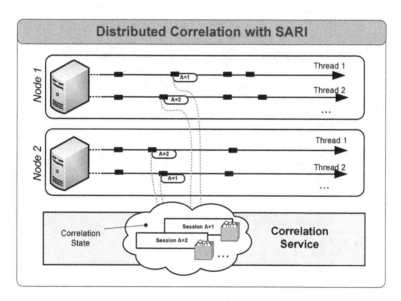

Fig. 8. Managing Correlations with Sessions

6 SARI Application Scenario

In this section, we present an example from the transportation and logistics domain. We show the modeling of SARI rules as well as how to validate and visualize rule processing results.

Fig. 9 shows the example's correlation set in the graphical rule editor of SARI. With the SARI rule editor, users can easy link attributes from a set of event types. Thereby, the rule editor automatically generates the resulting correlation band. Below the graphical editor, the resulting correlation bands are listed. For this example, we have separate correlation bands for both the transport and the shipment, which are combined by the correlation set. The modeled correlation set can be now assigned to an arbitrary number of SARI rules for detecting patterns from a correlated set of events. SARI rules allow defining complex event patterns, which use event conditions and timers for triggering response actions. Event conditions and timers can be arbitrarily combined with logical operators in order to model complex situations. Response actions are triggered when preconditions evaluate to true.

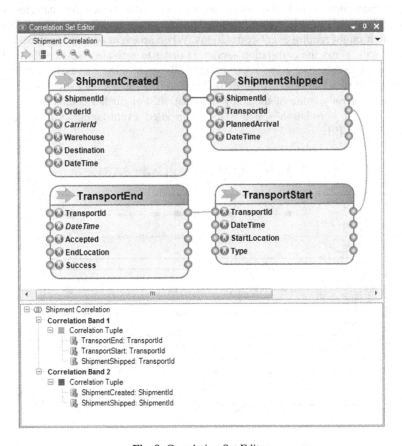

Fig. 9. Correlation Set Editor

Fig. 10 shows a SARI rule which monitors the transports of carrier 1200 airfare transports from Munich. If the shipment arrives more than 30 minutes late, two response actions are triggered: 1) an alert event is generated, and 2) a carrier satisfaction score is decreased. For correctly correlating the shipment and transport events, a user can refer to the correlation set from the previous section in the rule properties. A detailed discussion on the elements of SARI rules is given by Schiefer et. al in [13].

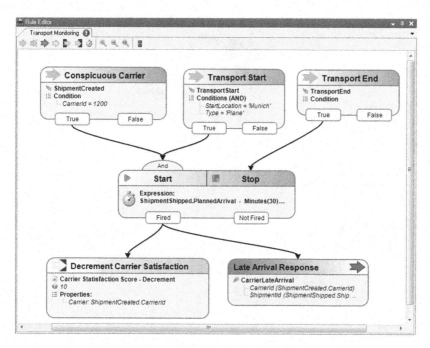

Fig. 10. SARI Rule – Transport Monitoring

6.1 Visualization of Rule Processing Results

SARI provides interactive visualizations of event streams to support business analysts in exploring business incidents. In the following, we show the event-tunnel visualization framework [15] for visualizing correlated events and validating rules. The event-tunnel is based on the metaphor of considering the event stream as a cylindrical tunnel, which is presented to the user from multiple perspectives. The visualization is able to display relationships between events which can be used for discovering root causes and causal dependencies of event patterns. The event-tunnel supports different types of placement policies for arranging events in the event tunnel. The centric event-sequence placement-policy (CESP policy), for instance, is focused on sequences of correlated events, such as the events of a business process instance. The policy plots event sequences over time (from the tunnel inside to the outer rings) and avoids overlapping events by clock-wise positional shifts. This technique results in characteristic patterns being abstract visual representations of business transactions.

Fig. 12 shows a screenshot of the 3D event-tunnel view for the events of a shipment process. While bullets represent events, the colored bands between them highlight correlation bands. The display of correlated events enables an abstract representation of underlying business transactions with characteristic patterns. If a rule generated some output (for instance an alert), it is displayed as part of the correlated event sequence. Business analysts can highlight events with characteristics of interest (e.g., delayed shipments, shipments with a triggered alert) for deeper analysis. Based on event correlations, it is possible to drill through the data, navigating to related

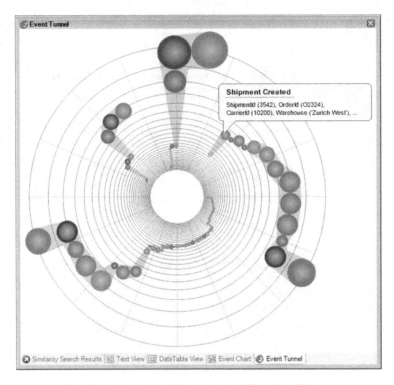

Fig. 11. Event-Tunnel Visualization of Correlated Events

(i.e., correlated) events which enables intuitive tracking of business incidents and root cause analyses. For more information on the event-tunnel, the CESP algorithm and evolving graphical patterns, readers may refer to Suntinger et. al [15].

7 Conclusion and Future Work

In large organizations, huge amounts of data are generated and consumed by business processes or human beings. Business managers need to respond with up-to-date information to make timely and sound business decisions. This paper described an approach for modeling relationships between events with correlation sets with the aim of correlating data from event streams for rule processing. We introduced various types of correlation sets and illustrated the usage of correlation sets with the SARI system and various types of applications. Our approach eases the separation of correlation concerns from rules, thereby making them more manageable and useful in a wide range of contexts. The work presented in this paper is part of a larger, long-term research effort aiming at developing an event stream management platform called SARI. A key focus of this future research work will be the automatic discovery of relationships and similarities between events and event sequences.

References

1. Abadi, D.J., Ahmad, Y., Balazinska, M., Çetintemel, U., Cherniack, M., Hwang, J.H., Lindner, W., Maskey, A.S., Rasin, A., Ryvkina, E., Tatbul, N., Xing, Y., Zdonik, S.: The Design of the Borealis Stream Processing Engine. In: Proc. of the Conf. on Innovative Data Systems Research, Asilomar, CA, USA, pp. 277–289 (2005)
2. Adi, A., Etzion, O.: AMIT - the situation manager. The VLDB Journal 13(2), 177–203 (2004)
3. Babcock, B., Babu, S., Datar, M., Motwani, R., Widom, J.: Models and Issues in Data Stream Systems. In: Proc. of 21st PODS, Madison, Wisconsin (May 2002)
4. Chen, S.K., Jeng, J.J., Chang, H.: Complex Event Processing using Simple Rule-based Event Correlation Engines for Business Performance Management. In: CEC/EEE (2006)
5. Esper, http://esper.sourceforge.net (2007-03-10)
6. Feldkuhn, L., Erickson, J.: Event Management as a Common Functional Area of Open Systems Management. In: Proc. IFIP Symposium on Integrated Network Management. North-Holland, Amsterdam (1989)
7. Houck, K., Calo, S., Finkel, A.: Towards a practical alarm correlation system. In: IEEE/IFIP Symposium on Integrated Network Management (1995)
8. Katzela, I., Schwartz, M.: Schemes for fault identification in communication networks. IEEE Transactions on Networking (1995)
9. McGregor, C., Schiefer, J.: Correlating Events for Monitoring Business Processes. In: 6th International Conference on Enterprise Information Systems (ICEIS), Porto (2004)
10. Moser, T., Roth, H., Rozsnyai, S., Biffl, S.: Semantic Event Correlation Using Ontologies. In: 3rd Central and East European Conference on Software Engineering Techniques (CEE-SET 2008), Brno (2008)
11. Nygate, Y.A.: Event correlation using rule and object base techniques. In: IEEE/IFIP Symposium on Integrated Network Management (1995)
12. RuleCore, http://www.rulecore.com/ (2007-03-10)
13. Schiefer, J., Szabolcs, R., Saurer, G., Rauscher, C.: Event-Driven Rules for Sensing and Responding to Business Situations. In: International Conference on Distributed Event-Based Systems, Toronto (2007)
14. Seirio, M., Berndtsson, M.: Design and Implementation of an ECA Rule Markup Language. In: RuleML. Springer, Heidelberg (2005)
15. Suntinger, M., Obweger, H., Schiefer, J., Gröller, E.: The Event Tunnel: Exploring Event-Driven Business Processes. IEEE Computer Graphics and Applications (October 2008)
16. Wu, P., Bhatnagar, R., Epshtein, L., Bhandaru, M., Shi, Z.: Alarm correlation engine (ACE). In: Proceedings of the IEEE/IFIP 1998 Network Operations and Management Symposium (NOMS), New Orleans (1998)
17. Zdonik, S., Stonebraker, M., Cherniack, M., Cetintemel, U., Balazinska, M., Balakrishnan, H.: The Aurora and Medusa Projects. IEEE Data Engineering Bulletin (2003)

Semantic Rule-Based Complex Event Processing

Kia Teymourian and Adrian Paschke

Freie Universität Berlin
Institut for Computer Science
AG Corporate Semantic Web
Königin-Luise-Str. 24/26, 14195 Berlin, Germany
{kia,paschke}@inf.fu-berlin.de
http://www.inf.fu-berlin.de/groups/ag-csw

Abstract. One of the critical success factors of event-driven systems is the capability of detecting complex events from simple and ordinary event notifications. Complex events which trigger or terminate actionable situations can be inferred from large event clouds or event streams based on their event instance sequence, their syntax and semantics. Using semantics of event algebra patterns defined on top of event instance sequences for event detection is one of the promising approaches for detection of complex events. The developments and successes in building standards and tools for semantic technologies such as declarative rules and ontologies are opening novel research and application areas in event processing. One of these promising application areas is semantic event processing. In this work we describe our research on semantic rule-based complex events processing.[1]

1 Introduction

Detection of events is one of the critical factors for the event-driven systems and e.g. event-driven business process management (edBPM). The permanent stream of low level events in business organizations, e.g. in Business Process Management (BPM) and Business Activity Monitoring (BAM), IT Service Management (ITSM), and IT Infrastructure Management (ITIM) needs an intelligent real-time event processor with declarative representation capabilities for defining the detection conditions of complex events and reaction rules. [15]

The promises of the combination of event processing and semantic technologies, such as rules and ontologies, which leads to semantic event processing (SCEP), is that the event processing rule engines now will **understand** what is happening in terms of events and (process) states and that they will **know** what reactions and processes they can invoke and what events it can signal. In this paper, we describe our ongoing research on semantic rule-based event processing . The use case of such declarative rule-based approaches can be for example in business process monitoring, fraud detection and many other fields which have a need for agile reactive behavior.

---

[1] This work has been partially supported by the "Inno-Profile Corporate Semantic Web" project funded by the German Federal Ministry of Education and Research (BMBF) http://www.corporate-semantic-web.de

G. Governatori, J. Hall, and A. Paschke (Eds.): RuleML 2009, LNCS 5858, pp. 82–92, 2009.

2 Semantic Events

Real-world occurrences can be defined as events that are happening over space and time. An event instance is an concrete semantic object containing data describing the event. An event pattern is a template that matches certain sets of events [7]. Event filters are constructive views on event sources such as an event cloud or event stream, which might be materialized in an event database. A filter F is a stateless truth-valued function that is applied to an event instance sequence. An event instance n matches a filter F if and only if it satisfies all attribute filters of F [2].

A common model for an event instance is e.g. a structure consisting of attribute/value pairs like {(type,StockQuote), (name,"Lehman Brothers"), (price,45)}. Events can be a compound of others events which build the complex events. Complex Event Processing (CEP) is about detection of complex events from a cloud of events which are partially temporal ordered by matching complex event patterns against event instance sequences.

There are some works like [19] and [1] trying to define a common top ontology for events using ontology languages such as RDFS [17] or OWL [16]. We investigated these ontologies and observed that they are trying to find a common set of attributes for events like: type, actors/agents, place references, time (start and end time, duration).

Instead, we propose that events should be describe by a modular and layered ontology model. A shared top ontology for events will capture the top concepts such as event, time, space, situation etc. Specific domain, application and task ontologies, which are modular sub-ontologies of the common top ontology, then represent more concrete specialized concepts. The domain-specific events are modeled as domain classes of the domain ontology and are in relationships with other classes of other domain ontologies. They can have object or data type properties. For example to describe the above example about the stock price change, an event class for this event can be named "stock_price_change", the names in name/value pairs can be mapped to new attributes for this event class. Each instance of this class is identified by an URI and is related with a rdf:type predicate to this class.

The Listing 1.1 shows a set of RDF triples which define an instance of this event:

```
{(event:e1002, rdf:type, event:stock_price_change),
(event:e1002, stock:Name, stock:Lehman_Brothers),
(event:e1002, stock:hasPrice, "45")}
```

Listing 1.1. Example of an event notification in RDF

A map of the attribute/value pairs to a set of RDF triples (RDF graph) can be used as event instance. The interesting part of this event data model is the linking of the existing knowledge (non-event concepts) to the event instances, for example the name of the stock which the event is about is not identified by a simply string name, but by an URI which links to the semantic knowledge about the stock. This knowledge can be used later for the processing of events, e.g. in the condition part of a Event-Condition-Action (ECA) reaction rule.

3 Semantic Event Detection

In our semantic event system an event instance is a set of RDF triples (RDF graph) and an event pattern is a graph pattern. More complex events are formed by combining smaller patterns in various ways. A complex conjunctive event filter is a complex graph pattern of the RDF graph. SPARQL [18] as a graph pattern matching language (RDF query language) can be exploited for event filtering by writing the event patterns in terms of SPARQL queries. As a result we get a constructive view over the events in the event system, which is a RDF database - a so called triple store.

We consider that in a semantic event processing system a part of the knowledge about the events is a static knowledge about the pre-defined event classes, i.e. the event types in an event ontology, the other part is as real-time data streams in form of RDF triple stream which notifies about the occurences of events. The system has to combine these knowledge references and generate new knowledge. Inferred triples are those triples which are not in the RDF triple stream but could be inferred from the (event) knowledge base of the system. Having such an semantic event pattern/filter makes it possible to detect complex event which is derived from the already happened events in combination with semantic knowledge of the processing system. For example, someone might be interested to be informed on any events about the oil companies which are related to the events of the Lehman Brothers stock prices.

Event notification triples are created by event detectors (or they are generated from the raw event data). After the generation of notification triples, they are sent to the event processing system which does the semantic filtering and rule-based inferencing. After the processing steps, the system decides if the triples should be stored persistently or simply dropped. This is decided based on a set of consumption policy rules which are pre-defined for each class of events. These rules can also be used to store the triples on (distributed) RDF storage clusters (RDF triple stores).

4 CEP Semantics - Interval-Based Event Calculus Event / Action Algebra

After having described how to semantically query and filter events we will now describe how to formalize complex event patterns based on a logical knowledge represenation (KR) interval-based event / action algebra, namely the interval-based Event Calculus [8,13,12,14,15].

In the interval-based Event Calculus (EC) all events are regarded to occur in a time interval, i.e. an event interval $[e1, e2]$ occurs during the time interval $[t1, t2]$ where $t1$ is the occurrence time of $e1$ and $t2$ is the occurrence time of $e2$. An atomic event occurs in the interval $[t, t]$, where t is the occurrence time of the atomic event. The interval-based EC axioms describe when events / actions *occur* (transient consumption view), *happen* (non-transient reasoning view) or are *planned* to happen (future abductive view) within the EC time structure, and which properties (*fluents* or *situations*) are *initiated* and/or *terminated* by these events under various circumstances.

For the syntax of the interval-based Event Calculus language we adopt a normal logic programming (LP) syntax [6] with an ISO Prolog like alphabet with functions, variables, negation etc, called Prova (`http://prova.ws`). For the interval-based EC we define a multi-sorted signature.

Definition *(Interval-based Event Calculus Signature). The interval-based EC signature is a multi-sorted signature S defined as a tuple $\langle E_S, F_S, T_S, X_S; \overline{E}, \overline{F}, \overline{T}, \overline{X}, \leq, arity, sort \rangle$ with sorts E_S for event type symbols, sorts F_S for fluent type symbols, sorts T_S for time interval type symbols, and sorts X_S for domain type symbols; $\overline{T}$ is the non-empty set of time interval function symbols, called time intervals, $\overline{E}$ is a non-empty set of event/action function symbols, called events, $\overline{F}$ is a non-empty set of fluent function symbols, called fluents, $\overline{X}$ is the signature of the used LP language including constant, function, predicate symbols. The function $arity(E_i)$, $arity(F_i)$, $arity(X_i)$ associates a non-zero natural number with each event E_i, fluent F_i, and X_i. The $arity(T_i)$ of time intervals is arity 2. But, for convenience reason and compliance with the classical time point based Event Calculus a time interval $[T_i, T_j]$ might reduced to a time point with arity 1 if $T_i = T_j$. $\leq$ is a partial ordering defined over the time intervals. The function sort associates with each k-ary event, fluent or time interval function symbol a $k + 1$-tuple of sorts. That is, if f is a event, fluent, or time interval function of arity k, then $sort(f)$ is a $k + 1$-tuple of sorts $sort(f) = (T_1, .., T_k, T_{k+1})$ where $(T_1, .., T_k)$ defines the sorts of the domain of f and T_{k+1} defines the sort of the range of f, where each T_i is some E_{S_j}, F_{S_j}, or T_{S_j} (repetitions are allowed). Similarly, $sort(X)$ gives the sort(s) of X.*

This multi-sorted signature allows us considering objects of different nature, such as events, fluents, time intervals or domain objects which are modeled as different sorts of individuals in a structure. That is, we assume not just a single universe of discourse, but several domains (the sorts) in a single multi-sorted structure.

Definition *(Interval-based Event Calculus Semantic Structure). A semantic structure is an interpretation I for the signature S. It is a tuple of the from $\langle D, \overline{E}^I, \overline{F}^I, \overline{T}^I, \overline{X}^I, TV^I \rangle$. Here D is a non-empty set called the domain universe of I and the (not necessarily disjoint) union of the sorts $\langle E_S^I \cup F_S^I \cup T_S^I \cup X_S^I \rangle$ is a subset of D. The members of D are called "individuals" of I. The other components of I are total mappings defined as follows:*

- $\overline{E}^I = \langle E_1^I, .., E_j^I \rangle$ is an interpretation of all event function symbols. For each function symbol E_i, of arity m and $sort(E_i) = (E_{S_1}, .., E_{S_m}, E_{S_{m+1}})$, E_i^I is a m-place function $E_i^I : D^m \to D$ with $E_i^I : E_{S_1}^I \times ... \times E_{S_m}^I \to E_{S_{m+1}}^I$.

- $\overline{F}^I = \langle F_1^I, .., F_j^I \rangle$ is an interpretation of all fluent function symbols. For each function symbol F_i, of arity m and $sort(F_i) = (F_{S_1}, .., F_{S_m}, F_{S_{m+1}})$, F_i^I is a m-place function $F_i^I : D^m \to D$ with $F_i^I : F_{S_1}^I \times ... \times F_{S_m}^I \to F_{S_{m+1}}^I$.

- $\overline{T}^I = \langle T_1^I, .., T_2^I \rangle$ is an interpretation of all time interval function symbols. For each function symbol T_i, of arity 2 and $sort(T_i) = (T_{S_1}, T_{S_2}, T_{S_3})$, T_i^I is a 2-place function $T_i^I : D^2 \to D$ with $T_i^I : T_{S_1}^I \times T_{S_2}^I \to T_{S_3}^I$.

- $\overline{X}^I$ is an interpretation of $\overline{X}$ in accordance with their sorts.

- I_{truth} is a mapping of the form $D \to \langle true, false \rangle$ used to define the truth valuation for well-formed formulas.

We now define how a semantic structure, I, determines the truth value $TVal(\varphi)$ of a formula φ. Due to space limitation we are focusing on the main axiom of the basic interval-based EC, where the basic $holdsAt$ axiom of the classical EC [5] is redefined to $holdsInterval$ for temporal reasoning about event intervals which hold between a time intervals, i.e., an interpretation of the $holdsInterval$ axiom is a mapping $I : [E1, E, l2] \times [T1, T2] \mapsto \{true, false\}$.

Definition (Interval-based Event Calculus Truth Valuation). *Truth valuation for the* $holdsInterval$ *axiom in the interval-based Event Calculus is determined using the function* $TVal(holdsInterval(F, T)) : I_{truth}(F^I \times T^I)$ *where F is an event interval* $[E1, E2]$ *and T is a time interval* $[T1, T2]$.

That is the declarative semantics of the main calculus axiom $holdsInterval$ is given by interpretations which map event intervals $[E1, E2]$ and time intervals $[T1, T2]$ to truth values.

We now define what it means for an event interval to be instantiated or terminated in an interval-based Event Calculus program.

Definition (Instantiation and Termination). *Let* Σ^{EC} *be an interval-based EC language,* D^{EC} *be a domain description (an EC program) in* Σ^{EC} *and I be an interpretation of* Σ^{EC}. *Then an event interval* $[E1, E2]$ *is instantiated at time point* $T1$ *in I iff there is an event E1 such that there is a statement in* D^{EC} *of the form* $occurs(E1, T1)$ *and a statement in* D^{EC} *of the form* $initiates(E1, [E1, E2], T)$. *A event interval* $[E1, E2]$ *is terminated at time point* $T2$ *in I iff there is an event E2 such that there is a statement in* D^{EC} *of the form* $occurs(E2, T2)$ *and a statement in* D^{EC} *of the form* $terminates(E2, [E1, E2], T)$.

Definition (Interval-based Event Calculus Satisfaction). *An interpretation I satisfies an event interval* $[E1, E2]$ *at a time interval* $[T1, T2]$ *if* $I([E1, E2], [T1, T2]) = true$ *and* $I(\neg[E1, E2], [T1, T2]) = false$.

An interpretation qualifies as a model for a given domain description, if:

Definition (Event Calculus Model). *Let* Σ^{EC} *be an interval-based EC language,* D^{EC} *be a domain description in* Σ^{EC}. *An interpretation I of* Σ^{EC} *is a model of* D^{EC} *iff* $\forall [E1, E2] \in \overline{F}$ *and* $T1 \leq T2 \leq T3$ *the following holds:*

1. If $[E1, E2]$ has not been instantiated or terminated at $T2$ in I wrt D^{EC} then $I([E1, E2], [T1, T1]) = I([E1, E2], [T3, T3])$
2. If $[E1, E2]$ is initiated at $T1$ in I wrt D^{EC}, and not terminated at $T2$ the $I([E1, E2], [T3, T3]) = true$
3. If $[E1, E2]$ is terminated at $T1$ in I wrt D^{EC} and not initiated at $T2$ then $I([E1, E2], [T3, T3]) = false$

The three conditions define the persistence of complex event intervals as time progresses. That is, only events/actions have an effect on the changeable event interval states (condition 1) and the truth value of a complex event state persists until it has been explicitly changed by another terminating event/action (condition 2 and 3). A domain description is consistent if it has a model. We now define entailment wrt to the main interval-based EC axiom of $holdsInterval$:

Definition (*Event Calculus Entailment*). *Let D^{EC} be an interval-based EC domain description. A event interval $[E1, E2]$ holds at a time interval $[1, T2]T$ wrt to D^{EC}, written $D^{EC} \models holdsInterval([E1, E2], [T1, T2])$, iff for every interpretation I of D^{EC}, $I([E1, E2], [T1, T2]) = true$. $D^{EC} \models neg(holdsInterval([E1, E2], [T1, T2]))$ iff $I([E1, E2], [T1, T2]) = false$.*

After having introduced the syntax and semantics of the interval-based Event Calculus we now outline the implementation of the calculus as a meta logic program. For a full description see [8,13,14].

```
holdsInterval([E1,E2],[T11,T22]):-
    event([E1],[T11,T12]), event([E2],[T21,T22]),
    [T11,T12]<=[T21,T22], not(broken(T12,[E1,E2],T21).
```

The event function $event([Event], [Interval])$ is a meta-function to translate instantaneous event occurrences into interval-based events: $event([E], [T, T])$: $-occurs(E, T)$. It is also used in the event algebra meta-program to compute complex events from occurred raw events according to their event type definitions. The *broken* function tests whether the event interval is not broken between the the initiator event and the terminator event by any other explicitly specified terminating event:

```
broken(T1,Interval,T2):-
            terminates(Terminator,Interval,[T1,T2]),
            event([Terminator],[T11,T12]), T1<T11, T12<T2.
```

Example

```
occurs(a,datetime(2009,1,1,0,0,1)).
occurs(b,datetime(2009,1,1,0,0,10)).
Query: holdsInterval([a,b],Interval)?
Result: Interval=
        [datetime(2009,1,1,0,0,1), datetime(2009,1,1,0,0,10)]
```

In the example an event a followed by an event b occurs. The $holdsInterval$ returns the occurrence time interval for the variable $Interval$. That is, time points/ time intervals, events/actions and fluents are n-ary literals L or $\neg L$ in the meta program. Remarkably, the logic programming rule engine Prova (http://prova.ws) which we are using supports a typed logic with Java and Description Logics support and provides query built-ins to external data sources, i.e. events can be e.g. stored and queried from databases such as an RDF triple store and for efficient time computations the Java Calendar API can be used. [10,9] For instance, the datetime formalization above is using the Prova ContractLog library which provides an efficient Java-based implementation of temporal reasoning in Prova logic program - see [10].

Based on this interval-based event logics formalism, we now implement typical event algebra operators, as can be found e.g. in SNOOP [3], and treat complex events as occurrences over an interval. In short, the basic idea is to split the occurrence interval of a complex event into smaller intervals in which all required component events occur, which leads to the definition of event type patterns in terms of interval-based event detection conditions, e.g. the SNOOP sequence operator (;) is formalized as follows $(A; B; C)$:

```
detect(e,[T1,T3]):-
    holdsInterval([a,b],[T1,T2],[a,b,c]),
    holdsInterval([b,c],[T2,T3],[a,b,c]),
    [T1,T2]<=[T2,T3].
```

In order to make definitions of complex events in terms of event algebra operators more comfortable and remove the burden of defining all interval conditions for a particular complex event type as described above, we have implemented a meta-program which implements an interval-based EC event algebra in terms of typical event operators:

```
Sequence:              sequence(E1,E2, .., En)
Disjunction:           or(E1,E2, .. , En)
Mutual exclusive:      xor(E1,E2,..,En)
Conjunction:           and(E1,E2,..,En)
Simultaneous:          concurrent(E1,E2,..,En)
Negation:              neg([ET1,..,ETn],[E1,E2])
Quantification:        any(n,E)
Aperiodic:             aperiodic(E,[E1,E2])
```

```
Example
Event Pattern is A ; (B ; C):    (Sequence)
EC algebra detection rule: detect(e,T) :- event(sequence(a, sequence(b,c)), T).
```

In this example under strict interpretation of the interval-based EC e.g. the event instance sequence (EIS) $\{abc\}$ does detect the complex event as defined by the event pattern sequence, $\{bac\}$ does not detect the event, $\{acb\}$ does not detect the event, $\{abac\}$ does not detect event and $\{abbabc\}$ does detect one complex event.

```
Example
Event Pattern is A (A, B,C):     (Aperiodic)
detect(e,T):- event(aperiodic(b,[a,c]),T).
```

Here the EIS $\{abc\}$ does detect the event, $\{abbc\}$ does detect the event, but $\{acbb\}$ does not detect the event.

For space reasons we have only described the formalization of complex events, but the interval-based Event Calculus can also used for the definition of complex actions, e.g. to define an ordered sequence of action executions or concurrent actions (actions which must be performed in parallel within a time interval), with a declarative semantics for possibly required rollbacks. For more information see [10,8,14].

We will now illustrate the rule-based event processing approach by a more extensive example typically found in industry:

Example: *A Manager node is responsible for holding housekeeping information about various servers playing different roles. When a server fails to send a heartbeat for a specified amount of time, the Manager assumes that the server failed and cooperates with the Agent component running on an unloaded node to resurrect it. A reaction rule for receiving and updating the latest heartbeat in event notification style is:*

```
rcvMsg(XID,Protocol,FromIP,inform,heartbeat(Role, RemoteTime)) :-
    time(LocalTime)
    update(key(FromIP,Role),"occurs(heartbeats(_0, _1, _2, _3)).",
            [ FromIP, Role, RemoteTime, LocalTime] ).
```

The rule responds to a message pattern matching the one specified in the *rcvMsg* arguments. *XID* is the conversation-id of the incoming message; *inform* is the performative representing the pragmatic context of the message, in this case, a one-way information passed between parties; *heartbeat(...)* is the payload of the message. The body of the rule enquires about the current local time and updates the record containing the latest heartbeat from the controller. This rule follows a push pattern where the event is pushed towards the rule systems and the latter reacts by updating the event information (in the RDF triple store - not shown here).

A pull-based ECA rule that is activated every second and for each server that fails to have sent heartbeats within the last second will detect server failures and respond to it by initiating failover to the first available unloaded server. The accompanying derivation rules *detect* and *respond* are used for specific purpose of detecting the failure and organizing the response.

```
eca(
   every('1S') ,
   detect(controller_failure(IP,Role,'1S')) ,
   respond(controller_failure(IP,Role,'1S')) ) .

every('1S'):-
   sysTime(T),
   interval(timespan(0,0,0,1),T).

detect(controller_failure(IP,Role,Timeout),LocalTimeNow) :-
   sysTime(LocalTimeNow),
   event(neg(heartbeat(IP,Role,RemoteTime,LocalTime)),holdsAt(status(IP,loaded),T)),
   LocalTimeNow-LocalTime > Timeout.

respond(controller_failure(IP,Role,Timeout)) :-
   sysTime(LocalTime),
   first(holdsAt(status(Server,unloaded),LocalTime)),
   add(key(Server),
       "happens(loading(_0),_1).",[ Server, Local-Time]),
   sendMsg(XID,loopback,self,initiate,failover(Role,IP,Server)).
```

The ECA logic involves possible backtracking so that all failed components will be resurrected. The detection conditions for the complex event *controller_failure* for a server with a particular *IP* are defined in terms of the interval-based Event Calculus algebra, where the event is detected if there is no *heartbeat* during the specified *timeout* while the server is loaded. The state of each server is managed via an event calculus formulation:

```
initiates(loading(Server),status(Server,loaded),T).
terminates(unloading(Server),status(Server,loaded),T).
initiates(unloading(Server),status(Server,unloaded),T).
terminates(loading(Server),status(Server, loaded),T).
```

The current state of each server is derived from the happened loading and unloading events and used in the ECA rule to detect the first server which is in state *unloaded*.

Note, as an event instance base for detecting complex events according to the detection patterns, which are formalized in terms of the event algebra operators as exemplified above, the (distributed) RDF storage clusters are used as active working memory (active knowledge base). The rule-based event algebra meta program queries event data via Prova's SPARQL query built-ins as event facts from the RDF clusters. Using triple store technology this dynamic integration approach is highly efficient. This approach is quite similar to using relational databases as event database, but with the major benefit of providing semantic event meta data and built-in ontological inference.

5   Rule-Based CEP Middleware

In this section we describe the main components of the rule-based CEP Enterprise Service Bus middleware which we have implemented. As in OMGs model driven architecture (MDA) the middleware distinguishes:

- a platform specific model (PSM) which encodes the rule statements in the language of a specific execution environment
- a platform independent model (PIM) which represents the rules in a common (standardized) interchange format (e.g. a markup language)
- a computational independent model (CIM) with rules in a natural or visual language.

The Prova (http://prova.ws) reaction rule language with the interval-based Event Calculus meta program formalization is used on the PSM layer as a concrete rule-based execution and event processing engine. Several Prova rule engines can be deployed as distributed web-based inference services on the ESB (http://responder.ruleml.org). The Prova rule engines have dynamic access to external data sources and object representations, e.g. using semantic triple stores as active event data stores. The quasi-standard RuleML/Reaction RuleML (http://reaction.ruleml.org) is used on the PIM layer as a standardized interchange format between the distributed rule inference engine on the ESB. The ESB

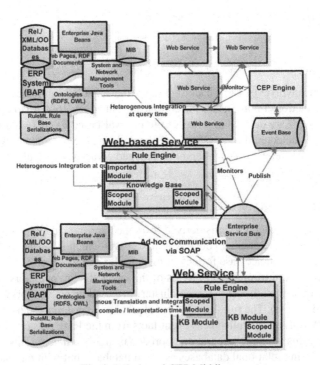

Fig. 1. Rule-based CEP Middleware

is used as object service broker for the rule inference services and as asynchronous messaging middleware supporting all common transport protocols such JMS, HTTP or SOAP (or Rest) (more than 30 protocols are supported) between the services. For an visual RuleML/Reaction RuleML editor on the CIM layer see [10]. Figure 1 illustrates the architecture of the middleware.

6 Conclusion and Future Works

First ideas for semantic event processing, like [4] or [1], are functionally inadequate from a event processing process perspective. This inadequacy is due to the lack of domain and application specific semantics for events, processes, states, actions, and other concepts that relate to change over time. Without such formal semantics rules or workflow like logic that react to events and govern e.g. (business) processes must remain at the level of procedural implementation - as it is in most current event processing applications - rather than declarative knowledge which makes the event-based applications, services and processes more integrative, adaptable and agile. Capturing domain-specific events, processes and other aspects of reality that occur and change over time is a fundamental challenge which we addressed in a practical manner in this paper.

In this paper we introduced a practical rule-based approach for semantic-enabled event processing which we implemented based on highly efficient industry-strength technologies, namely RDF storage clusters (using Triplestore technology), Enterprise Service Bus (using enterprise service technology), and Business Rules Managment System for Complex Event Processing (based on rule-based event processing technology). We have described syntax, semantics, implementation and integration of the components in the overall architecture. We have already successfully applied our rule-based event processing in the context of IT Service Management and Service Level Agreements. [10] Our next steps is to apply our semantic rule-based event processing approach in event-driven Semantic BPM, that is within executable BPEL orchestration flows as external semantic web services. [11]

References

1. Jans, A.: Unification of geospatial reasoning, temporal logic, & social network analysis in event-based systems. In: DEBS 2008: Proceedings of the second international conference on Distributed event-based systems, pp. 139–145. ACM, New York (2008)
2. Pietzuch, P., Mühl, G., Fiege, L.: Distributed Event-Based Systems. Springer, Heidelberg (2006)
3. Chakravarthy, S., Krishnaprasad, V., Anwar, E., Kim, S.K.: Composite events for active databases: Semantics contexts and detection. In: VLDB 1994, pp. 606–617 (1994)
4. Etzion, O.: Semantic approach to event processing. In: DEBS 2007: Proceedings of the 2007 inaugural international conference on Distributed event-based systems, pp. 139–139. ACM, New York (2007)
5. Kowalski, R.A., Sergot, M.J.: A logic-based calculus of events. New Generation Computing 4, 67–95 (1986)
6. Lloyd, J.W.: Foundations of logic programming, 2nd extended edn. Springer-Verlag New York, Inc., New York (1987)

92   K. Teymourian and A. Paschke

7. Luckham, D.: The Power of Events: An Introduction to Complex Event Processing in Distributed Enterprise Systems. Addison-Wesley Longman, Amsterdam (2002)
8. Paschke, A.: Eca-lp / eca-ruleml: A homogeneous event-condition-action logic programming language. In: RuleML 2006, Athens, Georgia, USA (2006)
9. Paschke, A.: A typed hybrid description logic programming language with polymorphic order-sorted dl-typed unification for semantic web type systems. In: OWL-2006 (OWLED 2006), Athens, Georgia, USA (2006)
10. Paschke, A.: Rule-Based Service Level Agreements - Knowledge Representation for Automated e-Contract, SLA and Policy Management. In: IDEA, Munich (2007)
11. Paschke, A., Teymourian, K.: Rule based business process execution with bpel+. In: Proc. 5th International International Conference on Semantic Systems (i-Semantics 2009) (2009)
12. Paschke, A.: Eca-lp / eca-ruleml: A homogeneous event-condition-action logic programming language. CoRR, abs/cs/0609143 (2006)
13. Paschke, A.: Eca-ruleml: An approach combining eca rules with temporal interval-based kr event/action logics and transactional update logics. CoRR, abs/cs/0610167 (2006)
14. Paschke, A.: A homogenous reaction rule language for complex event processing. In: Proc. 2nd International Workshop on Event Drive Architecture and Event Processing Systems (EDA-PS) (2007)
15. Paschke, A., Bichler, M.: Knowledge representation concepts for automated sla management. Decis. Support Syst. 46(1), 187–205 (2008)
16. W3C. Owl web ontology language, http://www.w3.org/TR/owl-features/
17. W3C. Rdf vocabulary description language, http://www.w3.org/TR/rdf-schema/
18. W3C. Sparql query language for rdf,
    http://www.w3.org/TR/rdf-sparql-query/
19. Abdallah, S., Raimond, Y.: The event ontology,
    http://motools.sourceforge.net/event/event.html

Generation of Rules from Ontologies for High-Level Scene Interpretation

Wilfried Bohlken and Bernd Neumann

Cognitive Systems Laboratory, Department Informatik, University of Hamburg
22527 Hamburg, Germany
{bohlken,neumann}@informatik.uni-hamburg.de

Abstract. In this paper, a novel architecture for high-level scene interpretation is introduced, which is based on the generation of rules from an OWL-DL ontology. It is shown that the object-centered structure of the ontology can be transformed into a rule-based system in a native and systematic way. Furthermore the integration of constraints - which are essential for scene interpretation - is demonstrated with a temporal constraint net, and it is shown how parallel computing of alternatives can be realised. First results are given using examples of airport activities.

1 Introduction

High-level scene interpretation can be roughly defined as understanding images or video streams at abstraction levels above single objects. Typical tasks are traffic scene interpretation in driver assistance systems, criminal acts recognition, and other monitoring tasks such as airport activity recognition, which is used as an example domain in this paper. Scene interpretation systems are typically conceived as knowledge-based systems where extensive high-level knowledge is modelled using declarative knowledge representation techniques. So far, no standard architecture has emerged. In his long-standing work on traffic scene interpretation [7,1,2], Nagel developed situation-graph trees, where hierarchically organized frame-based state descriptions of traffic situations are embedded in a state-transition structure. A similar structure was also realised by [3,4,8] in terms of scenarios for recognizing bank robberies or airport activities. Compositional and taxonomical hierarchies of structure-based configuration systems as a framework for flexible scene interpretation strategies realising both bottom-up and top-down interpretation steps are proposed in [10]. Using a hierarchical framework ranging from the pixel level to high-level semantic structures, a grammar-based scene interpretation system was developed [17].

The usefulness of high-level symbolic scene interpretation on top of low-level image processing for activity recognition in video streams was demonstrated by [6,16]. In this approach the activity models are represented in a self-made, non-standardized formalism. Scene interpretation for more complex activities, considered in this paper, calls for for well-founded knowledge representation and standardized inference procedures.

G. Governatori, J. Hall, and A. Paschke (Eds.): RuleML 2009, LNCS 5858, pp. 93–107, 2009.
© Springer-Verlag Berlin Heidelberg 2009

This was the motivation to investigate the use of Description Logics (DL) for scene interpretation [13] and multimedia interpretation and retrieval [11]. It was shown that a DL system can transparently represent compositional and taxonomical hierarchies (which provide the backbone for scene interpretation) in an object-centered manner, and that several DL inference services (such as inheritance and classification) can be exploited for the interpretation process. On the other hand, it is difficult in a DL system to represent constraints between objects, which are often decisive for defining and recognizing high-level entities. Furthermore, a DL system does not provide a framework for flexible, stepwise scene interpretation as required for complex applications. In this paper, we propose a novel architecture for high-level scene interpretation which exploits well-founded object-centered knowledge representation using OWL-DL, but avoids the limitations of DL systems by transforming knowledge structures into rules in the rule-based system Jess. This approach promises several advantages:

- High-level knowledge can be represented in OWL in an object-centered transparent manner, scaling well to large knowledge bases.
- The consistency of the knowledge base can be checked automatically using a DL reasoner connected to OWL (such as Pellet or Racer).
- Logic-based inferences implied by the OWL representation can be automatically translated into corresponding rules and inheritance mechanisms of Jess, realising the skeleton of a scene interpretation system.
- Constraint processing and other procedural components not representable in OWL can be realised in the Java background of Jess.
- Data-driven rule-based processing facilitates flexible interpretation as required for highly variable scenes and realistic image analysis results.

Note that in this architecture, as in most other approaches, high-level scene interpretation is conceived as a process, which takes low-level image analysis results in terms of primitive objects as input and delivers assertions about the scene as output, for example "Aircraft refueling has begun at 13:02:23" for airport activity monitoring. As in many applications objects cannot be recognized reliably, it remains the task of high-level interpretation to disambiguate or even correct low-level classifications. This must be kept in mind when devising high-level inference rules.

A basic interpretation step is recognizing an aggregate from its parts in accordance with the compositional hierarchy defined in the OWL knowledge base. In Section 2, we describe how OWL aggregates are transformed into Jess rules providing such interpretation steps. Our implementation differs from an automatic transformation of OWL to Jess described in [5] in several respects and promises advantages with respect to scalability and generality. We also show how the constraint checks required for aggregate instantiations are embedded into the rules to be executed in the Java background system.

In Section 3, we describe the basic bottom-up interpretation process based on the rules generated from the compositional hierarchy. Here, one of the problems is conflict resolution when several rules can fire. This must be expected throughout the interpretation process because of several possible lines of interpretation in the face of partial or unspecific evidence. It is shown how parallel interpretation threads can be generated automatically which terminate either with a valid interpretation or incomplete as dead ends.

In Section 4, we present an extended example from the airport activity domain. Section 5, finally, concludes with a discussion of the results and information about future work.

2   Rule Generation from Ontology

In the first part of this section, the usage of OWL-DL for an ontology for scene interpretation is motivated. Then aggregates are introduced as the main representational units of the ontology and finally the transformation of aggregates to Jess rules is presented. In the second part the integration of constraints into the rule-based system is described using a temporal constraint net.

2.1   Object-Centered Definition of Aggregates

Ontology and OWL DL. To describe scenes at a high conceptual level requires the expressiveness to define concepts like objects and events (e.g. for interpretation of video sequences), together with their properties and relations (e.g. temporal and spatial relations between parts of the scene) [13]. A description language for a conceptual knowledge base, which is used for scene interpretation, has to provide the expressiveness to satisfy these requirements. On the other hand, an ontology, which is developed for the purpose of scene interpretation tasks, is typically large and difficult to maintain manually. Therefore it is necessary that a reasoner (like Pellet or Racer) is available to perform automatic checks, e.g. with respect to class consistency, class equivalence, sub-class relation, disjunctiveness and global consistency. The description language OWL DL provides the maximum expressiveness without losing computational completeness (all entailments are guaranteed to be computed) and decidability (all computations will theoretically finish in finite time). Nevertheless the expressiveness is not sufficient for our purposes, particularly to specify n-ary constraints, but this gap can be closed with the OWL extension SWRL (*Semantic Web Rule Language*).[1] All in all, OWL DL has many desirable properties for our purposes, and in the following we assume that OWL DL is used for our ontology representation. It is also worth noting that OWL ontologies have become increasingly popular over the last years, primarily because of the idea of the *Semantic Web*, which increases the availability of tools for creating, publicising and distributing ontologies.

Aggregates. The main concepts in our OWL ontology are *physical* objects, for example `Vehicle`, `Person` or `Equipment`, and *conceptual* objects. These are more abstract objects, for example *events* as `Vehicle-Enters-Zone` or `Refueling`. Concepts are related to each other by super-class and sub-class relations, thus,

---

[1] http://www.w3.org/Submission/SWRL

forming a *taxonomy*. Other essential relations for a knowledge base for scene interpretation are the compositional relations, which express that a concept may have other concepts as parts inducing a compositional hierarchy. For example, the conceptual object `Vehicle-Enters-Zone` is composed of the conceptual objects `Vehicle-Outside-Zone` and `Vehicle-Inside-Zone`. Instances (or individuals in OWL terms) of these parts have to be in a specific temporal relation: an instance of `Vehicle-Outside-Zone` has to occur before the instance of `Vehicle-Inside-Zone`. Another constraint is that the respective instances of the physical objects `Vehicle` and `Zone` of both events are the same. These are the *conceptual constraints*. A concept and its parts together with the conceptual constraints form an *aggregate*, given by the following generic structure in a description logic setting [13]:

$$
\begin{aligned}
Aggregate\_Concept \; \equiv \; & Parent\_Concept_1 \sqcap ... \sqcap Parent\_Concept_n \sqcap \\
& \exists_{\geq m_1} hasPartRole.Part\_Concept_1 \sqcap \\
& ... \\
& \exists_{\geq m_k} hasPartRole.Part\_Concept_k \sqcap \\
& conceptual\ constraints
\end{aligned}
$$

In Figure 1, a simplified extract of the OWL ontology, used for modelling airport activities, is shown as a screenshot of Protégé, as pure OWL notation is not convenient to read. In the left frame the taxonomy is shown and in the right frame the properties of the selected conceptual object `Vehicle-Enters-Zone` are displayed.

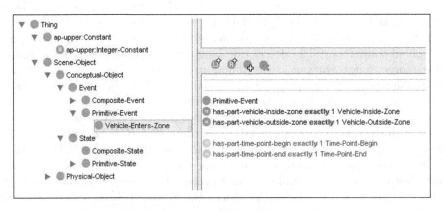

Fig. 1. OWL ontology in Protégé

For modelling the *conceptual constraints* we use SWRL. An example of a SWRL rule, which expresses the conceptual constraint that instances of the physical objects `Vehicle` and `Zone` have to be the same in both events, is given below (for simplification the temporal constraints are omitted here.

```
Vehicle-Enters-Zone(?vez) ^
has-part-vehicle-inside-zone(?vez, ?viz) ^
has-part-vehicle(?viz, ?v1) ^
has-part-zone(?viz, ?z1) ^
has-part-vehicle-outside-zone(?vez, ?voz) ^
has-part-vehicle(?voz, ?v2) ^
has-part-zone(?voz, ?z2) ^
->
swrlb:equal(?v1, ?v2) ^
swrlb:equal(?z1, ?z2)
```

In this way an aggregate hierarchy is modelled with primitive aggregates - like Vehicle-Inside-Zone - as leaves, which can be directly instantiated based on visual attributes of physical objects computed by the perceptual components of the scene interpretation system (see Section 3.1), and more complex aggregates, defined with aggregates as parts.

Transforming Aggregates to Jess Rules. In this paragraph it will be described, how aggregates can be transformed into rules for the rule-engine Jess in a systematic, automatable way, realising possible interpretation steps and sustaining the object-centered structure of aggregates.

Scene interpretation cannot be solely modelled as deduction. It has been shown in [14] that constructing a scene interpretation is essentially a search problem in the space of possible interpretations defined by the taxonomical and compositional relations by incrementally instantiating concepts while maintaining consistency. Four kinds of interpretations steps are necessary:

- Aggregate instantiation (moving up a compositional hierarchy).
- Aggregate expansion (moving down a compositional hierarchy).
- Instance specialisation (moving down a taxonomical hierarchy).
- Instance merging (unifying instances obtained separately).

In this paper, we will focus on the first step - aggregate instantiation - which is a bottom-up step and the backbone for scene interpretation.

A main structuring feature of the Jess rule language is a *template*, which can be seen as analogon to a Java class. A template is defined by a name and a number of *slots*, which are comparable to member variables of a Java class. In the first step of the transformation of aggregates to Jess, every concept is defined by a template with the name of the concept. The slots of the template are defined corresponding to the properties of the concept with an additional slot *name*, which holds the name of the instance (e.g. vehicle_17). Here our approach differs from the transformation of OWL and SWRL to Jess described in [5], where the properties are modelled as *ordered facts*. Ordered facts are simply Jess lists, which perform an implied template creation. This would mean to lose the object-centered structure of an aggregate, as the properties are decoupled from the concept template. The other significant difference is that we keep the

OWL taxonomy by defining the templates with `extends`, which is used to express inheritance in Jess. In this way the OWL subclass relation of class C and D

$$C \sqsubseteq D \tag{1}$$

is directly transformed into the template inheritance structure of Jess. In the realisation described in [5] the template structure is flat and the taxonomy is emulated by duplicating the facts along the taxonomical hierarchy, which could lead to problems with scalability.

In the second step of the transformation, a rule is defined for every aggregate of the OWL ontology. In the predicate part (LHS) of the rule, the parts of the aggregate are listed together with the slots which are needed to express the conceptual constraints, as far as possible. With `part-of` relations it can be checked that a part is not already integrated into another aggregate instance. Constraints that go beyond the scope of a single aggregate - for example temporal constraints - are processed procedurally in the Java part of the system and appear in the predicate part of the rule as a test function (*test conditional element*). This will be described in detail in Section 2.2. In the action part (RHS) of the rule, the aggregate is instantiated, properties are modified accordingly, and the temporal constraint net is updated.

An example for a Jess rule for the (simplified) aggregate `Refueling` is given below[2] (the temporal constraint processing is only sketched here).

```
(defrule Refueling
  ?tez-id <-
  (Tanker-Enters-Zone (name ?tez)
    (has-part-tanker ?v1)
    (has-part-zone ?z1)
    (part-of-refueling nil))
  ?dr-id <-
  (Do-Refuel (name ?dr)
    (has-part-tanker ?v1)
    (has-part-zone ?z1)
    (part-of-refueling nil)
  ?tlz-id <-
  (Tanker-Leaves-Zone (name ?tlz)
    (has-part-tanker ?v1)
    (has-part-zone ?z1)
    (part-of-refueling nil))
  ;; check temporal constraints in a test function
  =>
  ;; create new instance of Refueling
  (assert
    (Refueling (name ?rf-new)
      (has-part-tanker-enters-zone ?tez)
```

---

[2] `?x-id<-` is an identifier needed for modification of facts in the RHS part of a rule.

```
      (has-part-do-refuel ?dr)
      (has-part-tanker-leaves-zone ?tlz)))
  ;; modify properties of parts
  (modify ?tez-id (part-of-refueling ?rf-new))
  (modify ?dr-id (part-of-refueling ?rf-new))
  (modify ?tlz-id (part-of-refueling ?rf-new))
  ;; update temporal constraint net
)
```

2.2   Constraints

Spatial and temporal context play a special part in scene interpretation. But as already mentioned and shown in [13], it is difficult in a DL system to represent constraints between conceptual objects, and a DL system does not provide a framework for flexible, stepwise scene interpretation. In this section the integration of a global *temporal constraint net* is introduced which controls the activation of rules and stepwise aggregate instantiations, maintaining consistency of the temporal constraints.

Temporal Constraint Net. Temporal constraints are essential in a domain like airport activity monitoring. For the modelling of temporal relations, we use the convex time point algebra [15]. The *Allen temporal operators* used in the SWRLTemporalOntology[3] are not expressive enough for our purposes, because they only allow the modelling of qualitative relations, whereas the complexity of our domain requires quantitative models.

The basic format of a temporal relation in the convex time point algebra is

$$t_1 \geq t_2 + c_{12} \tag{2}$$

where t1 and t2 are interval-valued time points and c12 is an integer-valued constant. Using such inequalities, it is possible to model important features of the temporal structure of a scene model.

Figure 2 illustrates a more detailed aggregate of `Refueling`. In the OWL ontology every concept has two temporal data type properties: `has-start-time` for the beginning time point (`x-tb`) and `has-finish-time` for the ending time point (`x-te`). Other properties are not listed here. The temporal constraints are as follows. A `Tanker-Enters-Fuel-Access-Area` event has to occur before a `Tanker-Stopped-In-Fuel-Access-Area` event, which has to happen before a `Tanker-Leaves-Fuel-Access-Area` event. A `Handler-Plugged-Fuel` event has to occur before a `Handler-Unplugged-Fuel` event. Both events occur during the `Tanker-Stopped-In-Fuel-Access-Area` event. Every event has to fulfill a certain duration. Analog to the conceptual constraints in Section 2.1 also these temporal constraints can be expressed with SWRL rules in the form of inequalities, mentioned in (2). Part of the SWRL rule for the `Refueling` aggregate which

---

[3] http://protege.cim3.net/cgi-bin/wiki.pl?SWRLTemporalOntology

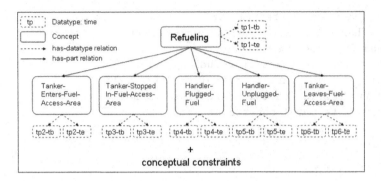

Fig. 2. Aggregate `Refueling` with time properties

concerns the temporal constraints of `Tanker-Stopped-In-Fuel-Access-Area` is given below.

```
Refueling(?rf) ^
has-part-tanker-stopped-in-fuel-access-area(?rf, ?tsifaa) ^
duration-of-tanker-stopped-in-fuel-access-area(?rf, ?tsifaa-dur) ^
has-Start-Time(?tsifaa, tsifaa-tb) ^
has-Finish-Time(?tsifaa, tsifaa-te) ^
swrlb:add(?sum-tb-dur, ?tsifaa-tb, ?tsifaa-dur) ^
swrlb:greaterThanOrEqual(?tsifaa-te, ?sum-tb-dur) ^
...
```

Beside the transformation of aggregates to Jess rules, the transformation process must also generate a *temporal constraint net* (TCN) out of the SWRL rules. The outcome is a single global TCN which includes all aggregates modelled in the OWL ontology. An extract of the TCN concerning the `Refueling` aggregate, is shown in Figure 3.

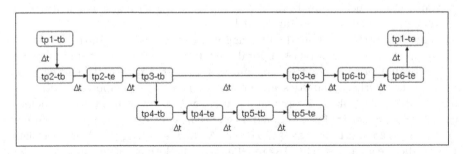

Fig. 3. Temporal constraint net for `Refueling`

The nodes are time points corresponding to the time marks given in the SWRL rules. The directed arcs represent inequalities, marked with an offset which represents the ideal value of the duration. Each node is interval-valued, where the interval denotes the range of the time points which is consistent with the constraints. Initially the intervals are open-ended, i.e. $[-\infty \ +\infty]$. When an aggregate is instantiated - that is when a rule fires - the corresponding nodes will receive concrete values. For example, if a `Tanker-Stopped-In-Fuel-Access-Area` event starts at 27, then the time point `tp3-tb` will receive the value [27 27]. New values are propagated through the constraint net as follows [12]:

- minima in edge direction: $\qquad t_{2min}' = \max(t_{2min}, t_{1min} + c_{12})$.
- maxima against edge direction: $\quad t_{1max}' = \min(t_{1max}, t_{2max} - c_{12})$.

A TCN is inconsistent, if for any node $t_{min} > t_{max}$ holds.

Implementation of Temporal Constraint Net with Shadow Facts. As the propagation of values through the constraint net has a procedural character and can effect all time points - i.e. not only the time points in one single aggregate - it is reasonable and efficient to implement this in Java, whereas a pure Jess implementation would be unnecessary complex and intransparent.

To establish a connection between the Java implementation and Jess objects, all time points and the TCN itself are implemented as *shadow facts* [9], i.e. every time point in Jess has a corresponding time point object in Java.[4] The usage of shadow facts enables Jess to perform reasoning about Java objects. Together with the integration of the TCN into the rules, a general structure of an aggregate bottom-up rule can be given schematically as follows (t_{xy} denote time points):

```
(defrule Aggregate-X-Rule
    - part1 with t11, t12
    - part2 with t21, t22
    ...
    (TCN (tMins $?tMins) (tMaxs $?tMaxs) (OBJECT ?tcn-obj))
    (test (call ?tcn-obj propagateAndCheckConsistency
        $?tMins
        $?tMaxs
        t11, t12, t21, t22,...))
    =>
    - instantiate aggregate X with t31, t32
    - modify part-of relations of parts
    (call ?tcn-obj update
        $?tMins
        $?tMaxs
        t11, t12, t21, t22,..., t31, t32))
```

When the TCN is initialised, all time points are "normal" objects (not connected to Jess facts). At the moment a rule is matched against the working memory, the `propagateAndCheckConsistency` function in the LHS part of the rule propagates

---

[4] Every shadow fact has a slot OBJECT which holds a reference to the Java object itself.

the time point values of the parts of the aggregate through a copy of the original TCN (because the original TCN must not be changed). If the TCN becomes inconsistent the function returns `false`, that means it cannot be satisfied with the currently checked instances, thus the rule must not be activated. If the function returns `true` (and all other constraints of the LHS are fulfilled), then the rule will be activated. If the rule fires, the `update` function in the RHS part of the rule integrates the time points into the original TCN (now these nodes are shadow facts) and propagates the values. In this way the instantiation of aggregates is achieved, while maintaining consistency of the temporal constraints.

Here it must be mentioned, that a pure Jess implementation of the temporal constraint net would be possible in principle with Jess functions. But the realisation of the TCN introduced here is only a preliminary stage for a more sophisticated component where probability distributions replace crisp time intervals so that a context-dependent certainty value can be generated for activated rules. This calls for more complex computations not easily realisable in Jess.

3   Interpretation Process

In this section, we describe the basic bottom-up interpretation process based on the rules generated from the OWL ontology. In the first part, a system overview of a general scene interpretation system is given and the interpretation process is described. In the second part, it is demonstrated, how parallel processing of interpretations can be realised.

3.1   Interpretation Process and System Overview

A basic framework for high-level scene interpretation can be subdivided into three main layers:

- The segmentation and tracking unit (low-level processing layer).
- The metric-symbolic interface (middle layer).
- The high-level interpretation layer.

In the segmentation and tracking unit, static or moving objects are detected by low-level image processing components. The objects are classified into *view* types. A view is a representation of the visual evidence of a physical object. Objects are tracked throughout image sequences, and object trajectories are computed for moving objects. In the middle layer, primitive aggregates are computed. In our domain of airport activities these are primitive states (which describe properties of physical objects that are true for a given time interval), like `Vehicle-Stopped-In-Zone` and primitive events (which describe one or several change(s) of properties of physical objects in a time interval), like `Vehicle-Enters-Zone`. These primitive aggregates serve as input for our rule-based high-level interpretation layer. With several interpretation steps, mentioned in Section 2.1, and the usage of conceptual knowledge, the interpretation layer performs the inference of high-level aggregates which represent assertions

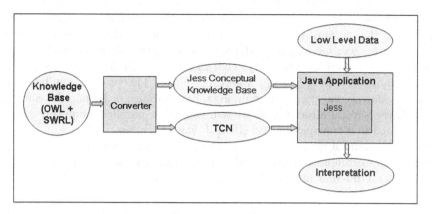

Fig. 4. Architecture of interpretation system

about complex activities in the scene. The usefulness of this architecture for scene interpretation was already demonstrated by [6].

In the initialisation (or offline) phase of the system, the concepts of the conceptual knowledge base are transformed to templates and aggregates are transformed to rules, all written to data files. An initialisation file for the temporal constraint net is generated out of the SWRL rules. These files together form the *Jess conceptual knowledge base*. In the working (or online) phase of the system, these data files are read by the Java application with the embedded Jess engine. The templates and rules are added to the engine, a temporal constraint net is initialised and also added to the Jess engine as a shadow fact. Now the system is ready to process the primitive aggregates provided by the middle layer. In the present stage of the project, the primitive aggregates are read from XML files, in the future they will be provided by a CORBA interface. Corresponding to the time marks given in the XML files, the primitive aggregates are added successively as facts to the working memory of the Jess engine, simulating an evolving scene. Then the *agenda*, i.e. the list of *activations* (rules that can fire when the engine is started) of the Jess engine is analysed. If the agenda is not empty, the command is given to run the engine. The rules fire and add new facts, representing instances of higher level aggregates, to the working memory. Continuing this in a loop, more and more aggregates - defined higher up in the hierarchy and representing more complex activities - are instantiated, consistent with the corresponding conceptual constraints (see Figure 4).

This way a framework for stepwise scene interpretation is realised. The consistency of the rules is guaranteed as far as possible, as they are generated from an OWL ontology which provides automatic consistency checks (except for SWRL rules).

3.2   Parallelisation

In this section the necessity of parallel computing in the scene interpretation process is motivated and the technical realisation is demonstrated.

As mentioned before, constructing a scene interpretation is essentially a search problem in the space of possible interpretations. In a real-time scene interpretation system, e.g. for airport activity monitoring, it cannot be avoided that evidence is processed incrementally. That means, early interpretation steps may be ambiguous because of lack of supporting context. This problem can be solved either by allowing backtracking to undo faulty decisions, or by parallel computing to follow several alternatives. We will show that parallel computing can be implemented in a transparent and efficient way, using Jess.

In our domain of airport activities it is not unusual that an instance of an aggregate, for example an instance of `Vehicle-Enters-Zone`, could be a part of one of several different instantiations (see Figure 5).

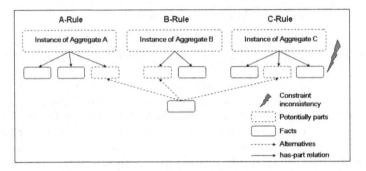

Fig. 5. Possible alternatives in an interpretation step

Assuming that the conceptual constraints are only fulfilled in rule A and rule B, both will be activated and put onto the agenda. Rule C will not be activated. Because the fact is exclusively part of either instance A or instance B, the order in which the rules fire is decisive for the result: if rule A fires, then rule B will be deactivated and vice versa (this is controlled by the `part-of` relation, mentioned above). To follow both alternative interpretation paths, the actual Jess engine is cloned in this situation. By using the Jess mechanism of serialisation and deserialisation, it is ensured that correct (deep) copies of Java objects, implemented as shadow facts, are created. After cloning the Jess engine, rule A and rule B are activated in both engines. Now, we want to fire rule A in clone_1 and rule B in clone_2. This can be achieved by using a special conflict strategy which can be easily set in the Jess engine to manipulate the execution engine accordingly. Concretely, to explicitly fire a certain rule, a strategy is set, which gives the priority to the activation with a certain activation name (this name is unique). Then both engines are executed in different threads. Directly after the first rule has fired in a clone, the strategy is reset to the original strategy (this can be done by an event handler). When the `setStrategy` function of the Jess engine returns, all remaining activations are re-ordered, according to the new (original) strategy.

In the next loop, new facts, provided by the middle layer, will be added to the working memory of every clone. If the TCN is not satisfiable anymore, then the thread dies. It can be assumed that in the beginning of the scene, the initial thread branches into several parallel threads very quickly, as there is less context information. But with preceding evolution of the scene the number of threads will decrease. For example a primitive event like Person-Enters-Zone can be a part of various events, whereas higher aggregates like Refueling are only part of one or two higher events. Future experience will show which maximal number of threads is useful.

4   Results

In this section a simple example of a scene interpretation process is demonstrated in the domain of airport activities which is the application domain in in the EU project *Co-Friend*.[5]

For our experiment we assume a simplified aggregate Refueling with the parts Tanker-Enters-Zone, Do-Refuel and Tanker-Leaves-Zone. Another aggregate Tanker-Enters-And-Leaves-Zone consists only of the parts Tanker-Enters-Zone and Tanker-Leaves-Zone, both with their respective conceptual constraints. The second aggregate is a model for the activity that a tanker enters and leaves the specific zone without refueling the aircraft for any reason. Normally an assured evidence for Do-Refuel should inhibit the activation of the rule Tanker-Enters-And-Leaves-Zone-Rule, but in future work other interpretation steps - beside bottom-up - will be realised which also include hypothesising facts, for example, in cases where evidence is missing because of occlusion. Hence, in general it could make sense to follow both alternatives.

An OWL-DL ontology, including the aggregates and physical objects described above, has been created with Protégé 3.4. The global consistency of the ontology has been checked with the OWL reasoner RacerPro 2.0. SWRL rules have been defined to express the conceptual constraints of the aggregates with the integrated SWRL functionality of Protégé 3.4.

The data files for the rules and the templates have been generated manually. The instances of the three aggregates are read from an XML data file (for simplification we assume Do-Refuel to be a primitive aggregate here) and added to the Jess engine one after another.

In Figure 6, an extract of the output of the experiment is shown. It can be seen that as soon as the instance of Tanker-Leaves-Zone is added to the working memory, the Refueling-Rule and the Tanker-Enters-And-Leaves-Zone-Rule are activated and put onto the agenda in engine_1. Then the engine is cloned, thus, a new engine_2 is created with the same status. Then both engines are executed. As desired, the Refueling-Rule fires in engine_1, and the Tanker-EntersAnd-Leaves-Zone-Rule fires in engine_2. Furthermore, the instantiation of Refueling results in adding a Refueling fact to the working

---

[5]   This work was partially supported by the EC, Grant 214975, Project Co-Friend.

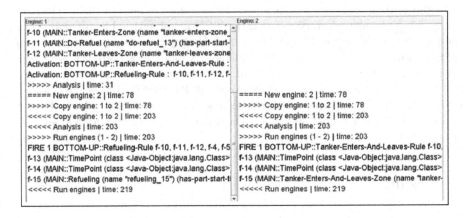

Fig. 6. Output for example of interpretation process

memory in engine_1, and a `Tanker-Enters-And-Leaves-Zone` fact is added in engine_2.

5   Conclusion and Future Work

In this paper we have presented a novel architecture for high-level scene interpretation, which is based on the generation of rules from an OWL-DL ontology. It has been shown how aggregates can be transformed into a rule base of Jess in a systematic way and how a global temporal constraint net can be integrated. A general rule pattern has been given for the transformation of aggregates into bottom-up interpretation rules which provide the backbone of scene interpretation. Furthermore, the usage of these rules in the scene interpretation process was explained with examples of airport activities. The technical functionality of parallel computing, with the intention to follow alternative interpretation steps, has been shown and a first simple experiment was demonstrated.

In ongoing work, rule generations for the remaining interpretation steps, i.e. aggregate expansion, instance specialisation and instance merging, are elaborated. All transformations from the OWL ontology into rules, templates, and the temporal constraint net will be fully automated.

As an advanced use of rule-based processing, the inclusion of *common sense* inferences will be investigated. The goal here is to conclude missing facts not provided by low-level image analysis by rules which reflect every-day human experiences, for example about natural motion of physical objects.

The original conflict strategy of Jess will be replaced by a probabilistic strategy to provide a preference measure for interpretations steps. In this way the most promising alternatives of interpretations will be traced in parallel as a beam search. Finally more complex experiments will be performed with real input data obtained from aircraft activities captured at Toulouse Airport in project Co-Friend.

References

1. Arens, M., Nagel, H.-H.: Behavioral knowledge representation for the understanding and creation of video sequences. In: Günter, A., Kruse, R., Neumann, B. (eds.) KI 2003. LNCS (LNAI), vol. 2821, pp. 149–163. Springer, Heidelberg (2003)
2. Arens, M., Ottlik, A., Nagel, H.-H.: Using behavioral knowledge for situated prediction of movements. In: Biundo, S., Frühwirth, T., Palm, G. (eds.) KI 2004. LNCS (LNAI), vol. 3238, pp. 141–155. Springer, Heidelberg (2004)
3. Borg, M., Thirde, D., Ferryman, J., Fusier, F., Valentin, V., Brémond, F., Thonnat, M.: A Real-Time Scene Understanding System for Airport Apron Monitoring. In: Proc. of IEEE International Conference on Computer Vision Systems (ICVS-06) (2006)
4. Bremond, F., Thonnat, M., Zuóniga, M.: Video Understanding Framework for Automatic Behavior Recognition. Behaviour Research Methods 3(38), 416–426 (2006)
5. Eriksson, H.: Using JessTab to Integrate Protégé and Jess. IEEE Intelligent Systems 18(2), 43–50 (2003)
6. Fusier, F., Valentin, V., Brémond, F., Thonnat, M., Borg, M., Thirde, D., Ferryman, J.: Video understanding for complex activity recognition. Machine Vision and Applications 18(3-4), 167–188 (2007)
7. Gerber, R., Nagel, H.-H.: Occurrence Extraction from Image Sequences of Road Traffic Scenes. In: van Gool, L., Schiele, B. (eds.) Proc. Workshop on Cognitive Vision, Switzerlan, pp. 1–8 (2002)
8. Georis, B., Maziére, M., Brémond, F., Thonnat, M.: Evaluation and Knowledge Representation Formalisms to Improve Video Understanding. In: Proc. ICVS 2006 (2006)
9. Friedman-Hill, E.: Jess in Action: Java Rule-Based Systems, Manning, Greenwich (2003)
10. Hotz, L., Neumann, B.: Scene Interpretation as a Configuration Task. In: Kuenstliche Intelligenz, vol. 3, pp. 59–65. BoettcherIT Verlag (2005)
11. Moeller, R., Neumann, B.: Ontology-based reasoning techniques for multimedia interpretation and retrieval. In: Kompatsiaris, Y., Hobson, P. (eds.) Semantic Multimedia and Ontologies: Theory and Applications, pp. 55–98. Springer, Heidelberg (2008)
12. Neumann, B.: Description of Time-Varying Scenes. In: Waltz, D. (ed.) Semantic Structures. Lawrence Erlbaum, Mahwah (1989)
13. Neumann, B., Möller, R.: On scene interpretation with description logics. In: Christensen, H.I., Nagel, H.-H. (eds.) Cognitive Vision Systems. LNCS, vol. 3948, pp. 247–275. Springer, Heidelberg (2006)
14. Neumann, B., Weiss, T.: Navigation through logic-based scene models for high-level scene interpretations. In: Proc. 3rd Int. Conf. on Computer Vision Systems (ICVS 2003), pp. 212–222 (2003)
15. Vila, L.: A survey on Temporal Reasoning in Artifical Intelligence. AI Communications 7(1), 4–28 (1994)
16. Van-Thinh, V., Brémond, F., Thonnat, M.: Automatic video interpretation: A recognition algorithm for temporal scenarios based on pre-compiled scenario models. In: Crowley, J.L., Piater, J.H., Vincze, M., Paletta, L. (eds.) ICVS 2003. LNCS, vol. 2626, pp. 523–533. Springer, Heidelberg (2003)
17. Zhu, S.-C., Mumford, D.: A Stochastic Grammar of Images. Now Publishers (2007)

RBDT-1: A New Rule-Based Decision Tree Generation Technique

Amany Abdelhalim, Issa Traore, and Bassam Sayed

Department of Electrical and Computer Engineering
University of Victoria, P.O. Box 3055 STN CSC,
Victoria, B.C., V8W 3P6, Canada
Ph.: (250) 721-6036; Fax: (250) 721-6052
{amany,itraore,bassam}@ece.uvic.ca

Abstract. Most of the methods that generate decision trees use examples of data instances in the decision tree generation process. This paper proposes a method called *"RBDT-1"*- rule based decision tree -for learning a decision tree from a set of decision rules that cover the data instances rather than from the data instances themselves. The method's goal is to create on-demand a short and accurate decision tree from a stable or dynamically changing set of rules. We conduct a comparative study of *RBDT-1* with three existing decision tree methods based on different problems. The outcome of the study shows that *RBDT-1* performs better than *AQDT-1* and *AQDT-2* which are rule-based decision tree methods in terms of tree complexity (number of nodes and leaves in the decision tree). It is also shown that *RBDT-1* performs equally well in terms of tree complexity compared with *C4.5*, which generates a decision tree from data examples.

Keywords: attribute selection criteria, decision rules, data-based decision tree, rule-based decision tree, tree complexity.

1 Introduction

The most common methods for creating decision trees are those that create decision trees from a set of examples (data records). We refer to these methods as data-based decision tree methods. The *attribute selection criteria* are the essential characteristics in all those methods [1]. These criteria are used to choose the best attributes to be assigned to the nodes of the decision tree. Examples of such criteria include the entropy reduction [2], the gini index of diversity [3], and others [4], [5].

On the other hand, to our knowledge there are only few approaches proposed in the literature that create decision trees from rules which we refer to as rule-based decision tree methods.

A Decision tree can be an effective tool for guiding a decision process as long as no changes occur in the dataset used to create the decision tree. Thus, for the data-based decision tree methods once there is a significant change in the data, restructuring the decision tree becomes a desirable task. However, it is difficult to manipulate or restructure decision trees once constructed. This is because a decision tree is a

G. Governatori, J. Hall, and A. Paschke (Eds.): RuleML 2009, LNCS 5858, pp. 108–121, 2009.

procedural knowledge representation, which imposes an evaluation order on the attributes in the tree. In contrast, rule-based decision tree methods handle manipulations in the data through the rules induced from the data not the decision tree itself. A declarative representation, such as a set of decision rules is much easier to modify and adapt to different situations than a procedural one. This easiness is due to the absence of constraints on the order of evaluating the rules [6].

On the other hand, in order to be able to make a decision for some situation using the set of rules we need to decide the order in which tests should be evaluated. In that case a decision structure (e.g. decision tree) will be created from the rules. So the methods that create decision trees from rules combine the best of both worlds. On one hand they easily allow changes to the data (when needed) by modifying the rules rather than the decision tree itself. On the other hand they take advantage of the structure of the decision tree to organize the rules in a concise and efficient way required to take the best decision. So knowledge can be stored in a declarative rule form and then be transformed (on the fly) into a decision tree only when needed for a decision making situation [6].

In addition to that, generating a decision structure from decision rules can potentially be performed faster than generating it from training examples because the number of decision rules per decision class is usually much smaller than the number of training examples per class. Thus, this process could be done on demand without any noticeable delay [2], [7]. Methods that create decision trees from examples or data require examining the complete tree to extract information about any single classification. Even converting the tree into a set of individual rules could also result in a large amount of rules if the tree is large, which could be the case when the tree is based on a large dataset. Otherwise, with methods that create decision trees from rules, extracting information about any single classification can be done directly from the declarative rules [8].

Although rule-based decision tree methods create decision trees from rules, they could be used also to create decision trees from examples by considering each example as a rule. Data-based decision tree methods create decision trees from data only. Thus, when generating a decision tree for problems were rules are provided e.g. by an expert and no data is available, rule-based decision tree methods are the only applicable solution.

This paper presents a new rule-based decision tree method called *RBDT-1*. To derive the tree, the *RBDT-1* method uses in sequence three different criteria to determine the fit (best) attribute for each node of the tree, which are referred to as the attribute effectiveness (AE), the attribute autonomy (AA), and the minimum value distribution (MVD).

The rest of the paper is structured as follows. Section 2 summarizes the related work. Section 3 discusses the rule generation approach and notations used in this work. Section 4 describes the *RBDT-1* method by illustrating, in particular, the preparation of the rules into a format that will be used by the method, the different criteria used in the attribute selection process, the pruning technique adopted by the method and also by providing an illustration of the method using a small dataset. Section 5 presents the results of an experiment in which, based on public datasets, the proposed method is compared to two existing methods for creating decision trees from declarative rules, namely the *AQDT-1* [6] and *AQDT-2* [8] methods, and to the *C4.5* algorithm [9] that creates decision trees from data examples. In Section 6, we make some concluding remarks and outline our future work.

2 Related Work

There are few published works on creating decision structures from declarative rules.

The *AQDT-1* method introduced in [6] is the first approach proposed in the litera-ture to create a decision tree from decision rules. The *AQDT-1* method uses four crite-ria for selecting the fit attribute that will be placed at each node of the tree. Those criteria are the *cost[1]*, the *disjointness*, the *dominance*, and the *extent*, which are ap-plied in the same specified order in the method's default settings.

The *AQDT-2* method introduced in [8] is a variant of *AQDT-1*. *AQDT-2* uses five criteria in selecting the fit attribute for each node of the tree. Those criteria are the *cost, disjointness, information importance, value distribution,* and *dominance,* which are also applied in the same specified sequence in the method's default settings. In both the *AQDT-1 & 2* methods, the order of each criterion expresses its level of im-portance in deciding the attribute that will be selected for a node in the decision tree. Although both *AQDT-1 & 2* are capable of generating a decision tree from a set of rules, experiments presented in this paper show that our proposed method *RBDT-1* produces a less complex tree in most of the cases.

Another point is that the calculation of the second criterion - the *information impor-tance* - in *AQDT-2* method depends on the training examples, which contradicts the method's fundamental idea of being a rule-based decision tree method rather than a data-based decision tree method. *AQDT-2* requires both the examples and the rules to calculate the *information importance* at certain nodes where the first criterion- *Dis-jointness* - is not enough in choosing the fit attribute. *AQDT-2* being both dependent on the examples as well as the rules results in an increase of the running time of the algorithm remarkably in large datasets especially those with large number of attributes.

In contrast the multi criteria calculations of the *RBDT-1* method, proposed in this work, for generating a decision tree require only the set of rules given to the method as an input, and does not require the presence of the examples used to induce those rules. The calculations of all the method's criteria are based on certain characteristics of the attributes intrinsic to the rules only.

Akiba et al. [10] proposed a rule-based decision tree method for learning a single decision tree that approximates the classification decision of a majority voting classi-fier. Their method was proposed as a possible solution to solve the issues of intelligi-bility, classification speed, and required space in majority voting classifiers. In their proposed method, if-then rules are extracted from each classifier which is a decision tree generated using the data-based decision tree method *C4.5*. They use the extracted rules used to learn a single decision tree. The goal of the method is to provide an approximation of the majority voting classifier rather than an exact matching behav-ior. The method that they propose depends both on the real examples used to create the classifiers (decision trees) and on a set of training examples that they create using the rules extracted from the classifiers. The procedure that they follow in selecting the best attribute at each node of the tree is based on the *C4.5* as well. The size of a deci-sion tree learned by Akiba et al. method while using rules extracted from multiple classifiers built by *C4.5* is about 1.2 to 4.2 times the size of a decision tree learned by *C4.5* from the data [10]. As will be shown in the experiments, when using rules ex-tracted from a *C4.5* decision tree, the *RBT-1* method generates a tree that is the same size as the decision tree learned by *C4.5* from data, even smaller in some of the cases with an equal accuracy.

In [11], the authors proposed a method called Associative Classification Tree (ACT) for building a decision tree from association rules rather than from data. They proposed two splitting algorithms for choosing attributes in the *ACT*. The first algorithm is based on the confidence gain criterion and the second is based on the entropy gain criterion. In both splitting algorithms the attribute selection process at each node relies on both the existence of rules and the data itself as well. Unlike our proposed method *RBDT-1*, *ACT* is not capable of building a decision tree from the rules in the absence of data, or from data (considering them as rules) in the absence of rules.

3 Rule Generation and Notations

In order for our proposed method to be capable of generating a decision tree for a certain dataset, it has to be presented with a set of rules that cover the dataset. The rules will be used as input to *RBDT-1* which will produce a decision tree as an output. The rules can either be provided up front, for instance, by an expert or can be generated algorithmically.

Let $a_1,...,a_n$ denote the attributes characterizing the data under consideration, and let $D_1,...,D_n$ denote the corresponding domains, respectively (i.e. D_i represents the set of values for attribute a_i). Let $c_1,...,c_m$ represent the decision classes associated with the dataset.

The datasets that we use in our experiments are based on classification problems where each example in the dataset belongs to only one class. Thus, a desirable form of a rule set would be a logically disjoint and complete family of rule sets. Thus, given a collection of rule sets, one for each class decision, no two rule sets for two different classes shall logically intersect and the union of all the rule sets shall cover the whole dataset. In such a case, each possible example in the dataset will belong to one of the predefined classes. So the decision classes induce a partition over the complete set of rules. Let P denote the complete set of rules and R_i denote the set of rules associated with decision class c_i. Hence, we have the following: $i \neq j \Rightarrow R_i \cap R_j = \varnothing$, where $1 \leq i, j \leq m$; $P = \bigcup_{1 \leq i \leq m} R_i$.

In our main experiment, we are comparing the decision tree generated by our proposed method to the *AQDT-1, AQDT-2*, and the *C4.5* methods. Since all the methods under comparison except the *C4.5* method are rule-based decision tree methods, one of the best ways to perform a fair comparison is to use the same rules extracted from *C4.5* decision tree itself as input to the other three rule-based methods. The method used to extract rules from the *C4.5* decision tree consists of converting each branch – from the root to a leaf – of the decision tree to an if-then rule whose condition part is a pure conjunction. This approach ensures that we will have a collection of disjoint rules.

We used *AQ19* [12] to generate the rule set used to illustrate *RBDT-1* method. *AQ19* is a rule induction program that belongs to the famous *AQ-type* family for machine learning and pattern discovery techniques, which are capable of creating logically disjoint rules.

4 RBDT-1 Method

In this section, we outline the format of the rules required for *RBDT-1* method, the attribute selection criteria of the method, and then summarize the main steps of the underlying decision tree building process. We also present the pruning technique adopted by the method. Finally, we illustrate the steps for generating a decision tree by the *RBDT-1* method using a small rule set.

4.1 Preparing the Rules

The decision rules must be prepared into the proper format used by the *RBDT-1* method. This is done by assigning a "don't care" value to all the attributes that were omitted in any of the rules. The "don't care" value is equivalent to listing all the values for that attribute.

For example, suppose that we have three attributes a_1, a_2 and a_3 with the same domain containing v_1, v_2 and v_3 as possible values.

Let us assume that the following rules correspond to class $c1$:

$r1$: $c1 \Leftarrow a_1=v_1 \& a_2=v_2,$ $r2: c1 \Leftarrow a_1=v_3$

The preparation of these two rules will result in the following formatted rules:

$r1: c1 \Leftarrow a_1=v_1 \& a_2=v_2 \& a_3="don't\ care",$ $r2: c1 \Leftarrow a_1=v_3 \& a_2="don't\ care" \& a_3="don't\ care"$

Each rule is submitted to *RBDT-1* in the form of an attribute-value vector. This vector will contain first the values of the attributes that appear in the rule followed by the class decision representing the last element of the vector. Thus, accordingly the previous two rules will be presented as follows:

$r1: (v_1, v_2, don't\ care, c1), r2: (v_3, don't\ care, don't\ care, c1)$

4.2 Attribute Selection Criteria

The *RBDT-1* method applies three criteria on the attributes to select the fittest attribute that will be assigned to each node of the decision tree. These criteria are *the Attribute Effectiveness, the Attribute Autonomy, and the Minimum Value Distribution.*

Attribute Effectiveness (AE). *AE* is the first criterion to be examined for the attributes. It prefers an attribute which has the most influence in determining the decision classes. In other words, it prefers the attribute that has the least number of "don't care" values for the class decisions in the rules, as this indicates its high relevance for discriminating among rule sets of given decision classes. On the other hand, an attribute which is omitted from all the rules (i.e. has a "don't care" value) for a certain class decision does not contribute in producing that corresponding decision. So it is considered less important than the other attributes which are mentioned in the rule for producing a decision of that class. Choosing attributes based on this criterion maximizes the chances of reaching leaf nodes faster which on its turn minimizes the branching process and leads to producing a smaller tree.

Using the notation provided above (see section 3), let V_{ij} denote the set of values for attribute a_j involved in the rules in R_i, which denote the set of rules associated with decision class c_i, $1 \le i \le m$. Let DC denote the 'don't care' value, we calculate $C_{ij}(DC)$ as shown in (1):

$$C_{ij}(DC) = \begin{cases} 1 & if \ DC \in V_{ij} \\ 0 & otherwise \end{cases} \tag{1}$$

Given an attribute a_j, where $1 \le j \le n$, the corresponding attribute effectiveness is given in (2).

$$AE(a_j) = \frac{m - \sum_{i=1}^{m} C_{ij}(DC)}{m} \tag{2}$$

(Where m is the total number of different classes in the set of rules).

The attribute with the highest AE is selected as the fit attribute. If more than one attribute achieve the highest AE score we will use the next criterion in our method, which is the *Attribute Autonomy* to determine the best attribute among them.

Attribute Autonomy (AA). *AA* is the second criterion to be examined for the attributes. This criterion is examined when the highest AE score is obtained by more than one attribute. This criterion prefers the attribute that will decrease the number of subsequent nodes required ahead in the branch before reaching a leaf node. Thus, it selects the attribute that is less dependent on the other attributes in deciding on the decision classes. We calculate the attribute autonomy for each attribute and the one with the highest score will be selected as the fit attribute. If more than one attribute achieve the highest AA score, we will use the next criterion in our method which is the *Minimum Value Distribution* to determine the best attribute among them.

For the sake of simplicity, let us assume that the set of attributes that achieved the highest AE score are $a_1, ..., a_s$, $2 \le s \le n$. Let $v_{j1}, ..., v_{jp_j}$ denote the set of possible values for attribute a_j including the "don't care", and R_{ji} denote the rule subset consisting of the rules that have a_j appearing with the value v_{ji}, where $1 \le j \le s$ and $1 \le i \le p_j$. Note that R_{ji} will include the rules that have don't care values for a_j as well.

The AA criterion is computed in terms of the Attribute Disjointness Score (ADS), which was introduced by [8]. For each rule subset R_{ji}, let $MaxADS_{ji}$ denote the maximum ADS value and let ADS_List_{ji} denote a list that contains the ADS score for each attribute a_k, where $1 \le k \le s, k \ne j$.

According to [8], given an attribute a_j and two decision classes c_i and c_k (where $1 \leq i, k \leq m; 1 \leq j \leq s$), the degree of disjointness between the rule set for c_i and the rule set for c_j with respect to attribute a_j is defined as shown in (3):

$$ADS(a_j, c_i, c_k) = \begin{cases} 0 & \text{if } V_{ij} \subseteq V_{kj} \\ 1 & \text{if } V_{ij} \supseteq V_{kj} \\ 2 & \text{if } V_{ij} \cap V_{kj} \neq (\emptyset \text{ or } V_{ij} \text{ or } V_{kj}) \\ 3 & \text{if } V_{ij} \cap V_{kj} = \emptyset \end{cases} \tag{3}$$

The *Attribute Disjointness* of the attribute a_j; $ADS(a_j)$ score is the summation of the degrees of class disjointness $ADS(a_j, c_i, c_k)$ given in (4):

$$ADS(a_j) = \sum_{i=1}^{m} \sum_{\substack{1 \leq k \leq s \\ i \neq k}} ADS(a_j, c_i, c_k) . \tag{4}$$

Thus, the number of *ADS_List* that will be created for each attribute a_j as well as the number of *MaxADS* values that are calculated will be equal to p_j. The *MaxADS*$_{ji}$ value as defined by [8] is $3 \times m \times (m-1)$ where m is the total number of classes in R_{ji}. We introduce the *AA* as a new criterion for attribute a_j as given in (5):

$$AA(a_j) = \frac{1}{\sum_{i=1}^{p_j} AA(a_j, i)} \tag{5}$$

Where $AA(a_j, i)$ is defined as shown in (6):

$$AA(a_j, i) = \begin{cases} 0 & \text{if } MaxADS_{ji} = 0 \\ \\ 1 & \text{if } \left((MaxADS_{ji} \neq 0) \wedge \left((s=2) \vee (\exists l : MaxADS_{ji} = ADS\_List_{ji}[l]) \right) \right) \\ \\ 1 + \left[(s-1) \times MaxADS_{ji} - \sum_{l=1, l \neq j}^{s} ADS\_List_{ji}[l] \right] & \text{otherwise} \end{cases}$$

$$\tag{6}$$

The *AA* for each of the attributes is calculated using the above formula and the attribute with the highest *AA* score is selected as the fit attribute. According to the above formula, $AA(a_j, i)$ equals zero when the class decisions for the rule subset examined corresponds to one class, in that case *MaxADS=0*, which indicates that a leaf node is reached (best case for a branch). $AA(a_j, i)$ equals 1 when *s* equals 2 or when one of the attributes in the *ADS_list* has an *ADS* score equal to *MaxADS* value (second best case). The second best case indicates that only one extra node will be required to reach a leaf node. Otherwise $AA(a_j, i)$ will be equal to 1 + (the difference between the *ADS* scores of the attributes in the *ADS_list* and the *MaxADS* value) which indicates that more than one node will be required until reaching a leaf node.

Minimum Value Distribution (MVD). The *MVD* criterion is concerned with the number of values that an attribute has in the current rules. When the highest *AA* score is obtained by more than one attribute, this criterion selects the attribute with the minimum number of values in the current rules. *MVD* criterion minimizes the size of the tree because the fewer the number of values of the attributes the fewer the number of branches involved and consequently the smaller the tree will become [8]. For the sake of simplicity, let us assume that the set of attributes that achieved the highest *AA* score are $a_1, ..., a_q$, $2 \leq q \leq s$. Given an attribute a_j (where $1 \leq j \leq q$), we compute corresponding *MVD* value as shown in (7).

$$MVD(a_j) = | \bigcup_{1 \leq i \leq m} V_{ij} | . \tag{7}$$

(Where |X| denote the cardinality of set *X*).

When the lowest *MVD* score is obtained by more than one attribute, any of these attributes can be selected randomly as the fit attribute.

4.3 Building the Decision Tree

We describe, in this section, the *RBDT-1* approach for building a decision structure from a set of decision rules. In our case the decision structure is a decision tree which is a single-parent decision structure. In the decision tree building process, we select the fit attribute that will be assigned to each node from the current set of rules *CR* based on the attribute selection criteria outlined in the previous section. *CR[1]* is a subset of the decision rules that satisfy the combination of attribute values assigned to the path from the root to the current node. From each node a number of branches are pulled out according to the total number of values available for the corresponding attribute in *CR*. Each branch is associated with a reduced set of rules *RR* which is a subset of *CR* that satisfies the value of the corresponding attribute. If *RR* is empty, then a single node will be returned with the value of the most frequent class found in the whole set of rules. Otherwise, if all the rules in *RR* assigned to the branch belong to the same decision class, a leaf node will be created and assigned a value of that

---

[1] *CR* will correspond to the whole set of rules at the root node.

decision class. The process continues until each branch from the root node is terminated with a leaf node and no more further branching is required.

4.4 Pruning Decision Rules

RBDT-1 is capable of handling the problem of generating a decision tree from noisy
training data. In *RBDT-1*, we handle noisy data by removing rules that cover only a
small portion of the data that could be considered noise [13]. The examples that were
covered by the truncated rules can often be covered by applying an analogical matching procedure. The analogical matching procedure determines the degree of similarity
between the examples to be classified and the rules of a given decision class, and
selects the best matching decision class [14]. In [15] experiments show that such a
rule truncation method not only simplifies decision rules which could lead to a simpler decision tree, but could also improve their prediction accuracy in some cases.

In *RBDT-1*, rules are pruned if their support level is less than or equal to a predefined threshold. The support level of a rule is the percentage of the total number of
examples covered by the rule (called *the t-weight*) to the total number of examples in
the given decision class.

4.5 The Weekend Problem

In this section, the steps for generating a decision tree by the *RBDT-1* method will be
explained in detail for a small dataset called the Weekend problem. The Weekend
problem is a dataset that consists of 10 data records obtained from [16]. We used the
AQ19 rule induction program to induce the rule set shown in Table 1 which will serve
as the input to our proposed method *RBDT-1* in this example. *AQ19* was used with the
mode of generating disjoint rules and with producing a complete set of rules without
truncating any of the rules.

Table 1. The weekend rule set induced by AQ19

Rule	Description
r1	*Cinema* ← *Parents-visiting="yes" & weather ="don't care" & Money ="rich"*
r2	*Tennis* ← *Parents-visiting="no" & weather ="sunny" & Money ="don't care"*
r3	*Shopping* ← *Parents-visiting="no" & weather ="windy" & Money ="rich"*
r4	*Cinema* ← *Parents-visiting="no" & weather ="windy" & Money ="poor"*
r5	*Stay-in* ← *Parents-visiting="no" & weather ="rainy" & Money ="poor"*

In order to choose the fit attribute for the root node of the tree, we first apply the
three criteria of the *RBDT-1* method on the attributes in the rules presented in Table 1.
The criteria are applied in the same order explained in the previous section. The *AE*
calculations for Parents-Visiting, Money and Weather attributes are {1, 0.75, 0.75}
respectively. Thus, the candidate attribute with the highest *AE* is the *parents-visiting*
attribute. Two branches will be pulled out from the *parents-visiting* attribute corresponding to its two values in the current rules, namely *yes and no*.

A subset of the weekend rule set will be assigned to the branch where
parents-visiting="yes". This subset of rules will consist of all the rules with

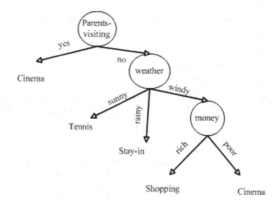

Fig. 1. The decision tree generated by RBDT-1 for the weekend problem before pruning

parents-visiting="yes" or *"don't care"*. The corresponding subset contains only one rule corresponding to rule {r1} with *"cinema"* as a decision. Thus, a leaf node *"cinema"* will be created and assigned to the branch where *parents-visiting="yes"*.

Another subset of the weekend rule set will be assigned to the branch where *parents-visiting="no"* corresponding to rules {r2, r3, r4, r5}. This subset of rules will consist of all the rules with *parents-visiting="no"* or *"don't care"*.

The *AE* calculations for Weather and Money attributes are {1, 0.75} respectively. Thus, the attribute with the highest *AE*; the *weather* attribute will be selected as the fit attribute for the branch *parents-visiting="no"*. Three branches will be pulled out from the *weather* attribute corresponding to its three values; *sunny, windy, and rainy* found in the *CR*. For the branch where *parents-visiting="no"* and *weather="sunny"*, there is only one rule that corresponds to that branch with a class decision *tennis* which is rule {r2}. So a leaf node with *tennis* as a decision will be created and assigned to that branch.

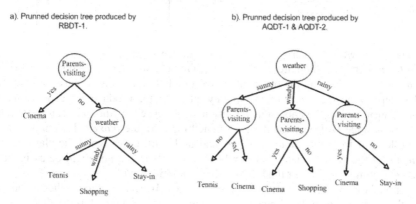

Fig. 2. The pruned decision tree generated by RBDT-1, AQDT-1 and AQDT-2 for the weekend problem

Table 2. Two sets of C4.5 Decision trees with prune option on and off along with the number of extracted rules and accuracy from each

Dataset	Prune option off			Prune option on		
	Tree Size (# nodes, # leaves)	# Extracted Rules	Acc[2]	Tree Size (# nodes, # leaves)	# Extracted Rules	Acc
Weekend	(2,4)	4	90 %	(2,4)	4	90 %
Lenses	(3,4)	4	91 %	(3,4)	4	91 %
Chess	(5, 10)	10	100 %	(5, 10)	10	100 %
Car	(52, 134)	134	96 %	(51,131)	131	96 %
Tic-Tac-Toe	(69, 139)	139	95 %	(47, 95)	95	93 %
Connect-4	(5317, 10635)	10635	91 %	(2142, 4285)	4285	87 %
Nursery	(264, 680)	680	99 %	(152, 359)	359	98 %
Balance	(22, 89)	89	80 %	(8,33)	33	75 %
MONK's 1	(29, 56)	56	97 %	(13,28)	28	100 %
MONK's 2	(88, 166)	166	85 %	(1,1)	1	67 %
MONK's 3	(5, 14)	14	100 %	(5,14)	14	100 %
Zoo	(8, 13)	13	99 %	(8, 13)	13	99 %
Breast-C	(27, 113)	113	85 %	(2,4)	4	74 %
Lung -C	(7, 12)	12	90 %	(6, 10)	10	87 %
Primary-T	(56, 67)	67	57 %	(41, 47)	47	58 %
Voting	(18, 19)	19	92 %	(5,6)	6	93 %

There are two rules corresponding to the branch where *parents-visiting="no" and weather="windy"* which are rules {r3, r4}. One of the rules corresponds to the decision class *shopping* and the other rule corresponds to the decision class *cinema*. Both rules depend on the value of the *money* attribute for producing the decision. The *money* attribute will be chosen as the fit attribute since it is the only candidate attribute left. A node will be created and assigned the attribute *money*. Two branches will be pulled out from the *money* node. A leaf node will be assigned the class decision *cinema* for the branch where *money= "poor"*. Another leaf node will be assigned the class decision *shopping* for the branch where *money="rich"*.

For the branch where *Parents-Visiting="no" and Weather="rainy"*, there is only one rule that corresponds to that branch with a class decision *stay-in,* thus it will be assigned

---

[2] *Acc* refers to the accuracy of the rules extracted and is calculated as the percentage of the number of examples correctly classified by the rules to the total number of examples.

to that branch. Overall, the corresponding decision tree created by the proposed *RBDT-1* method for the weekend problem is shown in Fig 1. It consists of 3 nodes and 5 leaves with 100% classification accuracy for the data. The decision tree created by *AQDT-1 & 2* using the same set of rules results in a tree of 5 nodes and 7 leaves - which is a bigger tree than that created by *RBDT-1* - with 100% classification accuracy for the data.

As indicated above, in *RBDT-1,* rules are pruned if their support level is less than or equal to a predefined threshold. Fig 2 shows a decision tree obtained after pruning the decision rules for the weekend problem in Table 1. In this case, we removed rules with the lowest support level. The produced decision tree by *RBDT-1* misclassified one example out of 10 giving a predictive accuracy of 90% and consists of 2 nodes and 4 leaves.

Fig 2 also shows the decision tree produced by *AQDT-1 & 2* using the same pruned rules; although the produced tree has the same accuracy as the tree produced with *RBDT-1,* it is bigger in size since it has 4 nodes and 6 leaves. When applying *C4.5* to the dataset of the weekend problem it resulted in a tree of the same size and accuracy as *RBDT-1* in Fig 2.

5   Experiment and Results

In order to evaluate the *RBDT-1* method, we conducted an experiment using 16 publicly available datasets, including the weekend dataset used in the previous section. Other than the weekend dataset, all the datasets appearing in Tables 2 and 3 were obtained from the UCI machine learning repository [17].

The evaluation consisted mainly of comparing the *RBDT-1* method with the *AQDT-1, AQDT-2,* and *C4.5* methods in terms of the complexity and accuracy of the decision trees produced. Since we were comparing our proposed method to *AQDT-1 & 2* methods which are all rule-based decision tree methods, it was a good idea to compare their performance with multiple rule sets.

Table 3. Comparison of the tree complexity of RBDT-1, AQDT-1, AQDT-2 & C4.5

Dataset	Experiment 1	Experiment 2	Dataset	Experiment 1	Experiment 2
Weekend	*RBDT-1, C4.5*	*RBDT-1, C4.5*	MONK's 1	*RBDT-1*	=
Lenses	*RBDT-1, C4.5*	*RBDT-1, C4.5*	MONK's 2	=	=
Chess	=	=	MONK's 3	*AQDT-1 & 2*	*AQDT-1 & 2*
Car	*RBDT-1, C4.5*	*RBDT-1, C4.5*	Zoo	*RBDT-1, C4.5*	*RBDT-1, C4.5*
Tic-Tac-Toe	=	=	Breast-C	=	=
Connect-4	*RBDT-1*	*RBDT-1, C4.5*	Lung-C	*RBDT-1, C4.5*	*RBDT-1, C4.5*
Nursery	*RBDT-1, AQDT-1 & 2*	=	Primary-T	*RBDT-1, C4.5*	*RBDT-1, C4.5*
Balance	=	=	Voting	*AQDT-1 &2*	=

The rules used for comparing the decision trees of *RBDT-1, AQDT-1, and AQDT-2* in the experiment are *C4.5-based rules*. Since *C4.5* is capable of handling datasets with missing values, we were capable of experimenting with both rules extracted from complete and incomplete datasets.

In our experiment, we used the C4.5 method under its default settings to generate two different decision trees for each dataset. One tree was generated with the pruning option turned on and the other with the pruning option turned off. Since the comparison was based on 16 datasets, 32 decision trees were generated. From each decision tree we extracted a rule set - as explained in section 3- which served as the input to the other three rule-based methods under comparison. So accordingly 32 rule sets were used in this experiment. The process is summarized in Table 2.

Results in Table 3 show that, in terms of tree size, *RBDT-1* performs better than *AQDT-1 & 2* in most of the rule sets. Based on the results in the two experiments, In Table 3, we illustrate the results of the comparison between *RBDT-1, AQDT-1, AQDT-2* and *C4.5* in two experiments, labeled experiment1 and experiment2. In each experiment the *C4.5* decision tree was generated from the whole set of examples of each dataset and the rules extracted from that tree served as input to the other three rule-based decision tree methods. In experiment1, the pruning option was turned off, while in experiment2 the pruning option was turned on. The name(s) appearing under each experiment correspond(s) to the method(s) that generated the smallest tree, while "=" indicates that all four methods produced the same tree.

AQDT-1 & 2 produce a larger tree by an average of 146.33 nodes with the exception of 3 rule sets where *RBDT-1*'s tree is larger by an average of 3 nodes. In addition, the results illustrate that *RBDT-1* is as effective as *C4.5* except in experiment1, where our method produced a slightly smaller tree for the *connect-4, MONK's 1* and *nursery* rule sets. In terms of accuracy, the four methods have equal performance.

6   Conclusion and Future Work

The *RBDT-1* method proposed in this work allows generating a decision tree from a set of rules rather than from the whole set of examples. Generating a decision structure from decision rules can potentially be performed much faster than by generating it from training examples.

Modifications to the data are handled easier in rule-based methods than in data-based methods. At the same time rule-based methods could transform the rules to a decision tree once we need to decide the order in which tests should be evaluated in those rules. Rule-based decision tree methods although designed to create decision trees from rules, could also generate decision trees from data examples. Rule-based decision tree methods are the only solution for generating a decision tree for applications where no data is available and only rules exist. The price of the *RBDT-1* advantages is the need to generate rules first before being capable of generating the tree. However, there are efficient rule learning systems available.

Experiments conducted in this work illustrates that in terms of tree complexity our proposed method *RBDT-1* performs better than *AQDT-1 & 2* in most of the rule sets, with an equal accuracy classification, while it is as effective as *C4.5* in terms of tree complexity and accuracy.

In our future work we will conduct more experiments using rule sets produced by different rule generation methods. We will also extend our method to address the problem of learning from rules that do not logically intersect. We intend to apply our method in application fraud detection where fraud data is not easily available and instead a rule-base could be created based on heuristics and expert knowledge. *RBDT-1* can summarize the rule-base into a decision tree for more readability and for obtaining the fastest decision.

References

1. Imam, I.F.: An Empirical Comparison between Learning Decision Trees From Examples and From Decision Rules. In: 9th International Symposium on Methodologies for Intelligent Systems, Zakopane (1996)
2. Quinlan, J.R.: Discovering rules by induction from large collections of examples. In: Michie, D. (ed.) Expert Systems in the Microelectronic Age, pp. 168–201. Edinburgh University Press (1979)
3. Breiman, L., Friedman, J.H., Oishen, R.A., Stone, C.J.: Classification and Regression Structures. Wadsworth Int. Group, Belmont (1984)
4. Cestnik, B., Karalie, A.: The Estimation of Probabilities in Attribute Selection Measures for Decision Structure Induction. In: Proceeding of the European Summer School on Machine Learning, pp. 22–31. Priory Corsendonk, Belgium (1991)
5. Mingers, J.: An Empirical Comparison of Selection Measures for Decision-Structure Induction. Machine Learning 3(3), 319–342 (1989)
6. Imam, I.F., Michalski, R.S.: Learning Decision Trees from Decision Rules: A Method and Initial Results from a Comparative Study. J. JIIS. 2(3), 279–304 (1993)
7. Witten, I.H., MacDonald, B.A.: Using Concept Learning for Knowledge Acquisition. J. IJMMS., 349–370 (1988)
8. Michalski, R.S., Imam, I.F.: Learning Problem-Oriented Decision Structures From Decision Rules: the AQDT-2 System. In: Raś, Z.W., Zemankova, M. (eds.) ISMIS 1994. LNCS (LNAI), vol. 869, pp. 416–426. Springer, Heidelberg (1994)
9. Quinlan, R.J.: C4.5: Programs for Machine Learning. Morgan Kaufmann, San Mateo (1993)
10. Akiba, Y., Kaneda, S., Almuallim, H.: Turning Majority Voting Classifiers Into A Single Decision Tree. In: 10th IEEE International Conference on Tools with Artificial Intelligence, pp. 224–230 (1998)
11. Chen, Y., Hung, L.T.: Using Decision Trees to Summarize Associative Classification Rules. Expert Syst. Appl. 36(2), 2338–2351 (2009)
12. Michalski, R.S., Kaufman, K.: The AQ19 System for Machine Learning And Pattern Discovery: A General Description And User's Guide. In: Reports of the Machine Learning and Inference Laboratory, MLI 01-2. George Mason University, Fairfax (2001)
13. Michalski, R.S., Imam, I.F.: On Learning Decision Structures. Fundamenta Informaticae 31(1), 49–64 (1997)
14. Michalski, R.S., Mozetic, I., Hong, J., Lavrac, N.: The Multi-Purpose Incremental Learning System AQ15 and its Testing Application to Three Medical Domains. In: Proceedings of AAAI 1986, Philadelphia, PA, pp. 1041–1045 (1986)
15. Bergadano, F., Matwin, S., Michalski, R.S., Zhang, J.: Learning Two-tiered Descriptions of Flexible Concepts: The POSEIDON System. Machine Learning 8(1), 5–43 (1992)
16. Colton, S.: Online Document (2004),
http://www.doc.ic.ac.uk/~sgc/teaching/v231/lecture11.html
17. Asuncion, A., Newman, D.J.: UCI Machine Learning Repository. University of California, School of Information and Computer Science, Irvine (2007),
http://www.ics.uci.edu/~mlearn/MLRepository.html

Process Materialization Using Templates and Rules to Design Flexible Process Models

Akhil Kumar[1] and Wen Yao[2]

[1] Smeal College of Business, Penn State University, University Park, PA 16802, USA
[2] College of Information Science and Technology, Penn State University, University Park,
PA 16802, USA
akhil@psu.edu, wxy119@psu.edu

Abstract. The main idea in this paper is to show how flexible processes can be designed by combining generic process templates and business rules. We instantiate a process by applying rules to specific case data, and running a materialization algorithm. The customized process instance is then executed in an existing workflow engine. We present an architecture and also give an algorithm for process materialization. The rules are written in a logic-based language like Prolog. Our focus is on capturing deeper process knowledge and achieving a holistic approach to robust process design that encompasses control flow, resources and data, as well as makes it easier to accommodate changes to business policy.

Keywords: flexible processes, rules, templates, materialization, algorithms.

1 Introduction

There are many approaches and frameworks for designing business workflows. Most of them are based on mapping a control flow that specifies the coordination of various activities (see, for instance, [1,2,8,15,21]). The control flow description of a process is also called a *process schema*. In general, there are a large number of process schemas in an organization. This occurs partly because many schemas are variants of one another with minor changes between them. Take for instance, an insurance company that writes policies for automobile, home and other kinds of insurance. When claim applications are made, then the company has to initiate a different process schema for an automobile accident claim as compared to a home damage claim. Moreover, even for a home damage claim, another different process must be enacted for a home whose value is less than $100,000 versus a home whose value is more than $250,000 because in the former case only one adjuster might be required to visit the home and appraise the damage, while in the second case two adjusters are required to submit independent reports of damage assessment. In general, if there are thousands of process variants it makes finding the correct process difficult and error prone.

Another complication may arise if the company changes its policy to require two independent assessments only when the value of the home is more than $500,000. Just this simple business policy change will necessitate a change in many process schema variants. It is time and effort consuming if every variant of the process

G. Governatori, J. Hall, and A. Paschke (Eds.): RuleML 2009, LNCS 5858, pp. 122–136, 2009.

schema affected by this change has to be modified. Thus process schema description gets tied into the business policy of the organization.

In this paper we propose a novel solution to process design based on the idea of process materialization. *Process materialization* means to generate on the fly a process schema (say, a BPMN [21] model in XML) that integrates the control flow, resource needs and data from a generic *process template*, which describes a very basic and general process schema, by applying business rules to the input data of a process. In general, these rules correspond to the business policy of an organization. Thus, this is a rule-based approach to process schema design so as to incorporate the business policy in a dynamic way. If policy changes, only the rules have to be modified while the template can remain unchanged. After the schema has been materialized, then the process would be executed by a workflow engine. The main advantage of this approach is that an end user does not have to create a large number of process schemas before hand and manually determine which schema to execute when a case (such as for an insurance claim) arrives. In addition, an end user does not need to modify a large number of process schemas when policy changes occur since the policy can be captured by rules.

Another advantage of this approach is that it leads to *holistic process design*. Workflow research has focused on the modeling of the control flow of a process, while other key aspects like data flow and resource needs of various tasks are neglected. In general, a holistic process model requires additional information like resource needs of each task, data values of parameters associated with a task, equipment and facilities needed for the completion of the task, etc. One of the goals in the proposed approach is to integrate the modeling of resource and data needs of various tasks as well into the process description. Thus, in the insurance example above, it should be possible to specify that: (1) if the damage exceeds $500,000 then the claim must be approved by a vice-president (a resource related constraint); and (2) at least one adjuster out of two should have more than 10 years of experience (a data related constraint). Modeling approaches that capture such requirements are needed.

Thus, the essence of our approach is: **process template + rules = materialized process**. Naturally, this leads to considerably more flexibility than conventional approaches and is suitable in scenarios involving variability and frequent changes in the environment, as well as resource intensive or ad hoc workflows. The organization of this paper is as follows. Section 2 provides an example to motivate our approach, presents an architecture to formalize this approach and also gives a formal representation for processes, while Section 3 describes our representation approach for rules and discusses rule processing. Then, Section 4 describes our materialization algorithm in detail. Finally, Section 5 provides a detailed discussion and Section 6 concludes the paper with directions for future work.

2 Preliminaries

2.1 Motivating Example

Figure 1 shows an example process template for an insurance process in BPMN notation [21]. In this template, after a claim is received by a customer representative, it is validated by a clerk to ensure that the customer has a valid policy that relates to this claim. The clerk also makes an assignment to two adjusters who will review and appraise the damage to the auto or the house, and then submit a report. The two adjusters may perform their jobs in parallel. This is indicated by a *parallel gateway* shown as a diamond with a cross. The first parallel gateway where multiple branches split has a corresponding parallel gateway where multiple paths merge. After the reports are received by the customer representative, they are checked for completeness and sent to an officer who will determine the settlement amount based on the reports. Subsequently, two approvals are required by a manager and a senior manager, and then the accounts manager will issue the payment to the customer.

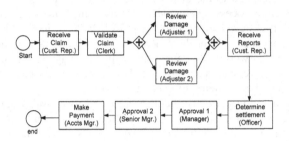

Fig. 1. Description of an insurance process in BPMN notation

Figure 1 shows the "normal" tasks, the roles that perform the tasks and the control flow relationships between tasks. However, the realized process may vary depending upon the actual data for a particular incident or case. Thus, the template can be customized for a specific case by applying rules to it. An example rule set is shown in Figure 2. The rules are written in plain English-like syntax. So, if the loss claimed is less than $500K, then only one adjuster is required (R1); however, if the loss is, more than $250K then the adjuster should have more than 10 years of experience (R7) and must fill the "long" form (R9). Furthermore, if the loss is more than $500K, the second adjuster should be classified as an expert (R8). After a settlement is assessed, either one or two approvals are required before payment is made. The number of approvals depends upon the amount of loss (R4, R5, R6). Finally, there are rules related to the urgency status of the case. If it is marked *expedite* then the approvals may be performed in parallel to save time (R2). On the other hand, if it is marked *urgent*, then the second approval may be deferred to after the payment is made (R3).

R1: If loss < $500K, **then** skip review by adjuster 2
R2: If application = expedite **then** perform approvals in parallel
R3: If application = urgent and loss < $500K **then**
                         defer second approval until after payment
R4: If loss < $100K, **then** need manager approval
R5: If $100K < loss < $500K, **then** need manager & senior manager approval
R6: If loss > $500K, **then** need manager + VP approval
R7: If loss > $250K, **then** need adjuster with minimum 10 years experience
R8: If loss > $500K, **then** need detailed assessment from an expert
R9: If loss > $250K, **then** adjustor must fill the "long" form

---

Fig. 2. Rules to be applied to the process template of Figure 1

Clearly, by applying the rules to the template on different case data, we obtain different processes. Two such materializations for two different cases are shown in Figure 3. In (a), the loss is $200 K and it is marked *expedite*, while in (b) the loss is $300K and it is marked *urgent*. We can see that different processes emerge.

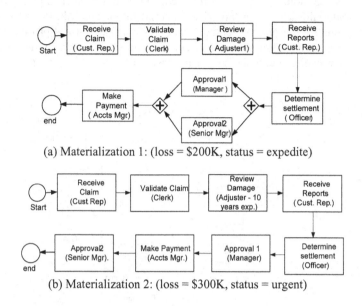

(a) Materialization 1: (loss = $200K, status = expedite)

(b) Materialization 2: (loss = $300K, status = urgent)

Fig. 3. Two materializations from process template and rules

The purpose of this example was to motivate the need for materialization. Our motivation in a nutshell is to reduce the number of processes, especially those which are minor variants of each other, and to make processes more adaptive and agile to business policy changes. Thus, our approach uses process templates to abstract similar processes and uses rules to separate business policy from process design to adjust to a constantly evolving environment.

Next, we give an architecture to show how to formalize this approach.

2.2 Architecture

A high level architecture for our approach is shown in Figure 4. A process designer can create, modify and delete process templates and rules using an editor. Each process template is associated with a number of rules that have the same process id. The editor checks the template for correctness and the rules for consistency. In the case where two rules conflict, the system will give a warning and ask the designer either to modify the rule or associate priorities with them. In addition, the editor checks the executability of rules on their associated template. For example, the editor will give an error warning to the process designer if delete(task t1) is contained in the rule while t1 does not exist in the process template. Detailed definition of semantics for the rule engine is discussed in the next section. As a result, valid process templates and rules are maintained respectively in the process template repository and shared rule repository. As the business policy changes over time, the process designer can easily change rules associated with a specific process while leaving the process template unchanged. When input data for a particular case is entered into the rule engine by the customer, the rule processing module determines the predicates that are true and passes them on to process materialization algorithm. The materialization algorithm uses these true predicates to modify the process template and create a materialized process instance schema for execution within the process engine. The rule editor can also check for data flow consistencies, i.e. make sure that a task will receive all its input data from the output of previous tasks.

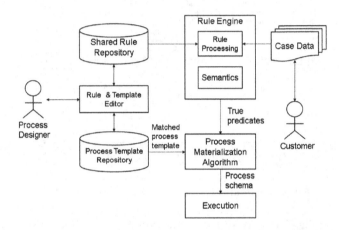

Fig. 4. An overall architecture of the materialization approach

2.3 Formal Representation of a Process

A business can be viewed as a collection of processes, and the robustness of these processes to a large extent is a crucial determinant of the success of the business. Business processes can be described using some simple constructs, and most workflow products provide support for these constructs. Four *basic constructs* that are used in designing processes are *sequence, parallel, decision structure* (or *choice*) and *loop*, as shown in BPMN notation in Figure 5.

In general, business processes can be composed by combining these four basic patterns as building blocks. They can be applied to atomic tasks, e.g. seq(A,B,...) to indicate that tasks A and B (and possibly other tasks) are combined in sequence, or to subprocesses, e.g. seq(SP1, SP2) to indicate that subprocess SP1 and subprocess SP2 are combined in sequence (see Figure 5(a)). Parallelism is introduced by using a parallel gateway to create two or more parallel branches which are synchronized by another parallel gateway as shown in Figure 5(b). We use 'P' or 'Par' to denote the parallel structure. Similarly, a choice structure, denoted as 'C' or Choice, is created with a pair of *exclusive OR (XOR) gateways* and denoted by C or Choice as shown in Figure 5(c). The first XOR node represents a choice or a decision point, where there is one incoming branch and it can activate any one of the two or more outgoing branches. Finally, a loop (denoted by 'L') is also drawn using a pair of XOR gateways but differently from a choice structure, as shown in Figure 5(d). The first XOR gateway takes only one out of all the incoming branches and the second XOR gateway represents a decision point that can activate any one outgoing branch.

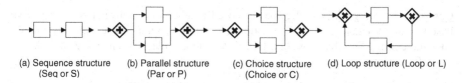

(a) Sequence structure   (b) Parallel structure   (c) Choice structure   (d) Loop structure (Loop or L)
(Seq or S)   (Par or P)   (Choice or C)

Fig. 5. Basic patterns to design processes in BPMN notation

The patterns are applied recursively to create complex processes. A process schema can also be described by rules in any rule-based language. We have chosen Prolog [6], but other languages can be used similarly. We first define base predicates for four aspects of process design and then use them to create process schemas and defining rules. The base predicates are shown in Table 1 along with a description. Figure 6 gives a description of the insurance claim process using these predicates. In general any structured process can be represented in this way. In the next section we discuss representation of rules in a formal way.

Table 1. Base predicates for defining a process and describing rules

Perspective	Base predicates	Description
General	task (t, name)	*t* is a task id and *name* is the task name
	status(proc_id, value)	Specify the status of a process [value=normal, expedite, urgent, OFF]
Control flow related	merge(t1, t2, blk1, S|P|C|L)	Merge task *t1* and *t2* into block *blk1* in relationship X, where X=S, P, C, or L
	delete(t)	Delete task t from the process
	insert(t, S_b|S_a|P|C|L,t1,[N])	Insert task *t* in sequence, parallel, choice or loop with task *t1* to create a node *N*. [S_b = before; S_a= after]
	replace(t1, t2)	Replace task *t1* with *t2*
	move(t, S_b|S_a|P|C|L, t1)	Move task *t* to a different place. The new place is defined in relation to task *t1*.
	change(t1, t2, S_b|S_a|P|C|L)	Change relationship between *t1* and *t2* to new relationship X; X = S_b, S_a, P, C, or L
Resource related	role(t, r)	Task *t* is performed by role *r*
Data related	data(attribute, value)	Assign a value to a data attribute
	prop(role, property_name, value)	Each role can have several properties and corresponding values with them
	data_in(t1, din)	*din* is an input data parameter for task *t1*
	data_out(t2, dout)	*dout* is an output data parameter for task *t2*

3 Rule Representation and Processing

3.1 Rule Categories

In general, we are interested in rules related to four aspects of process design:

Control flow rules: These rules may be used to alter the control flow of a process based on input data of a case. In addition to deleting or replacing a task, they can also alter the control flow by moving a task to a different place or changing the relationship of two tasks.

Resource related rules: These rules are concerned with resource assignment based on case data.

Data related rules: These rules are associated with properties or attributes of a resource related to a case or to other case data.

Other rules: These rules may relate to general reasoning, say related to the policies of the organization. An example of such a rule is: "if an insurance claim has not been completed in 7 days, the status is changed to expedite."

These categories can help to organize the rules and define them systematically. However, we show next all these types of rules can be represented in a common way.

```
Tasks                            Process structure
task(t1, receive_claim).          merge(t1,t2,blk1,seq).
task(t2, validate_claim).         merge(t3,t4,blk2,par).
task(t3, review_damage1).         merge(blk1,blk2,blk3,seq).
task(t4, review_damage2).         merge(blk3,t5,blk4,seq).
task(t5, receive_reports).        merge(blk4,t6,blk5,seq).
task(t6, det_settlement).         merge(blk5,t7,blk6,seq).
task(t7, approval1).              merge(blk6,t8,blk7,seq).
task(t8, approval2).              merge(blk7,t9,blk8,seq).
task(t9, payment).
Roles                            Data
role(t1,cust_rep)                 data_in(t1,loss)
role(t2,clerk)                    data_in(t1,policy_num)
...                              ...
```

Fig. 6. Describing the insurance claim process using predicates

3.2 Rule Representation

In Figure 2 the rules were represented informally. They can be written formally in a first order logic language like Prolog [6]. The rules are based on the predicates in Table 1. To illustrate, Figure 7 shows how the rules of Figure 2 will be expressed formally using the predicates from Table 1. The rules are based on case data. Case data is also expressed using the data predicate as follows:

```
data(loss, 10000).
data(date, 1-Jun-2009).
data(policy_num, sp34-098-765).
```

Rules R1, R2 and R3 are control flow related rules. R1 deletes task t4 from the process instance if the amount of loss is less than $500K since the second adjuster review is not required. R2 changes the relationship between tasks t7 and t8 from sequence to parallel if it is marked *expedite*. Similarly, R3 moves the second approval task (t8) to the end of the process if the process is marked *urgent*. Rules R4, R5 and R6 are resource related rules. They are all similar in that they make assignments of resources to tasks based on case data. Thus, in our example, the amount of loss determines the resource that is required to perform a task, such as an appraisal or an approval. R6 requires that if the loss amount is more than $500K, then a vice-president should perform the second approval task (t8). Finally, Rules R7, R8 and R9 are related to case data such as properties of the resources, or other data.

<u>control flow related</u>

R1: delete(t4)   :- data(loss, X), X < 500000.

R2: change(t7,t8,P) :- status(proc_id, expedite).

R3: move(t8, S_a, t9) :- status(proc_id, urgent),data(loss,X), X<500000.

<u>Resource related</u>

R4: role(t7, manager) :- data(loss, X), X > 0.

R4':delete(t8)   :- data(loss, X), X < 100000.

R5: role(t8, senior_manager) :- data(loss, X),
                                X > 100000, X < 500000.

R6: role(t8, vice_president) :- data(loss, X), X > 500000.

<u>Data related</u>

R7: prop(adjuster, min_exp, 10)   :-
                data (loss, X), X > 250000.

R8: prop(adjuster2, qualification, expert):-
                data(loss, X), X > 500000.

R9: prop(form, type, long)   :-
                data (loss, X), X > 250000.

Fig. 7. Different types of rules related to the insurance workflow process template

3.3 Rule Processing and Semantics for Conflict Resolution

When the above rule set is executed on case data, it will apply the predicates that are true to the process template. Assume the case data is as follows:

loss = $300k; status = expedite.

Now, on adding this data to the rule set we find that the rules that are applicable are: R1, R2, R5, R7 and R9. The corresponding predicates that are true as a result are:

```
Pred1:  delete(t4)
Pred2:  change(t7,t8,par)
Pred3:  role(t8, senior_manager)
Pred4:  prop(adjuster, min_exp, 10)
Pred5:  prop(form,type,long).
```

These predicates can be applied to the process template in order to instantiate a specific process for this case data. Although all rules for the template are valid and compatible with each other, true predicates generated by them may have conflicting results depending upon the order in which they are applied. For example, **insert**(t2, S_a, t1) and **insert**(t3, S_a, t1) applied to a process consisting of a single task 't1' can result in two processes S(t1, t2, t3) or S(t1, t3, t2) depending upon the order in which these operations are applied. Moreover, sometimes rules may fail. For instance **insert**(t1, S_a, t2) would fail if task 't2' has been deleted by a predicate in the previous step. Therefore, it is very important to specify the following semantics for handling such situations for process materialization. Some possible semantics are:

Arbitrary semantics: do not impose any order on the rules. Assume that all execution orders of rules are satisfactory from the user's point of view and are acceptable. If so, the execution priority of true predicates will follow the order of their corresponding rules in the rule repository.

Priority semantics: assign a priority to rules if the execution order is important. Higher priority rules will execute first, followed by lower priority rules in descending order. The user can assign priority based on the importance of the rules, so the most important predicates will be applied first. Besides, priorities may also be assigned based on timestamps with the more recent rules receiving higher priority.

Fail semantics: return failure when a rule cannot be executed due to different reasons. In this case a process cannot be materialized. Therefore, the user will have to intervene to modify the rules or assign new priorities to them.

It is highly recommended that the user should specify the priority semantics before applying the results of rule processing to materialize a new process. Otherwise, arbitrary semantics will be used automatically. If the materialization process fails, the user will be notified about the predicates that cause the failure.

4 Materialization Algorithm

The rule processing stage generates a list of true predicates that apply to the case data. These predicates relate to control flow, resources, and data of the case. The predicates that relate to the resources (such as role and prop predicates) and also those that relate to data (such as data predicate) are added to the case database directly. For example, prop(form,type,long) assigns the value long to the type attribute of the form object. Such data would be used as input data for task execution. However, the predicates that relate to the control flow are used as input for a materialization algorithm in order to generate a modified control flow schema for the process. The details of the materialization algorithm are discussed next.

4.1 Overview

Our materialization algorithm is based on creating a tree for the generic process template and then applying change operations to it. Each change operation will produce a corresponding operation on the tree. After all the changes are applied to the tree, the resulting tree reflects the control flow of the new process. This can be converted into Prolog rules or into any process description language to describe the process schema. A tree for the process template of Figure 1 is shown in Figure 8. There are several equivalent representations of such a tree. The tasks are at the leaf nodes, while internal nodes are control nodes that give relationships between tasks or blocks of tasks. Accordingly, the internal node labels are prefixed with the *node type* (S, P, C, or L). The child nodes of a sequence node are numbered in order from left to right. Thus, the leftmost child appears first in the sequence and the rightmost one is the last. For parallel and choice nodes, the order of appearance of the child nodes does not matter because of their execution semantics. A loop node has two child nodes, the first one for the forward path, and the second for the reverse, hence the order does matter. This tree can be stored in a tabular data structure as follows:

(node_id, type, child node, sequence#).

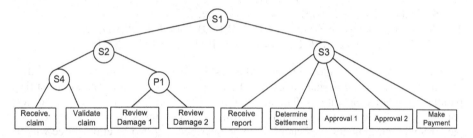

Fig. 8. A process tree for the process template of Figure 1

4.2 Algorithmic Details

The control flow modification predicates (or operations) were given in Table 1. They are summarized in Table 2 along with the pseudo-code for implementing them. We discuss these operations next in the context of Figure 8 which is a tree-like representation of the process of Figure 1. Although, this tree only presents the control flow perspective of the process, the resource and data related information can also be included in the nodes. This tree was drawn using the parent child relationships from the base predicates of Figure 6 after applying the rewriting rule 3 (see below). A **delete** operation simply removes the node corresponding to a task in the tree and if the parent of the deleted node has only one child left, then the child is moved up to take the place of the parent. Each non-leaf node should have at least two child nodes and a task is always a leaf node.

When a task is to be **inserted,** its position must be specified in the tree with respect to an already existing task node. Moreover, the relationship between the existing node and a new node should also be specified as sequence (S), parallel (P), choice (C) or loop (L). If it is a sequence it is necessary to state whether the new task is inserted before (S_b) or after (S_a) the current node. The insert procedure is to create a parent node P1 for the existing node (say, t1) and insert the new node t as a child of P1 in the tree. The parent node can optionally be given a new label N. The **replace** operation simply changes the label of a task node with its new name.

The **move** operation is like a delete followed by an insert. It removes a task from its current location in the tree and inserts it into a new position. This new position is defined with respect to an existing task node in the tree which serves as an anchor node. The **change** operation may be used to modify the relationship between two existing nodes t1 and t2 in the tree. However, this is possible only if a direct relationship (i.e., with a common parent node and no other siblings) exists between the two nodes. Otherwise, the operation would fail. In order to implement this operation, we first check if a direct relationship either exists already or can be found by rewriting the tree into an equivalent tree by means of rewriting rules. If it is possible to do so, then the parent node of t1 and t2 is changed to the new relationship. Otherwise, a failure message is given. The rewriting rules are as follows:

Rewriting rules (A, B, D are nodes of a tree)
1. $\mathbf{P}(\ldots, A,\ldots, B,\ldots) = \mathbf{P}(\ldots, B,\ldots, A,\ldots)$
2. $\mathbf{C}(\ldots, A,\ldots, B,\ldots) = \mathbf{C}(\ldots, B,\ldots, A,\ldots)$
3. $\mathbf{S}(A,B,D,\ldots) = \mathbf{S}(\mathbf{S}(A,B),D,\ldots) = \mathbf{S}(A, \mathbf{S}(B,D),\ldots)$
4. $\mathbf{P}(A,B,D,\ldots) = \mathbf{P}(\mathbf{P}(A,B),D,\ldots) = \mathbf{P}(A, \mathbf{P}(B,D),\ldots) = \mathbf{P}(B, \mathbf{P}(A,D),\ldots)$
5. $\mathbf{C}(A,B,D,\ldots) = \mathbf{C}(\mathbf{C}(A,B),D,\ldots) = \mathbf{C}(A, \mathbf{C}(B,D),\ldots) = \mathbf{C}(B, \mathbf{C}(A,D),\ldots)$

Nodes A, B, and D could represent either tasks if they are leaf nodes or root nodes of subtrees if they are internal nodes. The first two rules capture commutativity of the parallel and choice operations. The next three rules reflect associativity of sequence, parallel and choice operations. Thus, by Rule #3, three tasks in a sequence (i.e. a parent S with child nodes A, B and D) can be rewritten in a two level deep tree with a parent (say, S1) having child nodes S2 and D. The child node S2 in turn has two child nodes A and B. Clearly, both these structures are equivalent. The same argument applies to parallel and choice nodes.

Continuing with the running example, the result of rule processing (section 3.3) will send control flow predicates pred1 and pred2 to the process materialization algorithm. In the tree of Figure 8, the task Review Damage 1 and Review Damage 2 are executed in parallel. Applying Pred1 will remove Review Damage 2 so the parent node P1 only has one child node. Following the algorithm, we move task Review Damage 1 to replace P1. To apply Pred2, we consider the tasks Approval 1 and Approval 2. They have a sequence relationship between them; however, their parent node S3 has five child nodes. In order to change this relationship between only these two tasks to P, we can use rewriting rule #3 to rewrite the part of the tree under S3 as follows:

S3(rec., det., app1, app2, pay) = S3(S5(rec., det), S6(app1, app2), pay)

Now, since app1 and app2 have a common parent S3, we can rewrite as:

S3(S5(rec., det), S6(app1, app2), pay) → S3(S5(rec., det), P2(app1, app2), pay)

On the other hand, note that it would not be possible to change the tree such that tasks receive report and make payment are in parallel. On applying the true predicates Pred1 and Pred2 to the process template tree of Figure 8 by using the materialization

Table 2. Process Materialization algorithms

Operation	Algorithm

Delete (Node *t*, ProcessTree *p_tree*)	IF node *t* is NOT in *p_tree*
	    Report materialization failure;
	ELSE IF *parent(t)* has two child nodes
	    Move *t.sibling* to replace *parent(t)*;
	ELSE
	    Delete (t, *p_tree*);

Insert (Node *t*, Rel *X*, Node *t1*, [N], ProcessTree *p_tree*) Note: X = S, P, C, or L	IF node *t* or *t1* is NOT in *p_tree* Report materialization failure; ELSE {Create a new parent node *N* for *t1* && *N.node_type=X*; Add *t* as a new child of *N* ;}

Replace (Node *t1*, Node *t2*, ProcessTree *p_tree*)	IF node *t1* or *t2* is NOT in *p_tree*
	    Report materialization failure;
	    ELSE
	    Rename node *t1* with *t2*;

Move (Node *t*, Rel *X*, Node *t1*, ProcessTree *p_tree*):	IF node *t* or *t1* is NOT in *p_tree*
	    Report materialization failure;
	ELSE
	    Delete (t, *p_tree*) && Insert (t, X, t1, *p_tree*);

Change (Node *t1*, Node *t2*, Rel *X*, ProcessTree *p_tree*)	IF node *t1* or *t2* is NOT in *p_tree* Report materialization failure; ELSE {use rewriting rules to change *p_tree* to an equivalent tree *p_tree'* such that t1 and t2 have a common parent; Change parent (*t1*, *t2*).*node_type* to new relationship *X*;}

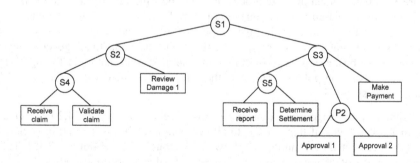

Fig. 9. Revised process tree after applying the materialization algorithm

algorithm, we obtain the new process tree of Figure 9. This tree can be converted into a process diagram or a description in, say, BPEL [7] or XPDL [27].

Although a formal proof is omitted for brevity, it is possible to show that one can translate any given correct process model described by a template into any other process model by applying the materialization operations.

5 Discussion and Related Work

Most process design techniques lead to *rigid* processes where policy is "hard-coded" into the process schema thus reducing flexibility. The motivation behind our approach in this paper is to overcome this drawback by integrating rules with generic process templates to materialize processes. The template captures the essence of the flow, while the rules allow modification based on case data, policy changes, resource availability, etc. The new approach is also more holistic since it can go beyond control flow and also capture case data and resource aspects of the workflow.

A preliminary proposal for process and rule integration is given in [18]. Other related work with similar objectives pertains to configurable models [25] and aggregate models [23] although they are not based on rules directly. Direct integration of the process and rule paradigms into a product is being made in the Drools Flow project [13]. Their goals are similar to ours, but they do not create a materialized process. Rather they provide a rule modeling construct or module as a way to include rules into their process. The rules module can be evaluated and a decision can be made accordingly. The advantage of our approach is that it can be integrated into existing process modeling paradigms. On materialization our process models can be expressed in any existing language (e.g., BPMN, BPEL and XPDL) and executed in an existing workflow engine. The generic process templates can also be written in any standard language.

There has been considerable amount of related work on execution of dynamic processes in the context of exception handling. The focus there is on modifying a running process when exceptions occur due to failed tasks, erroneous information, etc. However, this body of work is complementary to our work since it focuses on flexibility at run time. In contrast we are more interested in flexibility at design time.

The benefits from design time flexibility we foresee are:

(1) It leads to a cleaner process design. Thus, if the two materializations shown in Figure 3 were combined into one process it would become very difficult to read.

(2) It allows separation of organization policy from process flow. The basic template captures the essence of the main process flow by including the tasks and the normal order in which they are performed. However, variations to the generic process flow represent business policy, and these are captured more naturally through rules. When changes in the policy occur, only the rules are modified without affecting the generic process template.

Techniques for supporting dynamic change are discussed in [14,20,22,24] and elsewhere. The focus of this work is to allow operations like task insertion, deletion, etc. to be performed on running workflows in response to exceptions [5,7,12]. There have also been other approaches on designing flexible workflow models: based on deadline based escalation [3], satisfaction of constraints [19] or restrictions [11], availability of resources [17] and on graphs [26]. Another approach for designing workflows is centered on entities [4]. While all these methods try to reduce rigidity of a strict control flow approach, they lack the flexibility of rules. In [20], a rule based approach for dynamic modification in a medical domain is given with the focus on handling of exceptions. Rules can also be used to ensure that business processes comply with internal and external regulations. These rules could be specified in a first-order language [16] or in deontic logic [9].

Of course, on the downside additional cost is involved in managing the rules and ensuring consistency. However, if rules are created through a user friendly interface, then the burden on the user is minimized. Moreover, for the most part, we expect rules to be simple and their interactions few, thus reducing complexity. Another drawback with our approach is that it may give too much freedom to users to make ad hoc changes in business processes. This can be restricted by adding controls in the form of meta-rules that restrict use of modification operations only to certain users.

6 Conclusions

This paper described a novel proposal for designing flexible business processes based on combining process templates with business rules. We showed how the process materialization approach allows separation of basic process flow from business policy elements in the design of a process and also integrates resource and data needs of a process tightly. In future, we will incorporate "events", which are also an important business process design element, into our process template design approach. In Figure 1, we showed *start* and *end* events. Events add expressive power to a process model. We plan to add events of different types and provide formal definitions for them in future work.

We are building a prototype to test and evaluate this methodology following the proposed architecture. A further challenge lies in making an interface that allows users to describe the rules in an easy way and making the details of the language completely transparent to them. We intend to explore various solutions for this including the use of an English-like rule language such as SBVR [10] and providing a GUI interface to increase ease of use and prevent typing errors. More effort will also be devoted to the semantics for rule conflict resolution. Lastly, adding ontologies to the architecture would increase the expressive power of the framework.

References

1. van der Aalst, W.M.P.: The Application of Petri Nets to Workflow Management. The Journal of Circuits, Systems and Computers 8(1), 21–66 (1998)
2. van der Aalst, W.M.P., et al.: Workflow Patterns. Distributed and Parallel Databases 14(3), 5–51 (2003)
3. van der Aalst, W.M.P., Rosemann, M., Dumas, M.: Deadline-based escalation in process-aware information systems. Decis. Support Syst. 43(2), 492–511 (2007)
4. Bhattacharya, K., et al.: Towards Formal Analysis of Artifact-Centric Business Process Models. In: Business Process Management (BPM), Brisbane, Australia, pp. 288–304 (2007)
5. Chiu, D.K.W., Li, Q., Karlapalem, K.: Web interface-driven cooperative exception handling in ADOME workflow management system. Web Information Systems Engineering 26(2), 93–120 (2001)
6. Clocksin, W.F., Mellish, C.S.: Programming in Prolog. Springer, New York (1987)
7. Curbera, F., Khalaf, R., Leymann, F., Weerawarana, S.: Exception Handling in the BPEL4WS Language. In: van der Aalst, W.M.P., ter Hofstede, A.H.M., Weske, M. (eds.) BPM 2003. LNCS, vol. 2678, pp. 276–290. Springer, Heidelberg (2003)

8. Dumas, M., van der Aalst, W.M.P., Hofstede, A.H.M.: Process Aware Information Systems. Wiley Interscience, Hoboken (2005)
9. Goedertier, S., Vanthienen, J.: Designing Compliant Business Processes with Obligations and Permission. In: Proceedings of Workshop on Business Process Design, pp. 5–14 (2006)
10. Goedertier, S., Mues, C., Vanthienen, J.: Specifying process-aware access control rules in SBVR. In: Paschke, A., Biletskiy, Y. (eds.) RuleML 2007. LNCS, vol. 4824, pp. 39–52. Springer, Heidelberg (2007)
11. Halliday, J.J., et al.: Flexible Workflow Management in the OPENflow System. In: Proceedings of the Fifth IEEE International Enterprise Distributed Object Computing Conference (EDOC 2001), pp. 82–92 (2001)
12. Hwang, S.-Y., Tang, J.: Consulting past exceptions to facilitate workflow exception handling. Decision Support Systems 37(1), 49–69 (2004)
13. JBoss Community, Drools Flow,
    http://www.jboss.org/drools/drools-flow.html
14. Joeris, G.: Defining Flexible Workflow Execution Behaviors. In: Enterprise-wide and Cross-enterprise Workflow Management: Concepts, Systems, Applications, GI Workshop Proceedings – Informatik, pp. 49–55 (1999)
15. Kiepuszewski, B., ter Hofstede, A.H.M., Bussler, C.J.: On structured workflow modelling. In: Wangler, B., Bergman, L.D. (eds.) CAiSE 2000. LNCS, vol. 1789, pp. 431–445. Springer, Heidelberg (2000)
16. Kumar, A., Liu, R.: A rule-based framework using role patterns for business process compliance. In: Bassiliades, N., Governatori, G., Paschke, A. (eds.) RuleML 2008. LNCS, vol. 5321, pp. 58–72. Springer, Heidelberg (2008)
17. Kumar, A., Wang, J.: A framework for designing resource driven workflow systems. In: Rosemann, M., vom Brocke, J. (eds.) The International Handbook on Business Process Management, Springer, Heidelberg (2009) (forthcoming)
18. Lienhard, H., Künzi, U.-M.: Workflow and business rules: a common approach. BPTrends., http://www.bptrends.com/
19. Mangan, P., Sadiq, S.: On Building Workflow Models for Flexible Processes. In: Proceedings of the 13th Australasian Conference on Database Technologies (ADC), Melbourne, Victoria, Australia, vol. 5, pp. 103–109 (2002)
20. Müller, R., Rahm, E.: Rule-Based Dynamic Modification of Workflows in a Medical Domain. In: Buchmann, A.P. (ed.) BTW 1999, Freiburg im Breisgau, pp. 429–448. Springer, Berlin (1999)
21. OMG, Business Process Modeling Notation (BPMN) Version 1.0. OMG Final Adopted Specification, Object Management Group (2006)
22. Reichert, M., Dadam, P.: Adept_flex—Supporting Dynamic Changes of Workflows Without Losing Control. J. Intell. Inf. Syst. 10(2), 93–129 (1998)
23. Reijers, H., et al.: Improved Model Management with Aggregated Business Process Models. Data and Knowledge Engineering 68(2), 221–243 (2009)
24. Rinderle, S., Reichert, M., Dadam, P.: Correctness criteria for dynamic changes in workflow systems. Data and Knowledge Engineering 50(1), 9–34 (2004)
25. Rosemann, M., van der Aalst, W.M.P.: A configurable reference modeling language. Information Systems 32(1), 1–23 (2007)
26. Weske, M.: Flexible Modeling and Execution of Workflow Activities. In: Proceedings of the 31st Hawaii International Conference on System Sciences (HICSS), pp. 713–722 (1998)
27. XPDL. Workflow management coalition workflow standard,
    http://www.wfmc.org/xpdl.html

Introduction to "Rule Transformation and Extraction" Track

Mark H. Linehan[1] and Eric Putrycz[2]

[1] IBM T. J. Watson Research Center, 17 Skyline Drive, Hawthorne, NY 10532
mlinehan@us.ibm.com
[2] Apption Software, 290 Picton Ave., Suite 104, Ottawa, ON, K1Z 8P8
erik@apption.com

Abstract. In this short paper, we summarize the "Rule Transformation and Extraction" topic, defining the terms, describing some of the main approaches to the topic, and reviewing the current challenges for both rule transformation and extraction.

Keywords: rules, business rules, rule transformation, rule extraction.

1   Introduction

Rule *transformation* is the conversion of rules to other rule formats or other languages. For example, [1] describes a system that converts constraints given in the *Object Constraint Language* [2] to rules in the Object Management Group's (OMG) *Semantics of Business Vocabulary and Business Rules* [3] specification. Rules *extraction* is about discovering or recovering rules from sources such as natural language text or source code. The essential difference is that "transformation" involves conversions from rules, while "extraction" is about abstracting rules from sources that are not in some formal rules format.

In this short paper, we introduce this topic in some more detail, describing various kinds of rule transformation and extraction, some of the work that has been done in this area, and the challenges that should be addressed to make further progress on both forms of rule conversion.

2   Rule Transforamation

2.1   Rules and Model-Driven Architecture

Transformation of rules can be analyzed with respect to the OMG's *Model-Driven Architecture* (MDA) [4] stack, as shown in figure 1: MDA distinguishes three general architectural layers. The bottom "Platform Specific Modeling" (PSM) layer is about computing system models or implementations created to execute upon a specific computing platform, such as J2EE or .NET. The middle "Platform Independent Modeling" (PIM) layer considers implementation models explicitly designed to avoid commitments to particular execution environments.

G. Governatori, J. Hall, and A. Paschke (Eds.): RuleML 2009, LNCS 5858, pp. 137–143, 2009.

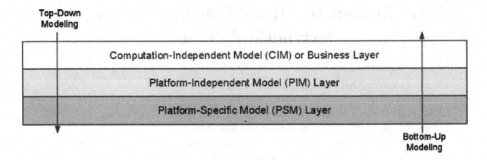

Fig. 1. Model-Driven Architecture (MDA) layers

The top "Computation Independent Modeling" (CIM) or business layer is about modeling businesses independent of implementation. For example, the OMG's Business Motivation Model "... provides a scheme or structure for developing, communicating, and managing business plans ..." [5, p. 1] at the CIM layer.

Most commonly, modeling practitioners apply the MDA model to categorize various types of information or processing models such as UML class models, entity-relation diagrams, Business Process Modeling Notation [6] process diagrams, or UML interaction diagrams. MDA can also classify rule languages, as shown in figure 2:

CIM or Business Layer	Semantics of Business Vocabulary and Business Rules (SBVR)		
PIM Layer	Rule Interchange Format (RIF)	Rule Markup Language (RuleML)	Object Constraint Language (OCL)
PIM Layer	JESS	ILOG JRules	Blaze Advisor

Fig. 2. Example rule languages positioned against the MDA layers

Most rule languages are vendor-specific and thus belong in the bottom, platform-specific (PSM) layer. Examples are the many commercial rule languages such as ILOG JRules and Blaze Advisor, as well as the non-commercial rule systems that are built for a specific programming platform such as Java. From the point-of-view of a rule system user, choosing any of these rule languages commits one to a particular vendor or computing environment. There also exist a number of PIM-layer rule languages, such as the World Wide Web Consortium's (W3C's) Rule Interchange Format [7], Rule Markup Language [8], OCL, and Common Logic [9]. These are PIM-level languages because they are vendor-and platform-independent. At the business or CIM layer, the SBVR specification defines a rule language intended to model business policies and rules in the way that business people think about them.

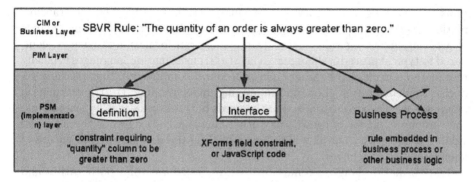

Fig. 3. One business-layer rule may transform to multiple implementation-layer rules or other artifacts

The MDA architecture promotes the idea of conversions among the three layers as shown in figure 1. Linehan [10] describes a top-down modeling implementation, that starts with business rules in SBVR, transforms them to the PIM layer using OCL, then further converts them to Java code at the PSM layer. In bottom-up modeling, one might start with rules in some vendor rule language, then convert the rules to a PIM-layer rule language such as RIF, then perhaps further abstract the key ideas into a business model. One reason for bottom-up modeling is to support the reimplementation of an older application in a more modern computing technology. Another reason is to help answer a question which many real businesses need or want to ask:"what rules are we actually operating under?" The OMG's Knowledge Discovery Metamodel [11] defines an overall approach to this type of bottom-up modeling for rules as well as other kinds of modeling artifacts.

Rule transformations are also possible within an MDA layer. For example, at the PSM level, one might want to transform rules from one vendor rule language to another.

Rule transformations within individual MDA layers, and between the PIM and PSM layers, are typically 1 : 1. That is, each source rule converts to one rule in the target language. Transformations from the business or CIM layer to either the PIM or PSM layer often is 1 : n: one rule at the business layer implies multiple rules in the implementation. Figure 3 shows an example of a business-layer rule that is transformed to multiple implementation aspects. The single business-layer rule affects the database design, the user interface, and the business logic. This might be accomplished by a single implementation-layer rule that is reused in multiple parts of the implementation, or more likely, by different runtime artifacts such as Database Definition Language (DDL) statements for the database design, Javascript code for the user interface, and perhaps a commercial rule system for the business logic.

2.2   Transformation Languages

Special-purpose languages exist for transforming elements between source and target languages. Examples include Extensible Schema Language Transformations [12], OMG's Query-View-Transformation [13], and the ATLAS Transformation Language [14]. These languages can be applied to transform many kinds of artifacts, including rules. The advantage of transformation languages, compared to typical procedural languages are: (a) transformation languages typically rely more on declarative rather than procedural components, thus possibly reducing implementation effort and improving understanding; (b) transformation languages usually are customized to the needs of transformations in general, and thus are more "fit for purpose" than other languages; (c) rigorous checking and validation of transformations may be enabled by the domain-specific nature of transformation languages.

Transformation languages may be used to transform rules both within and across the MDA layers shown in figure 1.

2.3   Challenges

A number of interesting technical challenges exist in the area of rules transformation:

- When transforming rules from the business layer to implementation layers, integration with other aspects of business solutions, such as business process models, is necessary but ad-hoc. Standards for rules, processes (e.g. BPMN, BPEL), and information models (e.g. OWL, entity-relation diagrams) exist in isolation from each other, yet complete solutions often require elements of each of these.
- Coping with the different functional and descriptive power of different rule languages. RuleML and RIF have made a start on this by defining different families of rule languages, categorized by semantic model and functional richness.
- Different rule languages have different objectives. Most rule languages are intended to support implementations and thus define an execution model. One language – SBVR – focusses on descriptive power and entirely ignores execution concerns such as tractability. It seems likely that some SBVR rules may not be efficiently executable in some implementation languages.
- It is highly desirable to provide traceability by linking the sources and targets of rule transformations. Such traceability documents where rules come from, aids in implementing or auditing future changes, and helps in debugging. As discussed above, there may be 1:n relationship between rules at the business versus the implementation layers. In bottom-up transformations, recognizing that multiple implementation-layer rules represent a single business-layer rule is difficult.
- Provably-correct transformations: providing proofs that transformations from one language to another do not lose or add semantics.

3 Rule Extraction

3.1 From Artifacts to Rules

Rule extraction consists of extracting or discovering rules or ontology from existing sources that are not normally thought of as rules and that are not expressed in any specific rule dialect.

The source artifacts can be various sources such as:

- source code [15, 16, 17] and the extraction process is based on reverse engineering techniques to obtain a data and execution flow;
- plain English text documents [18] or maintenance manuals [19], which require Natural Language Processing (NLP) tools such as tagging or morpho-lexical analysis.

The first step in the extraction process consists of parsing the source artifacts and locating the relevant information (step 1 in Figure 4). With source code, this step requires using a parser that extracts an Abstract Syntax Tree (AST) from the source code. Using this AST, it is possible to extract many knowledge elements such as the data structures and the execution flow that are necessary to build rules. Further analysis of the AST is also required to find all the operations which are potential candidates for rules, and build their context.

Once the knowledge elements are extracted, an ontology can be built from all the extracted knowledge elements (step 2 in Figure 4). For instance, with source code, the data declarations have to be connected with the identifiers used in calculation or other operations.

Because further transformations and other operations between all the extracted data are necessary, a repository is often created with all the knowledge elements and the ontology (step 3 in Figure 4).

The last step consists of extracting the rules from the repository into a formal or semi-formal dialect (step 4 in Figure 4), such as SBVR rules from all the relevant operations in the source code.

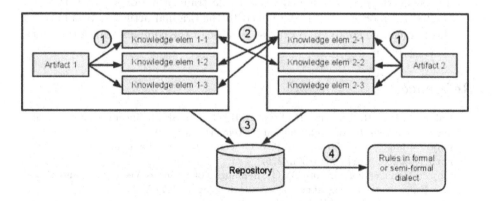

Fig. 4. Common steps of rule extraction

3.2 System Modernization

Rule extraction is often used in the context of system modernization. Legacy systems still represent an important share of today's IT. In 2008, 490 companies of the Fortune 500 are still using legacy systems to process more than 30 billion transactions or $1 trillion worth of business each and every day. In Canada, 10% of the total ICT employment are working with legacy systems. These legacy systems are often either replaced or integrated with a new system.

A common issue with legacy systems is that no documentation is available and the key people might be retired or have left. As a consequence, it is critical to extract rules from the legacy systems and simplify the modernization process.

Putrycz and Kark [15, 16] propose a business rule extraction technique which combines reverse engineering to find rules within the legacy code and plain English analysis with NLP to connect documents to data sources, and subsequently connect the business rules to the documents. This approach enables to extract rules that are expressed without technical identifiers, and are easily understandable by business analysts.

3.3 Challenges

- Rule extraction from documents lacks generic tools and techniques, and custom analysis is commonly required for each document model;
- To build high level business rules, multiple sources (e.g. source code and documentation) need to be combined. For example, the documentation on the data sources needs to be analyzed to translate all data elements from source code into business terms.
- In legacy systems, since the documentation can be partial or not existent, heuristics or other techniques need to be used to build high level rules from the basic rules extracted from source code operations.
- There is no generic benchmark for rule extraction, and thus, only manual validation can distinguish valid business rules from technical rules. This makes the detection of noise in the extracted rules a difficult problem.
- The same traceability challenge from rule transformation is valid, the extracted rules often need to be linked to the original artifacts. In addition, this link can be complex to express because of the complex transformations that can occur.

References

1. Cabot, J., Pau, R., Raventó, R.: From UML/OCL to SBVR Specifications: a Challenging Transformation. Information Systems (to appear)
2. Object Management Group (OMG): Object Constraint Language, Version 2.0., http://www.omg.org/spec/OCL/2.0/
3. Object Management Group (OMG): Semantics of Business Vocabulary and Business Rules, Version 1.0., http://www.omg.org/spec/SBVR/1.0/
4. Miller, J., Mukerji, J.: MDA Guide Version 1.0.1., http://www.omg.org/docs/omg/03-06-01.pdf

5. Object Management Group (OMG): Business Motivation Model, version 1.0., http://www.omg.org/spec/BMM/1.0/
6. Object Management Group (OMG): Business Process Modeling Notation, version 1.2., http://www.bpmn.org/
7. World Wide Web Consortium (W3C): Rule Interchange Format (RIF), http://www.w3.org/2005/rules/wiki/RIF_Working_Group
8. The Rule Markup Initiative, http://ruleml.org/
9. International Standards Organization (ISO): ISO/IEC 24707: 2007 - Information technology — Common Logic (CL): a framework for a family of logic-based languages, http://standards.iso.org/ittf/PubliclyAvailableStandards/c039175_ISO_IEC_24707_2007%28E%29.zip
10. Linehan, M.: SBVR Use Cases. In: Rule Representation, Interchange and Reasoning on the Web, Proceedings of the RuleML Interational Symposium, Orlando (October 2008)
11. Object Management Group (OMG): Knowledge Discovery Metamodel, http://www.omg.org/spec/KDM
12. World Wide Web Consortium (W3C): XSL Transformations, Version 1.0, http://www.w3.org/TR/xslt
13. Object Management Group (OMG): Meta Object Facility (MOF) 2.0 Query/View/Transform 1.0., http://www.omg.org/spec/QVT
14. Jouault, F., Kurtev, I.: On the Architectural Alignment of ATL and QVT. In: Proceedings of ACM Symposium on Applied Computing (SAC 2006) Model Transformation Track. Dijon, Bourgogne, FRA (April 2006)
15. Putrycz, E., Kark, A.W.: Recovering Business Rules from Legacy Source Code for System Modernization. In: Paschke, A., Biletskiy, Y. (eds.) RuleML 2007. LNCS, vol. 4824, pp. 107–118. Springer, Heidelberg (2007)
16. Putrycz, E., Kark, A.W.: Connecting Legacy Code, Business Rules and Documentation. In: Bassiliades, N., Governatori, G., Paschke, A. (eds.) RuleML 2008. LNCS, vol. 5321, pp. 17–30. Springer, Heidelberg (2008)
17. Wang, X., Sun, J., Yang, X., He, Z., Maddineni, S.: Business rules extraction from large legacy systems. In: Proceedings Eighth European Conference on Software Maintenance and Reengineering, CSMR 2004, pp. 249–258 (2004)
18. Ciravegna, F.: Adaptive Information Extraction from Text by Rule Induction and Generalisation. In: Proceedings 17th International Joint Conference on Artificial Intelligence, IJCAI 2001, Seattle (2001)
19. Yang, C., Orchard, R., Farley, B., Zaluski, M.: Authoring cases from Free-Text Maintenance Data. In: Perner, P., Rosenfeld, A. (eds.) MLDM 2003. LNCS, vol. 2734. Springer, Heidelberg (2003)

An SBVR Framework for RESTful Web Applications

Alexandros Marinos and Paul Krause

Department of Computing, FEPS, University of Surrey,
GU2 7XH, Guildford, Surrey, United Kingdom
{a.marinos,p.krause}@surrey.ac.uk

Abstract. We propose a framework that can be used to produce functioning web applications from SBVR models. To achieve this, we begin by discussing the concept of declarative application generation and examining the commonalities between SBVR and the RESTful architectural style of the web. We then show how a relational database schema and RESTful interface can be generated from an SBVR model. In this context, we discuss how SBVR can be used to semantically describe hypermedia on the Web and enhance its evolvability and loose coupling properties. Finally, we show that this system is capable of exhibiting process-like behaviour without requiring explicitly defined processes.

Keywords: SBVR, REST, SQL, Web-based Applications, Declarative Programming, Business Rules.

1 Introduction

Building an information system with the current methodologies is an uncertain proposition. Recent research indicates that only 35% of software development projects get completed in time and on budget [1]. This is a marked increase from 16.2% in 1995 [2], but even this has come at the expense of a longer and more complex development process. It is understandable then, that businesses tend to avoid modifying their production information systems until absolutely necessary, as any attempt at modification introduces further uncertainty.

An objective of modern digital ecosystems (DE) research is to help people, organizations and small and medium enterprises (SMEs) better dynamically integrate their activities, enabling them to utilize capabilities, access infrastructure, and compete in markets currently available only to large enterprises [3]. A large obstacle on the path towards realizing this vision is the inflexibility of information systems currently used by SMEs and other potential DE participants, which constitutes an internal barrier. Viewing the information system from the external perspective, the requirement to explicitly annotate provided services with semantics for exposition in a DE effectively limits the population of accurately described services available.

From a more general perspective, technologies can be seen as conforming to one of two different modes of use [4]. Sterile systems are systems whose function is limited by their design and will perform the same tasks for the duration of their lifespan. An example of a sterile system is the typewriter, the television or the telephone network. On the contrary, generative systems are built to enable novel and unplanned usage, far

G. Governatori, J. Hall, and A. Paschke (Eds.): RuleML 2009, LNCS 5858, pp. 144–158, 2009.
© Springer-Verlag Berlin Heidelberg 2009

beyond what their designers originally intended or could conceive. Typical examples of generative technologies are personal computers and the internet.

Examining information systems from this perspective, most of them are sterile. They have been built for specific tasks, contain a fixed set of processes, can handle specific data models and this functionality cannot be changed without significant reimplementation. Newer developments in the field offer some degree of flexibility but their core is still procedural, ultimately dependent on costly intervention by specialised intermediaries, and therefore resistant to rapid adaptation. This is in contrast with the inherently dynamic nature of business and human society within which these systems are applied and causes inefficiencies which hinder the fulfilment of the digital revolution's promise. We introduce the concept of the *generative information system*, built on declarative technologies, as a possible solution to these issues. In this paper, we use SBVR as a modelling language for such systems.

It is clear that SBVR was not intended [5] as a language from which to directly produce applications, at least to begin with. SBVR, as a declarative language, focuses on modelling the 'what' of a system, rather than defining the 'how' of its implementation. However, a given declarative model that is specified by its owner constrains the set of potential solutions that can implement it. Each new element of information added to the model reduces the number of compliant solutions. Within the set of compliant solutions, if two elements have a difference observable by the owner, such that one is acceptable and one is not, then the model is not completely expressing the owner's wishes. It must therefore be enriched with the additional information that retains the acceptable solution while excluding the unacceptable one. By iteratively repeating this process, we can arrive at a model that identifies only potential implementations that are acceptable to the owner. In practice however, the specification can only identify acceptable solutions at a level of granularity afforded by the expressivity of the language it is written in. With this caveat in mind, we use SBVR as the best balance between expressivity and user-accessibility (through SBVR Structured English), and explore the extent to which solutions can be automatically produced. Possible limitations that are encountered in expressing the specifications of the owners can act as feedback to the language design itself.

The subject of generating applications from business rules models has been first covered in the book 'What, not How: The Business Rules Approach to Application Development' by C.J. Date [6] which is also closely related to the Business Rules Manifesto by Ron Ross [7]. These works set the foundations for this paper, and we hope to extend the conceptual model they provide with discussions of business processes, extensibility and composable applications, ultimately aiming to produce a real-world production-capable framework based on SBVR, Relational Databases and the Architecture of the Web as expressed by REST. Section 2 discusses conceptual issues of producing applications from SBVR model and examines aspects of model checking and relational database schema inference. Section 3 examines how this information system can be made available on the web in a RESTful manner while Section 4 deals with issues of process-like behaviour. Section 5 gives some concluding remarks and discusses future work.

2 From SBVR Models to Applications

Current approaches to producing applications from SBVR models treat it as a code generation problem. This inevitably results in facing the tension between the declarative and imperative programming paradigms. When generating code, processes must be defined and programmed, however such processes are not natively defined by SBVR and this is so by design. Therefore attempts at code generation [8] must either arbitrarily select the processes that must be implemented, supplement SBVR with a workflow definition language such as XPDL and BPMN or extend SBVR itself to make it capable of specifying processes. The latter results in models that use SBVR as a verbose process language attempting to do what visual alternatives achieve far more concisely. Even then, the code that they orchestrate is still missing, at which point they revert to the need for a human programmer to fill in the gaps, a problem also faced by model-driven approaches like Executable UML in the past.

The alternative is to treat the model itself as the code to be executed and interpret it at run-time, possibly caching any decisions that can be reused. We view the static constraints of an SBVR model as defining the possible worlds that the data of an information system can describe. Additionally, dynamic constraints define the allowed transitions between these states. From this starting point, we explore how information systems can be generated, following the path set out by C.J. Date [6].

2.1 Validating an SBVR Model

SBVR supports constraints of two different modalities, Alethic and Deontic. The SBVR specification [5] describes their difference as follows:

"Alethic modal logic differs from deontic modal logic in that the former deals with people's estimate(s) of the possible truth of some proposition, whereas deontic modal logic deals with people's estimate(s) of the social desirability of some particular party's making some proposition true."

In this sense, it can be said that the alethic model defines the map of the territory that the deontic model navigates within. In order to analyse an SBVR model, we separate the rules into two models, alethic and deontic according to their modality. After checking each model for internal consistency, we infer the relationship between the sets of allowed states of the two models.

In business scenarios, constraints imposed from external sources (Nature, Government, Partner organisations) and therefore outside the jurisdiction of the model owner(s), are always alethic from the perspective of the business, whereas internal constraints can be either alethic or deontic. This difference also affects the enforcement of each constraint type, where altering an internal constraint is an option but not so for an external constraint.

The interaction between the alethic and deontic modalities is not discussed in the SBVR specification, but in order to produce executable code, it is an area we must examine. Following the notation in chapter 10 of [5], we want to ensure that situations do not occur where $OA \,\&\, {\sim}\lozenge A$ for any proposition A. A proposition cannot be made true if it is not possible for it to be true, so the propositions deemed obligatory by the deontic model must be possible in alethic model.

If some propositions obligatory in the deontic model are not possible in the alethic model and all the propositions possible are obligatory, we consider the deontic model to be superfluous as it adds no information about desirability beyond what is known to be possible. If the states allowed by the two models are disjointed, we consider the overall model to be invalid, since the deontic model only aims for states that are not deemed possible by the alethic model. If the two models partially overlap, the model can be executed, but the user should be warned that a subset of the states deemed desirable by the deontic model are unreachable.

After inferring the relationship between the two sub-models, we take their intersection as the effective model which we attempt to execute. Differences in modalities mean that there may or may not be recourse to the model owner in case a specific rule is violated depending on whether it is deontic or alethic respectively.

2.2 Inferring a Database Schema

While an SBVR model is an abstract construct, it defines the space which instances of terms and fact types are allowed to occupy when the model is itself materialized into an information system. This structure can be made explicit by extracting it and imprinting it onto a relational database. Relational databases, besides being the dominant persistence technology for information systems, are also an excellent candidate for persisting SBVR-based information systems because of their declarative nature. Relational databases are interfaced through SQL, a declarative language that defines the data structure and queries for a relational database. When discussing data structure, we focus on the SQL data definition language (SQL-DDL). While relational databases do not exhibit the expressivity that an SBVR model can, it is feasible to generate an SQL-DDL data model from an SBVR model. In this way, the maturity and performance of the many SQL databases can be harnessed while simultaneously using the integrity constraint checking as a basic model checker. The more advanced cases will of course still need to be checked against the SBVR model directly. Generation of a relational database schema has been referred to previously in [9].

To infer a database structure from an SBVR model, we begin by constructing a graph where each term and fact type is represented as a node. The edges link fact types with the terms they build on. Next, we need to define the relationship between each of the nodes.

For Date [10], defining a relationship requires integrating two aspects, one for each party in a relationship. The single-perspective relationships that Date considers are at most one (0..1), exactly one (1), one or more (1..*), and zero or more(0..*). While more detailed relations such as 0..5 etc. could be considered, there are diminishing returns to increasing levels of granularity, but also such relationships are also beyond the expressive capacity of the relational model.

In our graph, the edges initially connect fact types with terms. Since each fact type instance refers to exactly one term instance for each link, the relations of interest are those from the perspective of the terms. In the absence of constraints, a term instance can be referred to by multiple fact type instances (facts). The default edge label is therefore zero or more (0..*). So the fact type 'student is enrolled for course', in the absence of relevant constraints, would be represented as a node with 0..* edges to the student and course nodes. When rules exist that affect the cardinality of an edge, such as 'It is obligatory that each student is enrolled for exactly one course', the relation between the term student and the fact type 'student *is enrolled for* course' gets more constrained, in this case to an exactly one relationship (1).

Once the relationships are identified, we can begin to differentiate what will eventually become tables and what will be attributes for these tables. To begin, unary fact types, such as 'student *is under probation*' become Boolean attributes of the term they are connected to, in this case student. This is because they unambiguously say something about the term they are connected to, each term can only have one instance of that value, and this value can only be true or false. Similarly, we instantiate attributes from binary fact types which indicate that the one fact type role has a designation in an attributive namespace for the subject concept represented by the designation used for the other fact type role (e.g. student *has* name).

For other binary fact types, the simplest solution would be to represent them as tables on their own and leave enforcement of the relations between the data items to the SBVR model execution engine which will wrap the database. However, databases are highly optimized and their integrity constraints checking could take a lot of the burden off of our implementation, reusing the mature RDBMS software. So to represent different relationships we have five patterns for generating the equivalent SQL-DDL schema fragment for two nodes A and B. These generally include generating tables and using the Primary Key (PK) and Foreign Key (FK) as well as Nullable and Unique to express the relations specified in the graph.

The patterns, for two tables/nodes A and B are:

- Pattern I: PK of B is a FK in A with Uniqueness Constraint and is Nullable. If B has no other attributes, instead of an FK it becomes an attribute of A directly.
- Pattern II: PK of B is a FK in A with Uniqueness Constraint. If B has no other attributes instead of an FK, it becomes an attribute of A directly.
- Pattern III: PK of B is a FK in A and is Nullable. If B has no other attributes, instead of an FK it becomes an attribute of A directly.
- Pattern IV: PK of B is a FK in A. If B has no other attributes, instead of an FK it becomes an attribute of A directly.
- Pattern V: Intermediate Table A_B with PK of A and B as FKs and joint PK. If either A or B have no other attributes, instead of an FK they can become attributes of A_B.

We then apply these patterns as specified by Table 1. The cells that simply specify a pattern directly are those whose relationship semantics are exactly expressed by the results. The cells that merely specify 'Use' of a pattern are those whose semantics are not directly expressible in SQL, so a looser approximation needs to be used, with the rest of the input validation needing to be done by applying the SBVR constraints directly. We can observe that these cells are the ones related with the 1..* type of relationship which SQL cannot cover. Finally the cells that specify reverse use of a pattern are simply those where the appropriate pattern is the identified pattern with B and A substituted for each other. Due to the approximate nature of the relational model, an enclosing SBVR model execution engine must have sole write access to the database, in order to maintain consistency of the data. Alternatively, triggers could be implemented within the database itself for the more strict constraints, however their database-specific syntax and the difficulty of identifying the violated rule advise against this approach.

Table 1. Appropriate database patterns to express fact types as relations

Edge with A... Edge with B...	0..1	1	1..*	0..*
0..1	I	Reverse II	Use Reverse III	Reverse III
1	II	Same Table or Use V/II/Reverse II	Use V	Reverse IV
1..*	Use III	Use IV	Use V	Use V
0..*	III	IV	Use V	V

For n-ary fact types with n ≥ 2, such as Student *is marked with* grade *for* course, the relational model does not provide any way to represent this relation between terms, other than creating a new table. So for any combination of edge labels connected to the fact type, we use a separate table having the primary keys of the relevant terms as a foreign key. Another aspect of the data model that needs to be considered is the data types for the stored attributes. The 'SBVR Meaning and Representation Vocabulary' [5] gives us a number of data types such as (quantity, number, integer, text) that can be mapped to SQL primitives. This however puts a strict requirement on the terms that carry a value to belong to a concept type that specialises one of these data types such that it can be inferred. Terms that do not define a data type can still be represented as tables and defined in terms of their characteristics and connections. Terms that are defined to range over a fixed set of values, such as the term grade which could be defined as [A or B or C or D or F], can be translated as an SQL ENUM data type to avoid creating a new table. Finally, the issue of primary keys remains. While this could be inferred through attributes that have a uniqueness property or SBVR reference schemes, performance of the database may suffer when using textual keys. For this reason, each table gets an integer auto-incrementing id attribute added, which becomes the primary key of that table. This can be omitted when the table contains a unique integer, such as a code number.

With the steps discussed above, we can algorithmically infer a relational database schema from an SBVR model that uses most of SQL's expressivity to optimise access to the data. However, the expressivity gap must always be considered, and each new element of data that enters the database needs to be verified not only against the integrity constraints of the database schema, but also against the SBVR rules that are relevant to the terms related to the data item. For instance a rule that uses multiple fact types such as 'it is obligatory that each module that a student *is registered for, is available for* a course that the student *is enrolled in.*', cannot be expressed within the database schema. The process described in this section could potentially be implemented within a model transformation framework. QVT would be a candidate due to SBVR's serialisability in XMI, but the transformation would also require the existence of a suitable XMI target for relational databases.

2.3 Converting an SBVR Rule to an SQL Query

To verify whether or not a given state of relational database conforms to the more advanced constraints that SBVR imposes, we can convert the rules to SQL queries designed to verify the state of the dataset by essentially asking the question: "Is this rule constraint consistent with the state of the database?" Figure 1 shows the conversion of one rule from our example into an SQL query.

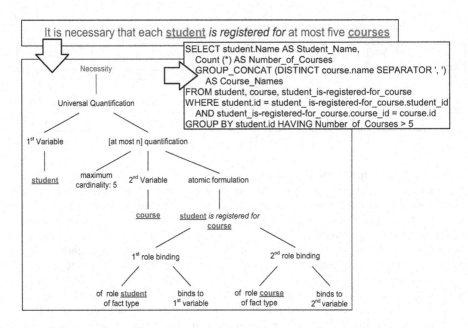

Fig. 1. Conversion of SBVR-SE to SBVR-LF to SQL Query

It should be noted that the above query is written in the dialect of MySQL 5.0. The result is a possibly empty set that contains a list of all the <Student_Name, Number_of_Courses, Course_Names> tuples that violate the rule. An empty set signifies that the rule is not violated throughout the dataset. In our example, if the user attempts to add a sixth course to the student 'John', the rule is violated and the query returns the values in table 2.

Table 2. Results of rule evaluation

Student_Name	Number_of_Courses	Course_Names
John	6	PY101, MA101, EN121, CS101, AF302, MG102

The logical formulation is transformed as follows: The FROM clause of the query includes all the variables of the formulation. In our example 'student', 'course', as well as the fact type table 'student_is-registered-for_course' are used. The WHERE clause connects the tables according to the model semantics. From there we transform the constructs of the logical formulation into SQL constructs. Specifically, the universal quantification becomes the 'GROUP BY' clause over the variable it introduces, whereas the at-most-n quantification combined with the maximum cardinality becomes the HAVING clause. The SELECT clause is formulated by importing the variables referenced according to their reference scheme. Since there exists a universal quantifier that groups courses, the name of the courses becomes a GROUP_CONCAT statement, and we also import the Number_of_Courses, which is needed in the

HAVING clause. Generalising this method is underway and its implementation is pending the implementation of the SBVR parser.

3 SBVR and the Architecture of the Web

When considering how SBVR models can be made executable, there is the issue of shifting between consistent states. The verbs that SBVR allows, in the form of fact types, are declaratives that describe a state and as such cannot be used to actively cause a shift from state to state. Even dynamic constraints which are not yet supported by SBVR can constrain but not cause transitions. What is needed is an architecture that allows users to express, in a standardized manner, the changes they want to affect on the data of the information system. This search led to the protocol and the architectural style that underpins the largest distributed system in the world, the Web. The protocol is HTTP, and the architectural style it instantiates is Representational State Transfer (REST). REST, was identified by fielding in 1999 [11], and has recently been popularised by works such as [12], [13]. In contrast to the less disciplined Remote Procedure Call (RPC) style used in most WS-* standards, REST is based on a number of explicit architectural constraints that govern interactions. While the constraints are abstract, it is important to state that they govern most of the daily interactions over the Web, both human-to-machine and machine-to-machine.

Fundamental constraints are that each significant entity should be named, uniquely identifiable and linkable. While this is reminiscent of the foundation of the business rules approach being on terms, it goes further to the level of instances, where each and every one should be identifiable also. While SBVR does include individual concepts, there is no explicit directive that each and every instantiation of a term should be named. In the case of the web, every instance, named a resource, should have a unique URI. The slight difference in approaches can be explained by the fact that while SBVR is concerned with the model-level description of a domain, REST is model-agnostic and only concerned with data. This distance can be covered by considering that each term identifies a collection of resources (individual concepts) and that this collection can be a resource by itself.

While linking URIs and terms brings SBVR closer to the web, the user is still without means to cause change in the state of the system. REST however indicates that resources should be manipulated through a uniform interface. In the case of HTTP this interface includes operations such as GET, PUT, POST and DELETE. The uniform interface is in fact the only way in which a client can interact with the resources that a service makes available. The rationale behind this constraint is that if the operations are insufficient to accomplish some functionality, then there are more resources that need to be identified, rather than overloading the interface with additional methods. By constraining the interface to a fixed set of methods, the interface designer is forced to extend the vocabulary of the application. This is an insight that can directly reflect on the modelling methodology and reveals the benefit of considering the run-time behaviour at design-time. By forcing the modeller to consider the model as an executable artefact, accessible from a constrained interface, they may discover entire new areas that need to be modelled that would have otherwise been overlooked, leaving room for ambiguity.

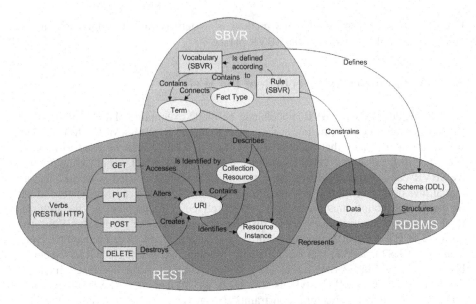

Fig. 2. Connections between REST, SBVR and Relational Databases

A number of other constraints are in effect when the REST architectural style is considered, including statelessness and the 'hypermedia as the engine of application state' (HEAS). Statelessness instructs that each request to the server contains all the information needed for the server to understand the request without need for session information on the server side. HEAS requires that resources use links to point to each other such that the clients can incrementally discover the API and any change on a resource URI will not trigger a catastrophic failure of the client but rather a recovery procedure during which the client will rediscover the new identifier the same way as the original identifier had been discovered. These constraints form an architectural style that is state-oriented rather than process oriented. By focusing around 'be' and not 'do' type interactions, REST can be considered a declarative architectural style, very well aligned with the design principles behind SBVR.

3.1 Constructing a RESTful Interface from an SBVR Model

A basic design principle of REST is that 'things' should be named. This gels perfectly with SBVR's term-orientation. Since vocabularies have a namespace URI, and terms are unique within a vocabulary namespace, it is trivial to assign a unique URI to each term. The following URI template [14] would be sufficient to accomplish this:

http://domain.org/{vocabulary}/{term}

However the question arises of what these URIs will return when requested from the application's server. Since the term can be seen as a collection containing instances, we can return this collection as a resource. The Atom Publishing Protocol [13] is being increasingly used as a general-purpose format for representing collections and can be used to grant our interface with significant standardised functionality, such as exposing collections as standard feeds and editing collections or instances. An aesthetic issue may be the need for pluralisation of terms when used for representing collections in URIs. This is one of the parts where the creation of SBVR for human-to-human interactions becomes apparent. While SBVR rules can freely pluralise terms, the exact plural of each term is not specified in the vocabulary. This issue also arises in the design of an SBVR parser. For English this can be partially solved by using an inflector library such as [15]. However this would leave terms with an incorrect pluralisation automatically inferred without a way of specifying the proper pluralisation. Ideally, an extension to the SBVR meta-model would allow for specification of the exact plural form of a term.

While we have discussed the issues of representing collections of instances, the issue of instances themselves remains. Following the unique identification patterns discussed in the relational schema generation, we can use a URI template such as the following to generate the URI for an instance of a term, like so:

http://domain.org/{vocabulary}/{term}/{identifier}

The serialisation of the content of a term can be subject to the content negotiation processes that HTTP specifies, but as a baseline, an XML serialisation can be assumed as a standard. Providing an XSLT stylesheet defining the transformation of that XML to HTML can also aid towards readability by human readers.

Another fundamental constraint of REST is that of 'Hypermedia as the Engine of Application State' (HEAS). This specifies that URI-named resources should link to each other so that they can be discovered by a new client with minimal initial information, or rediscovered in case of URI change. This also naturally overlaps with SBVR's assertion that fact types connect terms. In this sense we can use the fact types as links to instances of the terms that the fact type builds on. So, the representation of each term must also provide links to URIs that represent the set of instances of the fact type that concern the term in question. This can be reflected in the URI-space by a template such as:

http://domain.org/{vocabulary}/{term$_a$}/{identifier}/{fact type designation}/{term$_b$}

In the case of a binary fact type, presenting the subset of a term's instances that are connected by a fact type to the original term instance is acceptable. In the usual case where two terms are connected by only one binary fact type, the fact type designation can be optionally omitted. More complex queries including filters should also be possible, and work along these lines is underway but not strictly necessary for the operation of a system such as the one in this paper.

Table 3. Applying HTTP operations on Collections and Instances

	Student	
	Collection	Instance
GET	+	+
PUT		+
POST	+	
DELETE		+

Having defined identification of collections, instances and basic queries with URIs, we can now examine which HTTP operations apply on these URIs. Table 3 shows a basic application of HTTP operations on the resources defined by the student term.

Thus, knowing the URI of the student collection, we can not only see a representation via GET, but also create a new student via POST, identify a specific student via hyperlinks, and then modify the student instance through PUT or remove the student via DELETE. Access to data and ability to modify it is of course subject to authorisation and authentication of the user making the data retrieval and modification. The discussion up to this point has given enough information to present a unified view of SBVR, RESTful HTTP and Relational Databases as a foundation for a model-driven information system as seen in Figure 2, first published in [16] by the authors of this paper.

3.2 Using SBVR to Describe Resources

Along the way of aligning SBVR, RESTful HTTP and Relational Databases, we have also created a complementary way to use SBVR to describe resources, a use for SBVR that can potentially have an impact beyond systems natively described with SBVR and onto the mainstream web. For instance, having described the resource instance of student and accompanying relational representation, we now have a structured representation of a student. However, by separating all the elements (vocabulary and rules) of the model that mention 'student', and also adding all the terms and fact types necessary to express them, we can construct a subset of the model that describes the resource returned. This model can also give information about the data in the schema and the reaction that the server will have on various operations being applied to the resource as well as what can be expected by following specific links. So for instance, a request for a student instance may return a representation as seen in Figure 3.

```xml
<?xml version="1.0" encoding="UTF-8"?>
<student>
          <id>3465</id>
          <firstname>John</lastname>
          <lastname>Smith</lastname>
          <is-under-probation value="false" />

          <link rel="is-enrolled-in_modules"
              href="http://domain.org/school/student/3465/is-enrolled-
              in/modules" />

          <link rel="is-registered-for_course"
              href="http://domain.org/school/student/3465/is-
              registered-for/courses" />

          <link rel="is_marked_with-grade-for-course"
              href="http://domain.org/school/student/3465/is-marked-
              with/grade/for/courses" />
</student>
```

Fig. 3. Example XML Serialisation of model-derived resource

The server may internally have a model associated with it such as the one found in table 4. By publishing this model, the aware client can now start building a model of the application in general, and also specifically know what steps need to be taken for creating a new student resource. Something that must be taken into account is that since the new rules exposed by the system as a resource description are not changeable by a client system that may be using the resource, the modality published should become alethic as the rules now describe the environment that other systems operate within. The mechanism for publishing this SBVR description information in a web-friendly way has yet to be determined, although a direct link from the resource to an xml-serialised form of the model may be sufficient.

Table 4. Instance of an SBVR model subset describing a single resource

Terms	Fact Types	Rules
Student	Student *is under probation*	It is necessary that each **student** *is registered for* at most five **courses**.
Module	Student *is registered for* course	
Course	Student *is enrolled in* module	It is necessary that each **module** that a **student** *is registered for is available for* a **course** that the **student** *is enrolled in*.
Grade A or B or C or D or E	Student *has* first name	
First name	Student *has* last name	It is necessary that each **student** that *is under probation is registered for* at most three **courses**.
Last name	Student *is marked with* grade *for* course	
	Module *is available for* course	

4 Implicit Process Specification

The capabilities of the information system described so far are limited to satisfying sequential operations that satisfy the SBVR model over a basic RESTful API. However, fundamental to information systems is the ability to perform processes that make multiple state alterations as their result. In fact, for certain models, it may be impossible to move from one state to another without performing more than one operations, for instance in the case where a student must be registered for exactly five courses, a new course cannot be added if a course is not simultaneously removed. This calls for the execution of multiple operations simultaneously over HTTP, something the authors of this paper have already made progress in specifying RESTful Transactions [17]. Even with this capability, there is a fundamental tension between processes and declarative specifications, as processes focus on the 'how' rather than the 'what' which declarative models specify. To resolve this, we have been inspired by the motto of the logic programming community [18] which states that *Algorithm = Logic + Control*.

By approaching the process as a simplified algorithm and the SBVR model as the logic, we can see that perhaps processes can be dynamically generated through application of a control module to the SBVR model. Our solution is to introduce the meta-process as seen in Figure 4, which at its core examines the state resulting by each action of the (authenticated) user, and determines whether it will result in the system being in a consistent state, one where no rules are violated. In case of violations, we return the rule that has been violated as part of the response. This is an application of

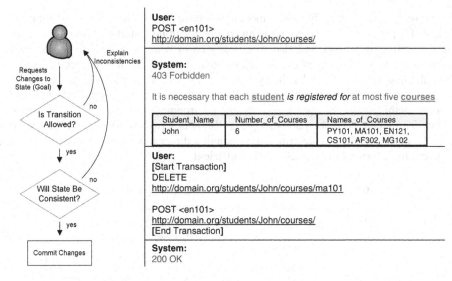

Fig. 4. The meta-process control structure

the business rules motto 'The rule is the error message' which seems to be quite effective in our case. Notice that this mechanism can also observe the violation of dynamic constraints such as the progression of marital status from 'married' to 'single' instead of 'divorced' or 'widowed'. The user can then amend their request with additional operations aimed at mitigating the violation. As the process iteratively continues, the user will either realise that their request is untenable within the constraints of the current model, or they will formulate a request that satisfied both their requirements and the system's.

This results in a declarative process-less system which can nevertheless exhibit process-like behaviour. Its main run-time difference with process-driven systems that is that it allows users to perform any allowable process instead of specifying at design-time the processes that the designers forecast will be useful to users. The design-time implication is that such systems are capable of naturally adapting their behaviour to changes in the model instead of requiring manual revision of their individual processes, which also risks inconsistencies in case of error as a rule may affect multiple processes and the designers have to infer which these processes are.

Also, by returning the violated rule in machine-to-machine interactions, the partner systems can update each other about changes in their models organically as violations occur and such changes can propagate through the network in the case of services composing other services.

As we have seen so far, SBVR can not only produce functional Web Applications, but also have them be self-describing by publishing the sub-model relevant to each resource. This, combined with transactional capabilities over HTTP opens the door for service composition by importing the sub-model that is published for each resource and using it as part of a new model. So a system could represent a service composition such as a travel arrangement and its model would be constructed by importing the sub-models used by other providers to describe the elementary resources the new system composes such as flights, car rentals, hotel bookings, etc. The resulting new resource (travel arrangement) can be made available to the users of the new service. Work has already been done in this area by the authors [19], however it remains to be naturally integrated with the meta-process presented in this paper.

5 Concluding Remarks and Future Work

In this paper we have described a framework that bridges the worlds of SBVR, Relational Databases and RESTful APIs to produce a functioning web application with an SBVR model as its starting point. Additionally, we have discussed the meta-process as a rule-driven control mechanism that makes the web application capable of exhibiting process-like behaviour without explicit processes defined. Finally, we discussed the possibility of composing such systems in a RESTful environment. Earlier work along these lines has been published in [16] however the present paper expands the work both in depth and scope.

An element that has not been discussed yet is that of a user interface to this system. The simplest approach to this would be to use templating systems to define custom interfaces for each resource and collection. While this is a good starting point, its lack of adaptability to model changes and imperative nature of templating languages make it only suitable as an intermediate measure. In the medium to long term a more flexible approach would be to add interface generation capabilities to the system with an accompanying rule-driven customisation mechanism such that the aesthetics of the system and the built in assumptions of the generator can be fully customised by the modeller. This would in effect turn SBVR to a declarative user interface modelling language. Additionally, the system as it stands is limited to information-driven tasks and cannot interface with complex algorithmic systems. This problem can partially be addressed through work in service composition where a system can be SBVR-described while not being generated from an SBVR model.

This description can act as a wrapper for the system to be included in other systems, giving them additional capabilities. Alternatively, a fact type can be introduced that instead of representing a class of facts given in a database, can instead represent facts that are only instantiated when requested, with the process of instantiation involving the execution of a processor-intensive algorithm. Finally, in implementing this system, an SBVR parser needs to be implemented that can convert SBVR-SE to SBVR logical formulation. While such parsers exist, none of them is open-source and freely available for use by the community. With this step completed, implementation of the system as described in this paper can commence in earnest.

References

1. Johnson, J.: CHAOS: the dollar drain of IT project failures. Application Development Trends, pp. 41–47 (January 1995)
2. Johnson, J.: CHAOS 2006 Research Project. CHAOS Activity News 2 (2007)
3. Nachira, F., et al.: Digital Business Ecosystems. Office for Official Publications of the European Communities (2007)
4. Zittrain, J.: The Future of the Internet–And How to Stop It. Yale University Press (2008)

5. Object Management Group, Semantics of Business Vocabulary and Rules Interim Specification (2006), http://www.omg.org/cgi-bin/doc?dtc/06-03-02 (accessed: 25/10/2007)
6. Date, C.J.: What Not How: The Business Rules Approach to Application Development. Addison-Wesley Professional, Reading (2000)
7. Ross, R.G.: The Business Rules Manifesto. Business Rules Group. Version 2 (2003)
8. Open Philosophies for Associative Autopoietic Digital Ecosystems (2008), Automatic code structure and workflow generation from natural language models, http://files.opaals.org/OPAALS/Year_2_Deliverables/WP02/D2.2.pdf
9. Linehan, M.H.: SBVR Use Cases. In: Proceedings of the International Symposium on Rule Representation, Interchange and Reasoning on the Web, pp. 182–196. Springer, Heidelberg (2008)
10. Date, C.J.: All for One, One for All, Part 2: How Many Cases Are There? Business Rules Journal 7(12) (December 2006), http://www.BRCommunity.com/a2006/b324.html
11. Fielding, R.T.: Architectural Styles and the Design of Network-based Software Architectures. University of California, Irvine (2000)
12. Richardson, L., Ruby, S.: RESTful Web Services. O'Reilly Media, Inc., Sebastopol (2007)
13. Gregorio, J., De Hora, B.: The Atom Publishing Protocol. Internet RFC 5023 (October 2007), http://www.ietf.org/rfc/rfc5023.txt
14. Gregorio, J.: URI Template. Internet Draft draft-gregorio-uritemplate-03 (September 2008)
15. Inflector, http://inflector.dev.java.net (accessed on June 28, 2009)
16. Marinos, A., Krause, P.: Using SBVR, REST and Relational Databases to develop Information Systems native to the Digital Ecosystem. In: IEEE Conference on Digital Ecosystems Technologies, DEST 2009 (to appear, 2009)
17. Marinos, A., Razavi, A., Moschoyiannis, S., Krause, P.: RETRO: A (hopefully) RESTful Transaction Model. Technical Report CS-09-01, University of Surrey, Guildford, Surrey (August 2009)
18. Kowalski, R.: Algorithm= logic+ control. Communications of the ACM 22, 424–436 (1979)
19. Marinos, P.K.: What, not How: A generative approach to service composition. In: IEEE Conference on Digital Ecosystems Technologies (DEST 2009) (to appear, 20090020039

Towards an Improvement of Software Development Processes through Standard Business Rules[*]

José L. Martínez-Fernández[1,2], Paloma Martínez[2], and José C. González-Cristóbal[1,3]

[1] DAEDALUS – Data, Decisions and Language S.A.
Avda. de la Albufera, 321
28031 Madrid, Spain
{jmartinez,jgonzalez}@daedalus.es
[2] Advanced Databases Group, Universidad Carlos III de Madrid
Avda. de la Universidad, 30
28911 Leganés, Spain
{joseluis.martinez,paloma.martinez}@uc3m.es
[3] DIT, Universidad Politécnica de Madrid
Avda. Complutense, 30
28040 Madrid, Spain
josecarlos.gonzalez@upm.es

Abstract. The automation of software development processes is a desirable goal of current software companies which would lead to a cost reduction in software production. This automation is the backbone of approaches such as Model Driven Architecture (MDA) or Software Factories. This paper proposes the use of *standard* Business Rules (using Rules Interchange Format, RIF) to specify application functionality along with a platform to produce automatic implementations for them. The novelty of this proposal is to introduce Business Rules at all levels of MDA architecture in a software development process, providing a supporting tool where production Business Rules are considered at every abstraction level. Production Business Rules are represented through standard languages, rule engine vendor independence is assured via automatic transformation between rule languages, and Business Rules reuse is made possible. The objective is to get the development of production Business Rules closer to non-technical people involved in the software development process through the use of natural language processing approaches, automatic transformations among models and semantic web languages such as Ontology Web Language (OWL).

Keywords: Business Rules, production rules, rule engines, Rules Interchange Format, RIF, Model Driven Architecture, MDA, OWL, Ontology Web Language.

[*] This paper has been partially supported by the Spanish Center for the Development of Industrial Technology (CDTI, Ministry of Industry, Tourism and Trade), through the project ITECBAN (Architecture for Core Banking Information Systems), INGENIO 2010 Programme. Other partners in ITECBAN are INDRA Sistemas, CajaMadrid, Sun Microsystems and Grid Systems. Special mention to our colleagues at INDRA must be done for their involvement in the specification of K-Site Rules: Fernando Alcántara, Pablo Leal, Juan Carlos Macho and Gonzalo Pando (in alphabetical order).

G. Governatori, J. Hall, and A. Paschke (Eds.): RuleML 2009, LNCS 5858, pp. 159–166, 2009.
© Springer-Verlag Berlin Heidelberg 2009

1 Introduction

In many software-based industries, the speed with which new products and services are efficiently developed and deployed is a definitive success factor [5][4]. This is the case, for example, in the insurance market, where the rules and conditions for establishing the scoring used to decide client fees are continuously changing. Suppose that an insurance company has to react to a new product from its competitor and decides to modify the conditions of one of its own products. An expert at the company decides the new conditions, which have to be notified to the Information Technology (IT) department. Then the corresponding software component is identified, re-coded, compiled, tested against errors and, if everything is all right, deployed. As can be seen, this process is usually hard, tedious and, most importantly, time-consuming. The business rules approach pursues the reduction of development time by putting governing business rules together and allowing their modification, without recoding or compiling them. Besides, the knowledge of the business resides in one place, in the form of business rules. At present, there are several Business Rules Management Systems (BRMSs) vendors such as ILOG JRules[1], Fair Isaac's Blaze Advisor[2] or open source alternatives like JBoss Rules[3], but companies integrating these products have to select one of them and this decision cannot be easily changed. So when a company decides to introduce a BRMS product as part of its IT systems, it becomes dependent on the provider of that product. Besides, phasing in a new BRMS product is, usually, an expensive and difficult process. On the other hand, it continues being difficult for non-technical business experts to define business rules without the intervention of technical people because there is no clear relation between business concepts and their implementation. This paper proposes a technical framework to address three basic features that current BRMS systems do not consider:

- To reduce the semantic gap between business concepts and business objects in rule-based environments by exploiting Semantic Web technologies.
- To provide independence from the BRMS implementing business rules.
- To allow easy integration of Business Rules into the software development process used in the target organization.

This approach has been put in practice in a tool called K-Site Rules. K-Site Rules is based on initiatives promoted by standardization organizations like W3C and OMG, who have defined different rule languages pursuing independence between rule definitions and implementations. Obviously, K-Site Rules focus on production rules [8] which are related with expressions having an *if-then* structure that can be supported by forward-chaining rule engines. In fact, any rule type which can be written using RIF language could be considered but, as an initial step, only production rules are taken into account.

The remainder of this paper is structured as follows: the next section describes the proposed framework, K-Site Rules, and the relation with MDA, which gives support to the automatic business rules development process. How K-Site Rules is integrated into the software development process is also explained in this section. The third

---

[1] http://www.ilog.com/products/jrules/
[2] http://www.fico.com/en/Products/DMTools/Pages/Fair-Isaac-Blaze-Advisor-System.aspx
[3] http://www.jboss.com/products/platforms/brms/

section provides an application example of the framework defined on the insurance industry domain. The final section includes some words on evaluation and presents preliminary conclusions.

2   K-Site Rules Framework Description

K-Site Rules framework pays special attention to capabilities facilitating business rule definition and management by non-technical users. Currently, most of rule development environments available allow business rules definition starting from a Unified Modeling Language (UML) model. This kind of model is not easily understandable by non-technical users because, usually, a business concept does not correspond to a unique UML class. For this reason, the use of ontologies to represent the business knowledge is proposed. Ontology, according to the meaning taken in computer science, is a knowledge representation paradigm containing entities and relations among them [3] People working on other knowledge disciplines are used to exploit ontology concepts or other related instruments so non-technical users can be closer to them than they are with respect to UML. Of course, these users are not intended to go through XML based ontology representations, a graphical view is provided instead. The idea is to define business rules over concepts represented in an ontology that can be matched against business objects, represented by UML classes. In this way, non-technical users can be provided with tools which help them to write business rules involving those business concepts. Following an MDA approach, automatic ways to transform these ontology concepts and rules definitions in the corresponding coded implementations must also be produced.

According to software engineering [11], four basic stages can be distinguished in a standard development process, these are: analysis, design, construction and test. K-Site Rules has been designed to allow automatic transformation (through MDA) from the design to the final implementation. K-Site Rules includes two different editors for the design stage, one for technical people, developed on a widely used IDE like Eclipse, and another one, for non-technical people, that is integrated with common web browsers. Taking into account the test stage, K-Site Rules provides tools for unit testing, allowing business rules builders to assure a correct logical behavior of their rules. With these unit testing tools, the non-technical user can provide some input values to check the behavior of the rule. If there is some undesired result, the user can go back, change the content of the rule and test again. For the business rules unit testing K-Site Rules generates default implementations for involved objects (if previous implementations are not available). Of course, only functional testing is possible, but nothing more is needed for non-technical users. Integration tests are not considered in K-Site Rules because they are part of the development of the system where rules are going to be used and no interference with processes already defined in the target organization is wanted. For the same reason K-Site Rules does not deal with the deployment of business rules once they have been built. Each target organization would have its own deployment policy and no interference in this process is desired, so K-Site Rules impose no restrictions on it.

An informal definition of the development process could be as follows: a user defines a business rule using an editor tool that guides rule writing; this editor restricts valid vocabulary to the concepts present in a given enterprise ontology about business data. The guiding editor produces an RIF compliant business rule definition, linked to ontology concepts (the RIF draft standard defines how to interact with ontology languages [10]). These expressions are automatically transformed into UML/OMG-PRR representations, which are independent from the rule engine that will be used to implement business rules. In fact, the last step is to obtain a specific model, once an available rule engine has been selected for implementation purposes; UML/OMG-PRR expressions are then automatically transformed to the rule engine language.

The work introduced in this paper defines and implements the three transformation steps that must be taken to complete the described business rule definition process:

- *UML to OWL transformation.* The OMG Group determined the necessity to establish ways to relate different knowledge representations using some kind of semantic model or ontology. This is the main goal of the Ontology Definition Metamodel, ODM, which is devoted to the definition of metamodels for ontology representations. Up to now there is a submitted specification under balloting [7]. It gives some guidance for mapping UML metamodels with OWL metamodels; so, ontology models could be obtained from UML models. The reverse step, from OWL to UML, is not covered; an UML element needs specific information that should not be expected in an ontology, so only some advice for matching ontology concepts with UML classes could be given.

    The transformation included in K-Site Rules considers conversions shown in Table 1, adapted from the ODM guidelines. Of course, other conversion alternatives are possible.

    Completeness of the transformation method included in K-Site Rules is not assured. For this reason, other approaches, like the one followed in [1] where a specific language, the ATLAS Transformation Language (ATL), is used to define the way transformations between models must be carried out in a Model-Driven Engineering (MDE) environment. One of the use cases defined in ATL provides an implementation for ODM specification. Some other examples on transformations among models applied to business rules scenarios can be found in [2].

    In a first approach, the authors have preferred to have complete control in UML/OWL transformation process, building a custom transformation.

    The final objective of this transformation is to build a simple graphic representation of the domain where business rules will be defined. This graphical representation will make it easier for non-technical users to know which concepts they can use to build business rules.

- *RIF to OMG-PRR transformation.* Another step to be given in this transformation is the translation between business rules in RIF and Production Rules Representations. The RIF format includes ways to represent business rules, the RIF Production Rules Dialect, (RIF-PRD) [12] that are easily mapped to OMG-PRR [8]. Several OMG members are working in both standardization groups. The work in [8] also points out that, although there is some overlap between these standards, the focus is different, OMG-PRR is oriented towards UML based tools and methods, while RIF is focused on web technologies and users. K-Site Rules includes a component in charge of this transformation. It is worth mentioning that, up to this point, standard expressions of business rules are provided so they are valid for any rule engine available. Taking a look at the MDA approach, this automatic transformation is needed.

Table 1. UML/OWL Conversion convention

UML component	OWL component
Package	Ontology
Class	Class
Attribute	DatatypeProperty or ObjectProperty
Method	DatatypeProperty or ObjectProperty
Association	ObjectProperty
Aggregation	ObjectProperty
Generalization	SubClass

- *OMG-PRR to specific platform transformation.* The proposed framework must also include a transformation module between Production Rules Representations and specific rules languages included in rule engine products. The standard, of course, does not include any guide for transforming PRR expressions to any rule engine. A shallow study of the standard has been made and a possible way to represent main PRR standard elements in JBoss Rules and ILOG JRules has been envisaged and implemented, although a deeper analysis is yet required. It is worth mentioning that, although the PRR standard is very recent, there are people from rule engine vendors involved in the definition of the standard, so mapping between PRR and the most relevant rule engine vendor languages should always be possible. This is the final automatic transformation needed according to the MDA approach.

3 Example in the Insurance Domain

To illustrate the business rules development process described a simple example is provided. Suppose a financial company where a domain expert has defined the following business rule to decide how much money a policy holder has to pay for his car insurance. The business rule will be called *insuranceFee*.

In this example there are really two business rules that, in case of necessity, could be stored in different groups of rules or *rulesets*. In the rest of this paper we are going to focus on the first business rule, defined in the first sentence of Fig. 1. It is a production rule as some action is taken when the conditions are fulfilled. In fact, there are two Business Rules that will be treated separately.

If the rated power of the target car of the insurance policy is less than 100 horsepower and its price is under 15000 euros then the annual base fee for this car must be decreased by 10%. On the other hand, if the policy holder is under 55 and over 30 the annual base fee must be increased by 2%, in any other case the annual base fee must be increased in 20%

Fig. 1. *InsuranceFee* rule, an example of business rule in the insurance domain

In this context a very simple business model is considered that could be represented by the OWL ontology. The ontology can be viewed through the ontology visualization tool included in K-Site Rules. The main goal of the supplied visualization tool is to allow non-technical users to have an idea of which concepts are available in the definition of business rules and which relationships between them are considered.

Using the OWL model obtained, the example business rule can be defined through K-Site Rules using the provided natural language editor, as shown in Fig. 2.

The RIF expression built will be the standard representation of the business rule, which must be transformed into the corresponding OMG-PRR expression. The final step in the rule generation process would be to obtain a representation of the business rules in a rule engine proprietary system.

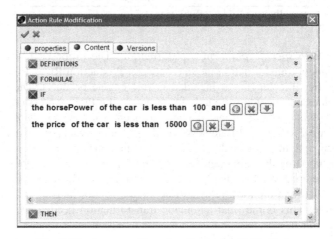

Fig. 2. The *insuranceFee* rule edited using K-Site Rules

In an ideal situation, if transformations have been correctly performed, there would be no need to modify the expression of business rules in rule engine proprietary languages, i.e. if a change must be made in the definition of the rule, it should be made at the natural language level through the corresponding editor, which will then automatically perform corresponding transformations. Of course, this editor is not only natural language based; it is also possible to define rules using decision tables or decision trees.

The example described is very simple in order to provide a clearer exposition of the business rule definition process. RIF includes the possibility of creating collections of business rules using <Group> and <sentence> statements which allows building business rules for more complex situations. Of course, K-Site Rules also includes this possibility, the user can define one or more rulesets when creating the flow diagram for a decision service and, into each ruleset, one or more elementary business rules can be added. K-Site Rules gives the name 'decision service' to the set of rulesets and conditions that are needed to provide some decision related functionality.

4   Conclusions and Future Work

Currently, there is a need in companies to define business domain knowledge in a declarative way to share a common understanding of subjects.

There are many user roles involved in the daily work in an enterprise information system, and the objective is to save costs as well as to improve business processes. The solutions have to be framed within a management initiative trying to align the technology existing in a company closely with its business strategy.

This paper has defined ongoing applied research on a new framework for business rules development with two main objectives in mind: to allow non-technical users to develop business rules with no or low intervention of developers and to reduce the development time needed to maintain or to produce new rule-based applications. The RIF and OMG PRR standard languages for representing business rules have been selected to conform a framework going through different abstraction levels from a business model to an application implementation, following an MDA approach in order to automate as much as possible the whole development process. The higher abstraction level uses ontology languages, such as OWL, and natural language to represent business models and business rules. These representations are transformed in a system description based on OWL, for business concepts, and RIF, for business rules. The following levels formed by a platform independent model described using UML and OMG-PRR languages and, finally, an implementation of the application over an available rule engine is obtained, constituting the last layer. It is important to notice that reusability has been taken into account when defining the framework, allowing an easy reuse of defined business rules.

The evaluation plan for the proposed business rule development framework follows now with a formal survey [5] for a set of developers and business analysts who will be confronted with the task of developing a rule-based application using K-Site Rules. The hypothesis to be tested at this stage of the work is: can developers and business analysts implement rule-based applications faster by using the proposed framework than by using traditional software development tools?

On the other hand, the business rule suffers two transformations in the development process with the possibility of obtaining more than one business rule in a level for a unique business rule in the upper level. Although the K-Site Rules platform gathers information about how different business rules are related at different abstraction levels, business rule transformations are unidirectional, from upper levels to lower levels, but it is not possible to make, in an automatic way, the inverse process. So, if a business rule must be modified, it would be more appropriate to change the standard definition, but not pieces of generated code. More effort has to be put in to allow complete correspondence among business rules at all levels.

References

[1] Cabot, J., et al.: From UML/OCL to SBVR specifications: A challenging transformation. Information Systems (2009), doi:10.1016/j.is.2008.12.002
[2] del Didonet Fabro, M., Albert, P., Bézivin, J., Jouault, F.: Achieving Rule Interoperability Using Chains of Model Transformations. In: Paige, R. (ed.) ICMT 2009. LNCS, vol. 5563, pp. 249–259. Springer, Heidelburg (2009)
[3] Gómez-Pérez, A., Fernández-López, M., Corcho, O.: Ontological Engineering. 1st edn., Springer, Heidelburg (2004); 2nd printing, vol. XII, p. 403 illus.159
[4] Greenfield, J., Short, K., Cook, S., Kent, S.: Software Factories, Assembling Applications with Patterns, Models, Frameworks and Tools, Editorial Wiley (2004)

[5] Lawrence Pfleeger, S.: Software Engineering: Theory and Practice. Prentice Hall PTR, Upper Saddle River (2001)

[6] Miller, J., Mukerji, J.: MDA Guide Version 1.0.1, OMG (2003), http://www.omg.org/docs/omg/03-06-01.pdf (last visit: 17/02/2009)

[7] OMG, Ontology Definition Metamodel (ODM). Available as ptc/2008-09-07 (2005), http://www.omg.org/docs/ptc/08-09-07.pdf (last visit: 17/02/2009)

[8] OMG, Production Rules Representation (PRR), Beta 1. Available as ptc/2008-09-07 (2006), http://www.omg.org/spec/PRR/1.0 (last visit: 17/02/2009)

[9] OMG, Semantics of Business Vocabulary and Business Rules (SBVR), First Interim Specification, Available as dtc/06-03-02 (March 2006), http://www.omg.org (2006)

[10] Paschke, A., Hirtle, D., Ginsberg, A., Patranjan, P., McCabe, F.: RIF Use Cases and Requirements, Working Draft. (2008), http://www.w3.org/TR/2008/WD-rif-ucr-20080730/ (last visit: 10/06/2009)

[11] Pressman, R.: Software Engineering: a practitioner's approach, 3rd edn. McGraw-Hill, Singapore (1992)

[12] Marie, S., de Christian: RIF Production Rules Dialect (RIF-PRD) (2008), http://www.w3.org/TR/2008/WD-rif-prd-20080730/ (last visit: 10/06/2009)

A Rule-Based System Implementing a Method for Translating FOL Formulas into NL Sentences

Aikaterini Mpagouli and Ioannis Hatzilygeroudis

University of Patras, School of Engineering
Department of Computer Engineering & Informatics, 26500 Patras, Hellas
{mpagouli,ihatz}@ceid.upatras.gr

Abstract. In this paper, we mainly present the implementation of a system that translates first order logic (FOL) formulas into natural language (NL) sentences. The motivation comes from an intelligent tutoring system teaching logic as a knowledge representation language, where it is used as a means for feedback to the students-users. FOL to NL conversion is achieved by using a rule-based approach, where we exploit the pattern matching capabilities of rules. So, the system consists of rule-based modules corresponding to the phases of our translation methodology. Facts are used in a lexicon providing lexical and grammatical information that helps in producing the NL sentences. The whole system is implemented in Jess, a java-implemented rule-based programming tool. Experimental results confirm the success of our choices.

Keywords: Rule-Based System, Natural Language Generation, First Order Logic.

1 Introduction

To help teaching the course of "Artificial Intelligence", in our Department a web-based intelligent tutoring system has been created. One of the topics that it deals with is first-order logic (FOL) as a knowledge representation language. One of the issues in the topic is the translation of natural language (NL) sentences into FOL formulas. Given that this is a non-automated process [1, 2], it is difficult to give some hints to the students-users during their effort to translate an "unknown" (to the system) NL sentence into a FOL formula. However, some kind of help would be provided, if the system could translate (after checking its syntax) the proposed by the student FOL formula into a NL sentence.

The solution to the above problem requires some kind of natural language generation (NLG). While NLG is a very active research domain [3], it seems that there is not any systematic effort for solving the above problem. There are however some efforts in the area of paraphrasing [4, 5], which translate rule-based software specification expressions into NL. Also, there are some efforts for the reverse problem: translating NL sentences into FOL formulas [6, 7, 8]. On the other hand, although there are efforts in the NLG domain related to other problems that exploit the advantages of the rule-based approach [9, 10, 11], there is no such approach used in the efforts related to paraphrasing and the inverse problem.

G. Governatori, J. Hall, and A. Paschke (Eds.): RuleML 2009, LNCS 5858, pp. 167–181, 2009.
© Springer-Verlag Berlin Heidelberg 2009

In this paper, we mainly present a rule-based system that implements a method for translating FOL formulas into NL sentences. The translation process is an extension of a previous one [13] and is informally presented here. The structure of the paper is as follows. Section 2 deals with related work. Section 3 presents the FOLtoNL conversion process. Section 4 deals with the use of rules for the implementation of the conversion method. Section 5 presents some examples of the system's application and finally, Section 6 concludes the paper.

2 Related Work

In the existing literature, we have traced only one work [12], which has the same objective as ours. However, it uses the object-oriented features of Java for implementing the corresponding method and lacks the generality of our system. For example, it does not allow for formulas with more than two variables and more than one implication.

On the other hand, works in paraphrasing could be considered as closely related. Paraphrasing is the process of describing the elements of a Conceptual Schema (CS) by means of NL expressions. Object Constraint Language (OCL) is the standard language to specify business rules on UML-based conceptual schemas. In [4] a system for automatically translating formal software specifications in OCL to NL is described. It is implemented using the Grammatical Framework (GF), a grammar formalism that is written from the perspective of linearization rather than parsing. In [5] a method that generates NL explanations for business rules expressed in OCL is presented. This method has as an intermediate step the translation of the OCL expression into a SBVR (Semantics of Business Vocabulary and Business Rules) representation.

Also, there are a number of efforts for translating NL sentences into FOL formulas, which can be considered as related to our work. In [6] an application of Natural Language Processing (NLP) is presented. It is an educational tool for translating Spanish text of certain types of sentences into FOL implemented in Prolog. In [7], ACE (Attempto Controlled English), a structured subset of the English language, is presented. ACE has been designed to substitute for formal symbolisms, like FOL, in the input of some systems in order to make the input easier to understand and to be written by the users. ACE expressions are automatically and unambiguously translated in the formal symbolism used as input language for such systems. Finally, in [8], a Controlled English to Logic Translation system, called CELT, allows users to give sentences of a restricted English grammar as input. The system analyses those sentences and turns them into FOL.

None of the above efforts uses a rule-based approach in implementing the corresponding method. However, there are efforts in the NLG domain that do that and benefit from the use of rules. For example, in [9] Jess, a rule-based expert system shell [14], is used to implement the representation and reasoning mechanism of an agent for storytelling. In [10] a rule-based approach is followed for overgenerating referring expressions. Finally, in [11], Jess capability of using modules of rules is exploited in implementing a framework, called ERIC, for real time commentary suitable for many domains. All of them stress the benefits of using rules to implement the corresponding systems/methods.

3 FOLtoNL Conversion Process

Our FOLtoNL conversion method takes as input FOL formulas [1] of the following syntax:

Expression → Atom | Expression Connective Expression | (Quantifier Variable) Expression | ~Expression | (Expression)
Atom → Predicate(Term[,Term])
Term → Function(Term) | Constant | Variable
Connective → & | V | =>
Quantifier → forall | exists

We distinguish FOL expressions in two different categories: *Sentences* and *Implications*. A Sentence can be either a single Atom or a set of FOL formulas connected by the connectives '&' and/or 'V'. An Implication consists of two FOL formulas that are connected by the connective '=>'. The first of the two formulas of an implication, which is before the connective '=>', is called the *antecedent*, whereas the second, which follows the symbol '=>', is called the *consequent* of the implication.

We also distinguish different cases of Sentences and Implications. One Sentence can be characterized as simple or complex. A *simple Sentence* does not contain any implications in it, whereas in a *complex Sentence*, one or more of the formulas connected by '&' or 'V' are implications. For example, the formula "(exists x) cat(x) & likes(Kate,x)" is a simple Sentence, whereas the formula "human(Helen) & ((forall x) human(x)=>mortal(x))" is a complex one.

Likewise, we divide Implications in three categories: simple, complex and special ones. A *simple Implication* consists of two simple Sentences connected by '=>'. In a *complex Implication*, the antecedent or the consequent is a complex Sentence (or both sides are complex Sentences). Finally, a *special Implication* has another Implication as antecedent, or as a consequent (or has Implications in both sides). A *consequent-special* Implication consists of an antecedent simple Sentence and a consequent Implication. In an *antecedent-special* Implication the antecedent is an Implication and the consequent can be either a Sentence or an Implication.

3.1 Basic Approach

The entire method of converting FOL formulas into NL is based on the translation of an Implication. Sentences are considered as Implications that consist of only a consequent (their antecedent is always True). As known, the *Universe of Discourse* (UoD), for a FOL formula that contains variable symbols, is the set of all the possible values of those variable symbols. In other words, the UoD represents the set of all the entities that participate in the FOL formula, i.e. the entities for which the formula counts. The antecedent of a FOL Implication defines the UoD of that Implication's consequent formula. Based on the latter fact, we start from the consequent of an Implication and get a *Primary Translation* that may contain variable symbols. Then, we specify NL phrases to substitute for the variable symbols of the Implication based on its antecedent and the quantifiers. These NL phrases are called *NL substitutes*. The last step is to use the appropriate NL substitutes in the place of variable symbols in the Primary Translation and get the *Final Translation*. In the absence of variable symbols, we get the translations of the two sides of the Implication and combine them into an if-then sentence.

In order to clarify the above, we present an example. Let us consider the three FOL formulas: "(forall x) lives(x, Patras)", "(forall x) human(x)=>lives(x,Patras)" and "(forall x) (human(x) & young(x) & loves(x,sea)) => lives(x,Patras)". The Primary Translation is "x lives in Patras" and is the same for all the three formulas, given that they have the same consequent. The next step is to find appropriate NL substitutes for x in each case. The first formula is a Sentence. Since there is no antecedent, x is determined by the quantifier and its NL substitute is "everything". In the second case, the antecedent restricts the UoD to the set of humans. The antecedent, along with the quantifier, result in the NL substitute "every human" for x. The more informative the antecedent, the more restricted becomes the UoD. Thus, in the third case, the NL substitute for x becomes "every young human that loves the sea". By substituting the appropriate NL substitutes for x, we get the final translations of the above formulas, which are: "Everything lives in Patras", "Every human lives in Patras" and "Every young human that loves the sea lives in Patras", respectively.

As already mentioned, a Sentence is translated as if it were an Implication without an antecedent. All variable symbols are translated based on their quantifiers. We distinguish four different cases of quantifiers: universal, existential, negated universal and negated existential. For example, let us consider the Sentence "(exists x) cat(x)". The NL substitute for x is "something" and the final translation of the Sentence is "Something is a cat". In case of a different quantifier, x could have been translated as "everything", "not everything" or "nothing". It is obvious that in the absence of variable symbols, the final translation is identical to the primary one.

So far, we have presented the translation of simple Implications and simple Sentences. When translating a complex Sentence, we first translate the Sentence ignoring all the Implications contained in it. Then we get the translations of the Implications that were ignored in the first step. Finally, we add the translations of the Implications to the end of the initial translation of the Sentence, using the words "and" or "or", according to the connectives. Thus, the complex Sentence "(human(Helen) & ((forall x) human(x)=>mortal(x)))" is translated as "Helen is a human and every human is a mortal".

In case of a complex Implication, we get the initial translation ignoring all the Implications contained in the antecedent. Those Implications are translated separately. The translations of the ignored Implications are added to the end of the initial translation, after the phrases "provided that" and connected with "and". As an example we present the formula "(human(Helen) & ((forall x) human(x)=>mortal(x))) => mortal(Helen)", which is translated as "if Helen is a human, then Helen is a mortal, provided that every human is a mortal". The phrase "provided that" declares that what follows is part of the antecedent of the Implication. The translation in the above example may seem somewhat awkward, but this is due to the fact that this particular example leads to an if-then sentence as a basic translation. In cases like this, the part after "provided that" can be appropriately embedded in the if-part of the if-then sentence: "if Helen is a human and every human is mortal, then Helen is mortal". However, in the presence of variable symbols in the basic Implication, things are different. As an example, consider the last formula in Table 5 of Section 5. The part "provided that" is necessary to provide a general way for translating such complex Implications.

What remains is the translation of Special Implications. A consequent-special Implication can easily be transformed into an equivalent simple Implication and be translated as such. For example, the expression "(forall x) (human(x) => (little(x) => (~dwarf(x) => child(x))))" is equivalent to "(forall x) (human(x) & little(x) & ~dwarf(x)) => child(x)" and is translated as "Every little human that is not a dwarf is a child". An antecedent-special Implication of the form (((a=>b)=>c)=>d) would be translated as "if (if (if a then b) then c) then d", which is rather odd and confusing. In order to avoid the previous translation we use the combinations "if-then" and "the fact that-means that" interchangeably. Thus, we get translations of the form "if the fact that if a then b means that c then d".

3.2 Conversion Process

The conversion of an input FOL formula into NL comprises a number of stages. The first stage is the *Input Analysis*. The FOL formula is analyzed to a tree structure, called FOLtoNL tree. A FOLtoNL tree has all the Atoms of the input formula as leaf nodes and all the connectives of the input formula as intermediate nodes. It does not contain quantifier information. The latter information concerns the order of the quantifiers in the FOL formula, their types and the variables they bind and it is kept in appropriate lists.

The next stage is the *Transformation Stage*. The tree structure produced by the previous stage undergoes a number of transformations in order to reach a much simpler but equivalent form. Such transformations may be the elimination of successive negations, transformations of AND nodes, elimination of tautologies etc. The tree is further processed in order to eliminate negation nodes with AND or OR nodes as children, changing the labels of the connective nodes (and the polarity of their children) according to DeMorgan rules.

After the Transformation Stage the *Translation Stage* comes. First of all, translate all the Atoms of the formula. After that we translate special Implications. When this is done, we proceed with the translation of the rest of the Implications that have not been translated yet and finally, we check the root of the FOLtoNL tree. In case it is a '=>', the input formula is an Implication and its translation is ready. Otherwise, the input formula is a Sentence and a further translation is done at this point, using the available translations of Implications appropriately.

The Translation Stage uses a *Lexicon* especially built for the system. It consists of a large number of facts concerning words, called *word-facts*. Each word-fact is an instance of the following template: (word ?type ?gen ?form ?past ?exp ?stem ?lem), where 'word' declares that we have a fact describing a word ?lem and it is followed by the fields that describe that word. Each time some information for a particular word is needed, the Lexicon provides this information to the system. ?type refers to the word's part of speech. ?gen refers to the gender of the word if it is a noun. ?form has a different role for each different type: for verbs it declares the person, for nouns the number and for adjectives the grade. ?past is the past participle form if the word is a verb. ?stem is not exactly the stem of the word but something common for all different words of the same stem. For example, the words "love", "loves" and "loved" will have the word "love" as ?stem. The rule is this: the stem for a verb is the verb in first person, for a noun is the noun in singular form and for an adjective is its base form.

Finally, ?exp is a field pointing to another fact, in case there is a special syntax for the word ?lem. For example, usually the word "married" must be followed by the preposition "to". Thus, the word-fact for the word "married" will be: (word d 0 1 0 a1 marry married), where ?exp has the value a1. This means that the Lexicon also contains an expression-fact: (expression a1 ". to"). When the system asks for information about "married", the Lexicon, apart from other characteristics, will also provide the expression "married to" in order to substitute for the predicate "married".

3.3 Translation of Atoms

The translation of an Atom is based on its predicate and the translations of its terms. A constant term is translated as the constant itself. A variable term is translated as the variable symbol at this stage, and a function term f(x) is translated as "the f of x". In Table 1 we present the translation of single-term Atoms according to the type of their predicates and their polarity. Polarity is determined by the existence or non-existence of the negation symbol before the predicate of an Atom. In Table 2, we present the translation of two-term Atoms. P denotes the predicate and t, t1 and t2 are terms. We use <t-nl> to denote the translation of the term t, and $\underline{P}$ to denote that the predicate is changed in order to be in the correct form (number for nouns, person for verbs etc).

Table 1. Translation of single-term Atoms

P	Translation P(t)	Translation ~P(t)
Noun	<t-nl> is a/an P	<t-nl> is not a/an P
Adjective	<t-nl> is [the] P	<t-nl> is not [the] P
Verb	<t-nl> does $\underline{P}$	<t-nl> does not $\underline{P}$

Table 2. Translation of two-term Atoms

P	Translation P(t1,t2)	Translation ~P(t1,t2)
Noun	<t1-nl> is the P of <t2-nl>	<t1-nl> is not the P of <t2-nl>
Adjective	<t1-nl> is [more] P than <t2-nl>	<t1-nl> is not [more] P than <t2-nl>
Verb	<t1-nl> P <t2-nl>	<t1-nl> does not $\underline{P}$ <t2-nl>

One important general fact about FOL formulas is that the order of quantifiers determines the subject and objects of the NL sentence that is the translation of a FOL formula. The first quantifier introduces the subject and the rest introduce the objects. So far, the translation of an Atom uses its first term as a subject and the second as an object. In case of variable terms, this is correct as long as the first term is a variable bound by the first quantifier of the formula. Otherwise, the first term must be used as an object in the translation.

In case a predicate has a special syntax, with a particular preposition for example, the translation of Atoms is simplified as shown in Table 3. We use *exp* to denote the special syntax of the predicate. In this way, instead of translating the Atom married(x, y) as "x is the married of y" or "x is more married than y", we translate it correctly as "x is married to y". Likewise, the Atom lives(x, Patras) is not translated as "x lives Patras" but as "x lives in Patras".

Table 3. Translation of Atoms with P special syntax *exp*

P	Translation P(t1,t2)	Translation ~P(t1,t2)
Noun	<t1-nl> is exp <t2-nl>	<t1-nl> is not exp <t2-nl>
Verb	<t1-nl> exp <t2-nl>	<t1-nl> does not exp <t2-nl>

3.4 Translation of Implications

After all the Atoms have been translated, we translate the Implications as follows. First, antecedent-special Implications are translated in the way presented in subsection 3.1. Then we translate consequent-special Implications. These Implications include other Implications. Thus, the contained Implications are translated and then their translations are appropriately combined. After the translation of special Implications, translation of the Implications that has not been translated so far comes. If the initial FOL formula is a Sentence, it is translated as described in subsection 3.1, after translation of all existing Implications. In the sequel, we present more analytically the Implications translation process.

Implication translation is realized in two phases. In the first phase, we get the *Primary Translation*, based on the elements of the Implication's consequent. In this phase, we have two steps. In the first step, if there are Implications or constant-term Atoms contained in the consequent, then they are translated independently and ignored until the end of the second step. What remains in the consequent is a set of Atoms connected by specific connectives. In the second step, we make the *Aggregation*. We have three possible ways of aggregation. The translations of Atoms having the same predicate (verb) and first term (subject) are aggregated to a sentence that has as subject their common subject and as a verb their common verb and that has as object their different objects connected by the words "and" or "or". Likewise, we aggregate Atoms with a common subject and Atoms with a common verb (predicate) and object (second term). Atoms that can not be aggregated result in sentences that are identical in their translations. When aggregation is done, all produced sentences are connected via appropriate connective words and are combined with the translations of Implications and constant-term Atoms, ignored so far. The result is a NL sentence that contains variable symbols.

The second phase is the specification of NL substitutes, based on the antecedent and the quantifiers. The Atoms of the antecedent are divided in different categories according to the place of a variable symbol in them. Thus, we have Atoms that have the variable as a single term, Atoms that have the variable as a first term etc. Based on this categorization, each variable symbol can be translated into its NL substitute. We first use single-term Atoms that describe entities represented by the variables, in order to create primary substitutes for those entities-variables. Two-term Atoms express relations between entities and are used to enrich primary substitutes and give the complete NL substitutes. Note that we use appropriate referring expressions whenever we encounter a variable symbol that has already been translated and its translation has already been used previously.

To clarify the above, we give an example. The Primary Translation of the formula "~(forall x) (human(x) & clever(x) & lives(x,Patras) & loves(father(x),x)) => happy(x)" is "x is happy". The primary substitute for x is "not every clever human" and the referring expression is "that human". In order to include the information of the two-term Atoms of the antecedent in the final translation, we enrich the primary substitute that becomes "not every clever human that lives in Patras and whose father loves them". Since the basic noun for x is "human" that is in singular number and is of unknown gender we change the word "them" to "him/her" and we get the final NL substitute "not every clever human that lives in Patras and whose father loves him/her". By substituting x in the Primary Translation we get the final translation of the Implication, which is: "Not every clever human that lives in Patras and whose father loves him/her is happy".

What is left is to use the NL substitutes in the place of the variable symbols in the Primary Translation and change appropriately the words "their", "them" and "themselves" to the correct form. We also change the verbs, when needed, in order to be in the correct form.

4 Using Rules for Implementing FOLtoNL Conversion

The FOLtoNL process has been implemented in Jess [14]. Jess is a rule-based expert system shell written in Java, which however offers adequate general programming capabilities, such as definition and use of functions. We have combined the pattern matching capabilities of rules with the general programming capabilities of Jess in order to achieve pure NL in a simple (as much as possible) way.

In [3], the basic stages of all applied NLG systems are presented. Most of such systems comprise a Linguistic Realisation stage, where NL is produced from semantic representations via a specific grammar and a parser-realiser for it. Our system is a different case. Realization is not useful when the output NL of a system is rather simple, using only Present Perfect tense for example. Moreover, in our case there is a correspondence between the input and the output of the system: most of the words that appear in the output exist in the input formula and the Atoms provide information as to the syntax of the NL sentence they represent. Thus, rules are more appropriate to implement this kind of conversion in a simpler way.

Jess rule engine uses an improved form of a well-known algorithm called Rete [15] (Latin for "net") to match rules against the working memory. Jess is actually faster than some popular rule engines written in C, especially on large problems, where performance is dominated by algorithm quality. Rete is an algorithm that explicitly trades space for speed, so Jess memory usage is not inconsiderable. Jess does contain some commands which allow us to sacrifice some performance to decrease memory usage. Nevertheless, moderate-sized programs fit easily into Java's default 16M heap. By putting most specific patterns near the top of each rule's LHS, patterns that will match the fewest facts near the top and the most transient patterns near the bottom, we can achieve better performance.

The Implication and Sentence translation methods, as well as the creation of FOL-toNL tree from the input formula, are implemented with functions. We use functions in the parts of the process, where the pattern matching capabilities of rules can not help and would make the implementation less effective. Rules make it much easier to implement the transformations on the tree structure in order to simplify it, since they provide the ability to detect much faster all the parts of the tree that need transformation and perform it independently from the rest of the tree. Moreover, in case a new form appears after a transformation, a form that did not exist before but can be further transformed, then the appropriate rule fires immediately.

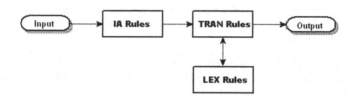

Fig. 1. FOLtoNL Inference Flow

Rules also help a lot during the translation of Atoms. All non-translated Atoms can be detected immediately due to pattern matching and be translated independently from all the rest. Rules also group the translation functions and impose a specific order on their execution. They ensure that all special Implications are translated after the translation of Atoms has finished and before the simple Implications are translated. They also ensure that a Sentence will be translated after all existing Implications have been translated. In a few words, rules and functions provide simplicity and effectiveness to our system.

In Fig. 1, the Inference Flow of our system is presented. Rules are divided into 3 groups: Input Analysis Rules (IA Rules), Translation Rules (TRAN Rules) and Lexicon Rules (LEX Rules). After all activated IA Rules have fired, TRAN Rules can be activated. TRAN rules pass the control to LEX rules when the characteristics of a specific word are needed in order to proceed with their translation. LEX rules update some global variables with appropriate values (the values of the characteristics of a word requested) and return the control to TRAN rules. When there are no more activated TRAN rules to be executed, the translation process has been completed and the NL sentence is returned as output.

The system's architecture is shown in Fig. 2. It includes three Jess modules: IA, TRAN and LEX, one module for each group of rules. Each Jess module has its own rule base and its own facts and can work independently from the rest of Jess modules. Focus is passed from one module to the other to execute its rules. IA and TRAN are the basic modules of the system, whereas LEX is the system's lexicon.

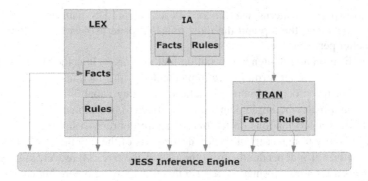

Fig. 2. FOLtoNL Architecture

```
(defrule unify-and-nodes
   (declare (salience 170))
   (ia mod)
   ?parent-node <- (Node (label and) (children $?children1) (level ?level1))
   ?node <- (Node (label and) (children $?children2) (level ?level2))
   (test (member$ ?node ?children1))
=>
   (foreach ?child ?children2
      (modify ?child (level ?level2)))
   (bind ?new-children (replace$ ?children1 (member$ ?node ?children1)
         (member$ ?node ?children1) ?children2))
   (modify ?parent-node (children ?new-children))
   (retract ?node) )
```

Fig. 3. Rule 'unify-and-nodes'

IA's most important function is 'readinput', which takes as an argument the user's input and processes it as a string. The result is to update the global lists with quantifier information and to build the FOLtoNL tree for the input FOL formula. This tree structure is the base of FOLtoNL conversion process. IA also contains rules that apply various transformations to the FOLtoNL tree, in order to simplify it. Such rules are 'unify-and-nodes', 'negated-sentence', 'negated-implication', 'negated-connective', 'remove-tautologies', "not-not" etc. After the execution of these rules, the produced FOLtoNL tree has been stored in the form of Node facts in TRAN module. We present a rule of this module in Fig. 3.

```
(defrule atom-interpretation
   (declare (salience 90))
   (main mod)
   ?atom-id <- (Atom (term1 ?term1) (term2 ?term2) (terms nil) (interpretation ""))
=>
      (atom-interpret ?atom-id) )
```

Fig. 4. Rule 'atom-interpret'

TRAN module contains all the rules and functions that implement the translation process. Some indicative rules of this module are: "atom-interpretation", 'ant-special-translation', 'cons-special-translation', 'implication-translation' and 'expression-translation'. The rules of this module call corresponding functions like: 'atom-interpret', 'implication-to-nl', 'sentence-to-nl' etc. We present the rule 'atom-interpretation' in Fig. 4.

```
(defrule part-of-speech
  ?w <- (stem ?lem)
  (word ?type ?gen ?form ?past ?exp ?stem ?lem)
=>
  (bind ?*type* ?type)
  (if (or (eq ?type N) (eq ?type n) (eq ?type d)) then
    (assert (nplural ?stem)))
  (if (eq ?type V) then
    (assert (vpast ?lem)))
  (if (or (eq ?type J) (eq ?type j)) then
    (bind ?*plural* ?lem)
    (bind ?*comparative-superlative* ?form)
    (assert (find-opposite ?lem)))
  (bind ?*stem* ?stem)
  (if (neq ?exp 0) then
    (assert (express ?past ?exp ?lem ?stem)))
  (retract ?w))
```

Fig. 5. Rule 'part-of-speech'

LEX consists of a large number of facts concerning words, called *word-facts*. Currently, the lexicon contains around 700 word facts. Apart from information about English words, LEX module also includes two basic rules for their treatment. The rule 'part-of-speech' is activated when a fact of the form (lemma ?lem) is inserted into LEX. This kind of fact declares that another module needs information for the English word '?lem' from the lexicon. When, subsequently, the focus is given to LEX the rule fires, looking for the specified word. Provided that the word is found, the rule updates the global variable '?*part*' with the word's type, i.e. part of speech. The rule 'expression' is used for words with a special syntax. This rule returns the syntax of such words to other modules and makes it possible for the word to appear in the correct from when being translated. There are also some rules that serve to provide other modules with information about the gender, the tense and/or the number of a specific word. In Fig. 5 we show the basic rule of LEX. Concerning issues of extendibility, the Lexicon could be extended to new application domains by mapping the word-facts to appropriate database records.

Jess uses the notion of templates, which are structured descriptors, consisted of slot-value pairs, for describing complex facts. We have defined six templates: Atom, Term, Function, Implication, Node, Variable, AtomSentence and SubSentence. Templates help in organizing the information of a complex fact in specific named slots and make very easy the access and modification of this information. In Fig. 6 we present some of these templates.

```
(deftemplate Atom
    (slot predicate)
    (slot term1)
    (slot term2)
    (slot positive (default 1))
    (slot constant-terms)
    (slot sentence)
    (slot passive)
    (slot passive-counts (default 0))
    (slot predicate-type)
    (slot interpretation (type STRING) (default "")))

(deftemplate Variable
    (slot id)
    (slot quantifier)
    (slot primary-substitute (type STRING) (default ""))
    (slot nl-substitute (type STRING) (default ""))
    (slot refer)
    (slot plural (default 0)))

 (deftemplate Implication
    (slot level)
    (slot antecedent)
    (slot consequent)
    (slot simple-ant)
    (slot simple-con)
    (slot special)
    (slot negated (default 0))
    (slot nl (default "")))
```

Fig. 6. Fact Templates

5 System Application

The system has been evaluated using a set of 100 formulas. The formulas were found in textbooks and exercises about converting NL to FOL. The 97% of the formulas were converted correctly (both syntactically and semantically) in NL. In Table 5, we present some examples of the system's application.

Table 5. Examples of FOLtoNL Application

Input (FOL Formula)	Output (NL Sentence)
(forall x) (bird(x)&~flies(x)&swims(x))=>penguin(x)	Every bird that swims and that does not fly is a penguin.
(forall x) bat(x) =>~feathered(x)	Every bat is not feathered.
~(forall x)(exists y) cares(x,y)	Not everything cares about something.
(forall x) (exists y) (exists z) (exists w) (human(x) & name(y) & age(z) & birthday(w)) => (has(x,y) & has(x,z) & has(x,w))	Every human has some name, some age and some birthday.
(forall x) human(x) => (loves(father_of(x),x) & loves(mother_of(x),x))	Every human is loved by his/her father and his/her father.
~((forall x) human(x))	It is not true that everything is a human.
(forall x) ~(exists y) (forall z) (person(x) & person(y) & person(z) & ~respects(x,y)) => ~hires(z,x)	Every person that does not respect no person is not hired by any person.
(forall x) (human(x) & lives(father_of(x)) & lives(mother_of(x))) => ~orphan(x)	Every human whose father and mother do live is not an orphan.
(exists y) (forall x) (student(x) & tutor(y) & smarter(x,y))	All are students and are smarter than some tutor. (There is some tutor such that all are students and are smarter than that tutor.)
(forall x1) (forall x2) (exists y) (boy(x1) & girl(x2) & human(y) & son(x1,y) & daughter(x2,y)) => (brother(x1,x2) & sister(x2,x1))	Every boy that is the son of some human is the brother of every girl that is the daughter of that human and has that girl as a sister.
(forall x) ~((human(x)=>mortal(x)) => (mortal(x) => human(x)))	It is not true that if every human is a mortal then every mortal is a human.
(forall x) (forall y) (student(x) & course(y) & (studies(x,y) V lucky(x))) => passes(x,y)	Every student that studies every course or is lucky passes every course.
(forall x) (forall y) ((human(x) & place(y)) => ((warm(y) & old(x) & (warm(y) => protects(y,health)) &loves(x,life)) => prefers(x,y)))	Every old human that loves life prefers every warm place, provided that every warm thing protects health.

6   Discussion and Conclusions

In this paper, we mainly present the implementation of an approach for translating FOL formulas into NL sentences, called the FOLtoNL method. The whole system is implemented in Jess and combines the pattern matching and the general programming capabilities of this language. We are not aware of other similar efforts, except one, although there are efforts to deal with related problems. Neither of them, however, uses a rule-based approach.

The system was evaluated using 100 FOL formulas. Those formulas emerged from exercises about converting NL sentences into FOL. Each exercise consists of a NL sentence that must be converted to FOL. The correct answer for each such exercise is a FOL formula that is known. Some of the exercises were selected from the intelligent tutoring system exercises database and others were found in textbooks that teach Logic as a knowledge representation language. The selected subset of formulas covers all the possible cases that can appear in the tutoring system for which the FOLtoNL method has been implemented, and generally almost all possible cases of exercises faced by someone who learns Logic. FOLtoNL was used to translate each one of the test formulas (with known answers) and the results were compared with the appropriate NL sentences.

97 formulas were translated in a quite natural and satisfactory way. The formulas not satisfactory translated included or were of the form "((exists x) apple(x)) & ((exists x) orange(x))" which was translated as "Something is an apple and an orange", instead of "Something is an apple and something is an orange" which is the correct translation. This is due to the fact that the system identifies entities by their variable symbols uniquely. It does not realize that the same variable symbol in different scopes refers to different entities. This is a problem easily solved by renaming a variable symbol that is the same with another one which belongs to a different scope. The other solution is to write it as two separate formulas, since it is a compound one.

Apart from the above problem, the general attitude of the system was satisfactory. There is no constraint in the number of variables and quantifiers of the input FOL formula, implications are handled in more than one levels and the order of quantifiers is taken into consideration. Choosing rules to implement our approach has been proven absolutely successful.

Adapting the system for human languages other than English requires much additional work, since in that case the problem of conversion becomes more complex. English has a much simpler syntax and grammar than most of other human languages. In order to translate FOL formulas of a different language in that language the Lexicon and the Translation Stage of the system should be designed again taking into consideration the characteristics of the particular language. However, we believe that our system could be more easily extended (or used as a basis) for translating software specifications written in Z language [16], which is based on FOL to a large degree.

A restriction of our system is that it cannot handle formulas containing predicates of arity greater than 2, i.e. having more than two terms. This is one of the concerns of our feature work.

Regarding the usefulness of our system, the first reactions of the students were encouraging. We need, however, a more systematic user study confirming our intuition about the system's usefulness and this is one of our future goals.

References

1. Genesereth, M.R., Nilsson, N.J.: Logical foundations of AI. Morgan Kaufmann, San Francisco (1988)
2. Brachman, R.J., Levesque, H.J.: Knowledge representation and reasoning. Morgan Kaufmann, San Francisco (2004)
3. Reiter, E., Dale, R.: Building natural language generation systems. Cambridge University Press, Cambridge (2006)

4. Burke, D.A., Johannisson, K.: Translating Formal Software Specifications to Natural Language. In: Blache, P., Stabler, E.P., Busquets, J.V., Moot, R. (eds.) LACL 2005. LNCS (LNAI), vol. 3492, pp. 51–66. Springer, Heidelberg (2005)
5. Pau, R., Cabot, J.: Paraphrasing OCL expressions with SBVR. In: Kapetanios, E., Sugumaran, V., Spiliopoulou, M. (eds.) NLDB 2008. LNCS, vol. 5039, pp. 311–316. Springer, Heidelberg (2008)
6. de Aldana, E.R.V. An application for translation of Spanish sentences into first order logic implemented in prolog (1999), http://aracne.usal.es/congress/PDF/
7. Fuchs, N.E., Schwertel, U., Torge, S.: Controlled natural language can replace first order logic. In: Proceedings of the 14th IEEE International Conference on Automated Software Engineering (ASE 1999), pp. 295–298 (1999),
   http://www.ifi.unizh.ch/groups/req/ftp/papers/ASE99.pdf
8. Pease, A., Fellbaum, C.: Language to logic translation with PhraseBank. In: Proceedings of the Second Global Conference (GWC 2004), pp. 187–192 (2004)
9. Theune, M., Faas, S., Nijholt, A., Heylen, D.: The Virtual Storyteller: Story Creation by Intelligent Agents. In: Gobel, S., Braum, N., Spierling, U., Dechau, J., Diener, H. (eds.) Proceedings of TIDSE 2003: Technologies for Interactive Digital Storytelling and Entertainment, pp. 204–215. Fraunhofer IRB Verlag (2003)
10. Varges, S.: Overgenerating Referring Expressions Involving Relations and Booleans. In: Belz, A., Evans, R., Piwek, P. (eds.) INLG 2004. LNCS (LNAI), vol. 3123, pp. 171–181. Springer, Heidelberg (2004)
11. Strauss, M., Kipp, M.: ERIC: A Generic Rule-based Framework for an Affective Embodied Commentary Agent. In: Proceedings of the 7th International Conference on Autonomous Agents and Multiagent Systems (AA-MAS 2008), pp. 97–104 (2008)
12. Zhu, J.: An NLG System Generates English From First-order Predicate Logic, MSc Project, Computer Science Department Old Dominion University Norfolk, VA (2006),
    http://www.cs.odu.edu/~gpd/msprojects/jzhu.1/
    ms_project_report.doc
13. Mpagouli, A., Hatzilygeroudis, I.: A Knowledge-based System for Translating FOL Formulas into NL Sentences. In: Iliadis, L., Vlahavas, I., Bramer, M. (eds.) Artificial Intelligence Applications and Innovations III, Proceedings of the 5th IFIP Conference on Artificial Intelligence Applications and Innovations (AIAI 2009), pp. 157–163. Springer, Heidelberg (2009)
14. Friedman Hill, E.: Jess in action: rule-based systems in Java. Manning Publishing (2003)
15. Forgy, C.L.: Rete: A Fast Algorithm for the Many Pattern/ Many Object Pattern Match Problem. Artificial Intelligence 19, 17–37 (1982)
16. Spivey, J.M.: Understanding Z: A Specification Language and its Formal Semantics. Cambridge University Press, Cambridge (2008)

An Empirical Study of Unsupervised Rule Set Extraction of Clustered Categorical Data Using a Simulated Bee Colony Algorithm

James D. McCaffrey and Howard Dierking

Volt VTE / Microsoft MSDN
One Microsoft Way
Redmond, WA 98052 USA
v-jammc@microsoft.com, howard@microsoft.com

Abstract. This study investigates the use of a biologically inspired meta-heuristic algorithm to extract rule sets from clustered categorical data. A computer program which implemented the algorithm was executed against six benchmark data sets and successfully discovered the underlying generation rules in all cases. Compared to existing approaches, the simulated bee colony (SBC) algorithm used in this study has the advantage of allowing full customization of the characteristics of the extracted rule set, and allowing arbitrarily large data sets to be analyzed. The primary disadvantages of the SBC algorithm for rule set extraction are that the approach requires a relatively large number of input parameters, and that the approach does not guarantee convergence to an optimal solution. The results demonstrate that an SBC algorithm for rule set extraction of clustered categorical data is feasible, and suggest that the approach may have the ability to outperform existing algorithms in certain scenarios.

Keywords: Association rules, data mining, pattern mining, rule set extraction, simulated bee colony algorithm.

1 Introduction

This study investigates the use of a biologically inspired simulated bee colony meta-heuristic algorithm to extract rules from clustered categorical data. Consider the general problem of analyzing a large set of clustered categorical data in order to extract a set of rules which efficiently describe the cluster membership of the existing data and which can be used to predict cluster membership of new data. Such problems have great practical importance. Examples include analyzing sales data to forecast consumer purchasing behavior, analyzing telecommunications data for possible terror-related activity, and analyzing genetic DNA sequences to discover biological system regulatory information. For the sake of concreteness, imagine a set of clustered categorical data which has three attributes: color, size, and temperature. Further, suppose that the color attribute can take on one of four categorical values: red, blue, green or yellow. The size attribute can be small, medium, or large. The temperature attribute value can be hot, cold, or warm. Each 3-tuple of color, size, and temperature is

G. Governatori, J. Hall, and A. Paschke (Eds.): RuleML 2009, LNCS 5858, pp. 182–192, 2009.

assigned to one of three clusters with cluster ID values of c0, c1, and c2 via some unspecified mechanism yielding the following artificially small data set:

```
Color    Size     Temp    ClusterID
===================================
Red      Small    Hot        c0
Red      Small    Cold       c0
Blue     Medium   Hot        c1
Green    Large    Cold       c1
Yellow   Large    Warm       c2
Blue     Small    Hot        c2
```

A human observer might conclude that one of the many possible rule sets which describe the data set is:

(Red, Small) => c0
(Blue, Medium) => c1
(Green) => c1
(Yellow) => c2
(Blue, Small) => c2

The rule set correctly categorizes all six tuples in the data set but requires nearly as many (five) rules as data tuples. A second human observer might be willing to sacrifice a certain amount of categorization coverage or rule set accuracy for a decrease in rule set size and conclude that a good rule set is:

(Red) => c0
(Green) => c1
(Yellow) => c2

This second, smaller, rule set correctly categorizes four of the six tuples in the data set. These examples point out one of the two main issues with rule set extraction, namely, there is no inherent definition of what an optimal rule set is and therefore any rule set extraction algorithm must a priori specify characteristics of the target rule set which define a goodness metric, such as tuple coverage, rule set size, and rule set accuracy. The second main issue with rule set extraction is the combinatorial explosion problem. Exhaustively analyzing all possible rule sets for a given data set of clustered data is not feasible in general. If the data set under analysis has n tuples, then rule set size can vary from 1 (in the degenerate case where one rule captures all tuples) to n (in the worst case where each tuple requires a distinct rule). If a data set contains c cluster ID values, and k attributes where each attribute Ai can take on one of Aij distinct values (possibly including a don't-care value), and if the rule set is structured in disjunctive normal form, then the total number of possible rule sets is given by:

$$\sum_{t=1}^{n} \left[t * \prod_{i=1}^{k} (A_{ij} * c) \right]$$ (1)

For example, for a data set consisting of n = 10,000 tuples, with k = 10 attributes, where all attributes can take on one of Aij = 20 distinct values, and c = 30 cluster ID values, the total number of disjunctive-form rule sets is $2.79 * 10^{15}$ sets. Even if rule sets could be evaluated at the rate of 1,000 sets per second, a complete examination of all possible rule sets would require roughly 88,000 years.

Rule set extraction of clustered categorical data can be considered a variation of a class of problems normally called association rule learning. The introduction of the term association rule learning is often attributed to a 1993 paper by Agrawal, Imielinski, and Swami [1]. That paper examined the problem of finding association rules based on frequency regularities between products in large scale transaction data recorded by point-of-sale systems in supermarkets. For example, one rule might indicate that customers who buy milk and eggs are also likely to buy cheese. Because the root study investigated this particular supermarket problem, this general class of problems is sometimes called market basket analysis. Association rules were defined by Agrawal et al. along the lines of the following. Let $I = I_1, I_2, \ldots, I_n$ be a set of n attribute values called items. Let P be a set of physical transactions where each physical transaction is a subset of the items in I consisting of those items purchased by a shopper. Let T be a database of transactions where each transaction t is represented by a binary vector where t[i] = 1 if the item I_i is in P and t[i] = 0 otherwise. There is one tuple in T which corresponds to each physical transaction. Let X be a set of some or all of the items in I. A transaction t satisfies X if t[i] = 1 for all items I_i in X. An association rule is an implication in the form $X => I_j$ where X is a set of items in I and I_j is a single item in I that is not in X.

Central concepts of standard association rule learning include the notions of support and confidence. The support, s(r), of a rule r is defined as the fraction of transactions in T that satisfy the union of items in the consequent and antecedent of the rule [2]. The confidence, c(r), of a rule r is a value between 0 and 1 inclusive such that $X => I_j$ is satisfied in the set of transactions T iff at least a fraction c of the transactions in T that satisfy X also satisfy I [2]. Support can be thought of as a measure of a rule's statistical significance, or importance. Association rule support metrics are used to deal with the combinatorial explosion problem by reducing the search space of all possible rule sets. Confidence can be thought of as a measure of a rule's strength or accuracy.

Association rules have been the subject of a large amount of research and many algorithms for association rule learning have been proposed and studied [3]. These algorithms differ from one another primarily in how they deal with searching the domain space of all possible rule sets and in how they define interesting/optimal rules. Example association rule learning algorithms include the Apriori algorithm, the Eclat algorithm, the FP-Growth algorithm, and the OneR algorithm [4]. One of the primary differences between the standard association rule learning problem and the rule set extraction problem addressed by this study is that in this study there is a clear distinction between attribute values, which can only be part of a rule antecedent, and cluster ID values, which can only be in the consequent part of a rule. Additionally, the assumption that the data set to be analyzed is a result of a clustering algorithm means each unique tuple in the data set is unambiguously associated with a single cluster ID value. Another significant difference between standard association rule learning and the rule set extraction problem of this study is that this study does not require the use of an a priori support metric to deterministically reduce the rule set search space. Instead the simulated bee colony algorithm used in this study explicitly does not constrain the search space of all possible rule sets for a given data set. Finally, association rule learning is designed to discover separate, possibly unrelated rules which may describe only part of a data set while the problem addressed by this study is to find a set of related rules which describe a data set as a whole.

2 Algorithms Inspired by Bee Behavior

Algorithms inspired by the behavior of natural systems have been studied for decades. Examples include algorithms inspired by ants, biological immune systems, metallurgic annealing, and genetic recombination. These algorithms are sometimes called meta-heuristic algorithms because they provide a high-level framework which can be adapted to solve optimization, search, and similar problems, as opposed to providing a stringent set of guidelines for solving a particular problem. A review of the literature on algorithms inspired by bee behavior suggests that the topic is evolving and that there is no consensus on a single descriptive title for meta-heuristics based on bee behavior. Algorithm names in the literature include Bee System, BeeHive, Virtual Bee Algorithm, Bee Swarm Optimization, Bee Colony Optimization, Artificial Bee Colony, Bees Algorithm, and Simulated Bee Colony.

Common honey bees such as Apis mellifera take on different roles within their colony over time [5]. A typical hive may have 5,000 to 20,000 individuals. Young bees (2 to 20 days old) nurse larvae, construct and repair the hive, guard the entrance to the hive, and so on. Mature bees (20 to 40 days old) typically become foragers. Foraging bees typically occupy one of three roles: active forgers, scout foragers, and inactive foragers. Active foraging bees travel to a food source, gather food, and return to the hive. Roughly 10% of foraging bees in a hive are employed as scouts. These scout bees investigate the area surrounding the hive, often a region of up to 50 square miles, looking for attractive new food sources. At any given time some foraging bees are inactive. These inactive foraging bees wait near the hive entrance. When active foragers and scouts return to the hive, depending on the quality of the food source they are returning from, they may perform a waggle dance to an audience of inactive foraging bees. This waggle dance is believed to convey information to the inactive foragers about the location and quality of the associated food source. Inactive foragers receive this food source information from the waggle dance and may become active foragers. In general, an active foraging bee continues gathering food from a particular food source until that food source is exhausted, at which time the bee becomes an inactive forager.

A 1997 study by Sato and Hagiwara used a model of honey bee behavior named Bee System to create a variation of the genetic algorithm meta-heuristic [6]. The algorithm essentially added a model of the behavior of scout bees to introduce new potential solutions and avoid premature convergence to local minima solutions. A 2002 study by Lucic and Teodorvic used a variation of the Bee System model to investigate solving complex traffic and transportation problems [7]. The study successfully used Bee System to solve eight benchmark versions of the traveling salesman problem. A

2004 paper by Nakrani and Tovey presented a honey bee inspired algorithm for dynamic allocation of Internet services [8]. The study concluded that bee inspired algorithms outperformed deterministic greedy algorithms in some situations. A 2004 study by Wedde et al. used a bee-inspired algorithm named BeeHive to solve classic routing problems [9]. The paper concluded that BeeHive achieved similar or better performance compared to other common algorithms. In 2005 Yang presented an algorithm named Virtual Bee Algorithm to solve general optimization problems [10]. The study concluded the bee-inspired algorithm was significantly more efficient than a genetic algorithm approach. A 2005 study by Drias et al. used a meta-heuristic named Bee Swarm Optimization to study instances of the Maximum Satisfiability problem [11]. The study concluded that Bee Swarm Optimization outperformed other evolutionary algorithms, in particular an ant colony algorithm. In 2005 Teodorovic and Dell'Orco presented an algorithm named Bee Colony Optimization [12]. The study examined different versions of the traveling salesman problem including those with fuzzy information. A 2006 paper by Basturk and Karaboga investigated a bee-inspired algorithm named Artificial Bee Colony to solve five multi-dimensional numerical problems [13]. The paper concluded that the performance of the bee algorithm was roughly comparable to solutions by differential evolution, particle swarm optimization, and evolutionary algorithms. A 2006 paper by Pham et al. used a meta-heuristic named Bees Algorithm to investigate a collection of ten benchmark optimization problems [14]. The paper concluded that the Bees Algorithm was comparable or superior to approaches including deterministic simplex method, stochastic simulated annealing optimization procedure, genetic algorithm, and ant colony system. A 2009 study by McCaffrey demonstrated that an algorithm named Simulated Bee Colony outperformed existing deterministic algorithms for generating pairwise test sets, for six out of seven benchmark problems [15].

3   Simulated Bee Colony Algorithm Implementation

There are many ways to map honey bee foraging behavior to a specific algorithm which extracts a descriptive rule set from a set of clustered categorical data. The three primary design features which must be addressed are 1.) configuration of a problem-specific data structure that simulates a foraging bee's memory and which represents a the location of a food source, which in turn represents a rule set, 2.) creation of a problem-specific function which measures the goodness, or quality, of a candidate rule set, and 3.) specification of generic algorithm parameters such as the numbers of foraging, scout, and inactive bees in the colony, and the maximum number of times a bee will visit a particular food source. The simulated bee colony (SBC) algorithm used in this study is perhaps best explained by example. Suppose the data set to be analyzed contains the data described in the Introduction section of this paper, with attributes of color (red, blue, green, yellow), size (small, medium, large), and temperature (hot, warm, cold), and where each tuple has been clustered into one of three categories (c0, c1, c2). The screenshot shown in Figure 1 shows the result of a sample program run and illustrates many of the implementation details.

```
C:\RuleExtraction\Run\bin\Debug> Run.exe

The input clustered tuples are:

Red     Small   Hot     c0
Red     Small   Cold    c0
Blue    Medium  Hot     c1
Green   Large   Cold    c1
Yellow  Large   Warm    c2
Blue    Small   Hot     c2

Initializing Hive

Number Active bees = 60
Number Inactive bees = 20
Number Scout bees = 20
Maximum number of cycles = 10,000
Maximum cycles without improvement = 10,000
Maximum visits to a food source = 100
Probability waggle dance will convince inactive bee = 0.9000
Probability a bee accepts a worse food source = 0.0100

Tuple coverage weight = 1.00
Rule accuracy weight = 5.00
Rule set size weight = 1.00
Attribute efficiency weight = 0.00
Cluster coverage weight = 1.00

Entering simulated bee colony algorithm main processing loop

Progress: |=========================================|
          ......................................

All cycles completed
Best rule set / memory matrix found is:

Red     Small   x       c0
Blue    Medium  x       c1
x       Large   Cold    c1
x       x       x       c2

Corresponding rule set quality = 0.9688

The memory matrix in if..then form is:

IF Color = Red AND Size = Small THEN
  ClusterID = c0
ELSE IF Color = Blue AND Size = Medium THEN
  ClusterID = c1
ELSE IF Size = Large AND Temp = Cold THEN
  ClusterID = c1
ELSE
  ClusterID = c2

End run
```

Fig. 1. Screenshot of an example test run of the SBC implementation

The SBC algorithm implementation used in this study models a bee as an object with four data members as illustrated in Figure 2. The primary data member is a two-dimensional string array named MemoryMatrix which corresponds to a bee's memory of the location of a food source, which in turn represents a rule set. The first row of the memory matrix in Figure 1 is equivalent to the implication (Red, Cold) => c0. The Status field identifies the bee's role (1 = an active forager). The RuleSetQuality field is a value in the range [0.00, 1.00] which is a measure of the goodness of the memory matrix. The NumberVisits field is a counter that tracks the number of times the bee object has visited a particular food. The honey bee colony as a whole is modeled as an array of bee objects. The SBC algorithm iterates through each bee in the colony and examines the current bee's Status field. If the current bee is an active forager, the algorithm simulates the action of the bee leaving the hive to go to the current food source in memory. Once there, the bee examines a single neighbor food source. A neighbor food source is one which, relative to the current food source, a.) has a single attribute value (including a don't-care value) in one rule randomly changed, or b.) has a different cluster ID value in one rule randomly changed, or c.) has a new random rule added, or d.) has a randomly selected existing rule removed. If the quality of the neighbor food source is superior to the current food source, the foraging bee's memory is updated with the neighbor location and the NumberVisits counter is reset to 0. The SBC algorithm also contains a condition where a foraging bee may accept a neighbor food source with a lower measure of goodness, with probability = 0.01. Otherwise the bee's memory does not change and the NumberVisits counter is incremented.

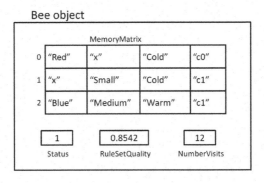

Fig. 2. Implementation representation of an SBC bee object

After examining a neighbor food source, an active bee returns to the hive. If the returning bee has reached a threshold for the maximum number of visits to its food source in memory, that bee becomes inactive and a randomly selected inactive bee is converted to an active forager. Otherwise the returning bee performs a simulated waggle dance to all inactive bees in the hive. This dance conveys the goodness of the current food source / rule set in the dancing bee's memory. Inactive bees with food sources in memory which have lower quality than the returning bee's food source, will update their memories to the returning bee's memory with probability = 0.90. Scout bees are not affected by the waggle dances of returning foragers. Instead, scouts leave the hive, examine a randomly selected food source, return to the hive, and perform a waggle dance to the audience of currently inactive bees.

The function which measures the quality of a particular rule set is composed of five components, and five associated weights. The first component measures the percentage of tuples in the data set which are touched by the rule set. This factor corresponds to the support metric of standard association rule learning. The second component measures the percentage of tuples which are correctly categorized by the rule set. This factor corresponds to the confidence metric of standard association rule learning. The third component measures rule set size efficiency and is the ratio of the number of rules in the rule set to the number of tuples in the data set. The premise is that, other factors being equal, a rule set with fewer rules is superior to a rule set with more rules. The fourth component of the goodness function measures attribute efficiency and is the ratio of the number of attributes in the rule set which have don't-care values to the total number of attribute values in the rule set. The premise here is that, other factors being equal, more don't-care values in a rule set is better than fewer don't care values. For example, a rule r1: (Red, Small) => c0 is better than a rule r2: (Red, Small, Hot) because fewer Boolean comparisons are needed to resolve r1 than r2. The fifth component of the goodness function measures the percentage of all possible cluster ID values which are contained in at least one rule in the rule set. The premise is that in most situations, it is extremely important that all cluster ID values be considered by a rule set. Each of the five components of the goodness function can take on values in the range [0.00, 1.00]. The goodness function assigns to each component value a weight in the range [0, 10], and returns the simple weighted average. These component weights must be specified as input parameters to the SBC algorithm.

4 Results

In order to evaluate the effectiveness of using a simulated bee colony algorithm to extract rule sets from clustered categorical data, the algorithm was implemented with a programming language and then executed against six benchmark data sets. The six test data sets were generated from hidden rule sets and varied across five dimensions: the number of attributes, the number of attribute values for each attribute, the number of cluster ID values, the number of rules in the hidden generator rule set, and the number of tuples in the data set. For each test data set, the algorithm was allowed to run until the hidden generating rule (or an isomorphic form of the generating rule) was found, or 10^8 iterations of the main processing loop had been performed. The characteristics of the six data sets, and the results are summarized in Table 1.

The number of attribute values in the third column in Table 1 include the possibility of a don't-care value for all six data sets. The hidden generating rules for each data set did not contain any don't-care values. The number of possible rule sets in column seven of Table 1 was computed based on the maximum number of possible tuples for a particular data set, which is the product of the number of attribute values less 1 (to account for the don't-care values) and the number of cluster ID values, rather than the actual number of tuples in the test data sets given in column six because the actual numbers of tuples selected for each data set were arbitrarily selected.

The SBC algorithm successfully discovered the underlying, hidden rule set generator for all six test data sets. In order to partially validate the test results, a program which generated purely pseudo-random rule sets (within the constraints of the data set parameters) was implemented and executed against each of the data sets. The random rule set generation program uncovered the underlying rule set generator for data set D0 but did not uncover the generators for data sets D1, D2, D3, D4, or D5. This result is not surprising given that the total number of possible rule sets for the small data set D0 is less than the maximum number of iterations allowed for the random rule set generator. The fact that the SBC algorithm successfully discovered the underlying rule generator for data set D5 is particularly noteworthy given the huge number of possible rule sets for that data set.

Table 1. Characteristics of benchmark data sets and empirical results

Data Set	Number Attributes	Attribute Values	Cluster ID Values	Generating Rules	Number Tuples	Possible Rule Sets	Generator Found?
DS0	3	(5,4,4)	3	4	6	$6.9 * 10^5$	yes
DS1	4	(4,4,4,4)	4	10	20	$5.4 * 10^8$	yes
DS2	4	(3,6,3,6)	5	162	300	$2.1 * 10^9$	yes
DS3	6	(3,4,5,5,4,3)	8	200	600	$1.1 * 10^{13}$	yes
DS4	10	$(3, \ldots, 3)$	10	300	1000	$1.0 * 10^{17}$	yes
DS5	10	$(5, \ldots, 5)$	10	400	2000	$4.7 * 10^{23}$	yes

The rule sets which are programmatically extracted from clustered data by the SBC algorithm can easily be expressed in multiple formats. For example, the program which produced the output shown in Figure 1, was easily reconfigured to emit a rule set using the RuleML language, as shown in Figure 3. RuleML is a markup language designed to provide a rich framework for expressing rules using XML. RuleML is a product of the Rule Markup Initiative, a collection of groups and individuals from industry and academia. Note that the rule shown in Figure 3 expressed in RuleML format is more structured than the same rule expressed in if. then format, and therefore RuleML rule format lends itself to efficient parsing and is particularly appropriate in situations where the SBC rule set output is intended to serve as input to another software system.

Because this study is primarily empirical, it is not possible to draw definitive conclusions from the results. However, the results do demonstrate that rule set extraction of clustered categorical data is feasible. A review of the literature did not reveal any studies which address the identical problem scenario as the one investigated here. However the association rule learning algorithms mentioned in the Introduction section of this paper could be adapted to produce a rule set which describes an entire data set. Compared to those approaches, the simulated bee colony (SBC) algorithm used in this study has the advantage of allowing full customizationof the characteristics of the rule set extracted from a particular data set. Additionally the SBC algorithm can be applied to arbitrarily large data set as opposed to some association rule learning

```
Best rule set / memory matrix found is:

Red      Small   x      c0
Blue     Medium  x      c1
x        Large   Cold   c1
x        x       x      c2

Corresponding rule set quality = 0.9688

The first rule in RuleML form is:

<Implies>
 <head>
  <Atom>
   <Rel>Membership</Rel>
   <Var>ClusterID</Var>
   <Ind>c0</Ind>
  </Atom>
 </head>
 <body>
  <And>
   <Atom>
    <Var>Color</Var>
    <Ind>Red</Ind>
   </Atom>
   <Atom>
    <Var>Size</Var>
    <Ind>Small</Ind>
   </Atom>
  </And>
 </body>
</Implies>

End of run
```

Fig. 3. Rule from rule set expressed in RuleML format

algorithms which require exponentially increasing processing time as the size of the data set increases. One disadvantage of using the SBC algorithm for rule set extraction is that the approach requires a relatively large number of input parameters, including both problem-specific parameters such as the definition of a function which computes the goodness of a rule set, as well as generic algorithm parameters such as the number of bee objects and the probability that an inactive forager bee will be convinced to alter memory based on a waggle dance of an active foraging bee with a better rule set solution. Because algorithms based on bee behavior are relatively unexplored, there are few if any guidelines for selecting input parameters and trial and error must be used to tune these parameter values. Another disadvantage of using an SBC algorithm is that because the technique is probabilistic, there is no guarantee that the approach will produce an optimal solution for a given set of inputs. However, when taken as a whole, the results of this study suggest that the use of a simulated bee colony algorithm for unsupervised rule set extraction of clustered categorical data is a promising new approach which merits further investigation, and that the use of an SBC algorithm may have the potential to outperform existing deterministic algorithms in certain problem scenarios.

References

1. Agrawal, R., Imielinski, T., Swami, A.: Mining Association Rules between Sets of Items in Large Databases. In: Proceedings of the International Conference on Management of Data, pp. 207–216 (1993)
2. Jochen, H., Ulrich, G., Nakhaeizadeh, G.: Algorithms for Association Rule Mining - A General Survey and Comparison. SIGKDD Explorations 2(2), 1–58 (2000)
3. Furnkranz, J., Flach, P.: An Analysis of Rule Evaluation Metrics. In: Proceedings of the 20th International Conference on Machine Learning, pp. 202–209 (2003)
4. Song, M., Rajasekaran, S.: A Transaction Mapping Algorithm for Frequent Itemsets Mining. IEEE Transactions on Knowledge and Data Engineering 18(4), 472–481 (2006)
5. Seeley, T.D.: The Wisdom of the Hive: The Social Physiology of Honey Bee Colonies. Harvard University Press, Boston (1995)
6. Sato, T., Hagiwara, M.: Bee System: Finding Solution by a Concentrated Search. In: Proceedings of the IEEE International Conference on Systems, Man, and Cybernetics, vol. 4, pp. 3954–3959 (1997)
7. Lucic, P., Teodorovic, D.: Transportation Modeling: An Artificial Life Approach. In: Proceedings of the 14th IEEE International Conference on Tools with Artificial Intelligence, pp. 216–223 (2002)
8. Nakrani, S., Tovey, C.: On Honey Bees and Dynamic Server Allocation in Internet Hosting Centers. Adaptive Behavior - Animals, Animats, Software Agents, Robots, Adaptive Systems 12(3-4), 223–240 (2004)
9. Wedde, H.F., Farooq, M., Zhang, Y.: BeeHive: An efficient fault-tolerant routing algorithm inspired by honey bee behavior. In: Dorigo, M., Birattari, M., Blum, C., Gambardella, L.M., Mondada, F., Stützle, T. (eds.) ANTS 2004. LNCS, vol. 3172, pp. 83–94. Springer, Heidelberg (2004)
10. Yang, X.S.: Engineering Optimizations via Nature-Inspired Virtual Bee Algorithms. In: Mira, J., Álvarez, J.R. (eds.) IWINAC 2005. LNCS, vol. 3562, pp. 317–323. Springer, Heidelberg (2005)
11. Drias, H., Sadeg, S., Yahi, S.: Cooperative Bees Swarm for Solving the Maximum Weighted Satisfiability Problem. In: Cabestany, J., Prieto, A.G., Sandoval, F. (eds.) IWANN 2005. LNCS, vol. 3512, pp. 318–325. Springer, Heidelberg (2005)
12. Teodorovic, D., Dell'Orco, M.: Bee Colony Optimization - A Cooperative Learning Approach to Complex Transportation Problems. In: Advanced OR and AI Methods in Transportation, pp. 51–60 (2005)
13. Basturk, B., Karaboga, D.: An Artificial Bee Colony (ABC) Algorithm for Numeric Function Optimization. In: Proceedings of the IEEE Swarm Intelligence Symposium, pp. 687–697 (2006)
14. Pham, D., Kog, E., Ghanbarzadeh, A., Otri, S., Rahim, S., Zaidi, M.: The Bees Algorithm – A Novel Tool for Complex Optimisation Problems. In: Proceedings of the 2nd International Virtual Conference on Production Machines and Systems, pp. 454–461 (2006)
15. McCaffrey, J.D.: Generation of Pairwise Test Sets using a Simulated Bee Colony Algorithm. In: Proceedings of the 10th IEEE International Conference on Information Reuse and Integration (2009)

Transformation of Graphical ECA Policies into Executable PonderTalk Code

Raphael Romeikat, Markus Sinsel, and Bernhard Bauer

Programming Distributed Systems, University of Augsburg, Germany
{romeikat,sinsel,bauer}@ds-lab.org

Abstract. Rules are becoming more and more important in business modeling and systems engineering and are recognized as a high-level programming paradigma. For the effective development of rules it is desired to start at a high level, e.g. with graphical rules, and to refine them into code of a particular rule language for implementation purposes later. An model-driven approach is presented in this paper to transform graphical rules into executable code in a fully automated way. The focus is on event-condition-action policies as a special rule type. These are modeled graphically and translated into the PonderTalk language. The approach may be extended to integrate other rule types and languages as well.

1 Introduction

Increasing complexity of information systems complicates their development, maintenance, and usage. Due to this evolution, the Autonomic Computing Initiative by IBM [1] proposes self-manageable systems that reduce human intervention necessary for performing administrative tasks. For realizing autonomic capabilities within managed objects, policies are a promising technique. The idea behind policy-based management is allowing administrators to control and manage a system on a high level of automation and abstraction. According to [2], policies are an appropriate means for modifying the behavior of a complex system according to externally imposed constraints.

The focus of this paper is on a certain type of policy called Event-Condition-Action (ECA) policies. ECA policies are considered as reaction rules that allow for specifying which actions must be performed in a certain situation. They specify the reactive behavior of a system in response to events and consist of a triggering event, an optional condition, and an action term.

Policy-based management is also a layered approach where policies exist on different levels of abstraction. Wagner et al. consider three different abstraction levels [3]. The business domain level typically uses a natural or a visual language to define terms and constrain operations. The platform-independent level defines formal statements expressed in some formalism or computational paradigm, which can be directly mapped to executable statements of a software platform. The platform-specific level expresses statements in a specific executable language. Strassner defines a flexible number of abstraction layers as the Policy Continuum [4]. The idea is to define and manage policies on each level in a

G. Governatori, J. Hall, and A. Paschke (Eds.): RuleML 2009, LNCS 5858, pp. 193–207, 2009.

domain-specific terminology, and to refine them e.g. from a business level down to a technical level.

An approach is presented that allows to graphically model ECA policies and transform those policy models into executable code. It uses techniques from model-driven engineering (MDE) to model policies in a language-independent way and to automatically generate code. Models are used to represent ECA policies based on common policy concepts that are represented in a generic metamodel. The policy language PonderTalk is also represented by a respective metamodel. Model transformations allow for generating executable PonderTalk code from an initial policy model. A full implementation of the approach exists as plugin for the software development platform Eclipse.

There have been other approaches for modeling information about policies and policy-based systems. The authors of [5] present a General Policy Modeling Language (GPML) as a means to design policies and map them to existing policy languages. This approach is also based on MDE concepts, but uses an UML profile for visualization and is based on the rule interchange language R2ML with a focus on logical concepts to map GPML policies onto existing policy languages. The Common Information Model (CIM) [6] by the Distributed Management Task Force (DMTF) represents a conceptual framework for describing a system architecture and the system entities to be managed. An extension to CIM to describe policies and to define policy control is provided by the CIM Policy Model [7]. Another type of information model is the Directory Enabled Networks next generation standard (DEN-ng) [8] by the TeleManagement Forum (TMF). DEN-ng is based on the Policy Continuum and considers different levels of abstraction. Policies are directly integrated into the models. Similar to the approach presented here, the CIM Policy Model and DEN-ng are independent of any policy language. They are as metamodels that enable the developer to describe a system and the enclosed policies in an implementation-independent way. Policies are specified in a declarative way while omitting technical details. However, only specification of policies is regarded in both approaches. They do not offer a possibility to transform a policy model to a particular language that can be executed by some engine. It remains an open issue to what extent PonderTalk and other policy languages are compatible with those policy models.

This paper is structured as follows. Section 2 gives an introduction to Model-driven Engineering and to the policy language PonderTalk. Section 3 describes the model-driven approach to transform graphical ECA policies into executable code. Section 4 describes the implementation of the approach. The paper concludes with related work and a summary in section 5.

2   Basics

This section presents a short introduction into model-driven engineering, which represents the foundation of the approach, and into the policy language PonderTalk, which is the target of the transformation.

2.1 Model-Driven Engineering

In software engineering one can observe a paradigm shift from object-orientation as a specific type of model towards generic model-driven approaches, which has important consequences on the way information systems are built and maintained. The model-driven engineering approach follows multiple objectives: apply models and model-based technologies to raise the level of abstraction, reduce complexity by separating concerns and aspects of a system under development, use models as primary artifacts from which implementations are generated, and use transformations to generate code with input from modelling and domain experts [9,10]. Model-driven solutions consist of an arbitrary number of automated transformations that refine abstract models to more concrete models (vertical model transformations) or simply describe mappings between models of the same level of abstraction (horizontal model transformations). Finally, code is generated from lower-level models. Models are more than abstract descriptions of systems as they are used for model and code generation. They are the key part of the definition of a system.

2.2 PonderTalk

Ponder2 [11] is a policy framework developed at Imperial College over a number of years. A set of tools and services were developed for the specification and enforcement of policies. Ponder2 offers a general-purpose object management system and includes components that are specific to policies.

Everything in Ponder2 is a managed object. Managed objects generate events and policies are triggered by those events to perform management actions on a subset of managed objects. This is also called local closed-loop adaptation of the system. There are managed objects that are available by default to interact with the basic Ponder2 system, i.e. factory objects to create events and policies. Besides that, user-defined managed objects are implemented as Java classes and used within Ponder2. Managed objects can send messages to other managed objects and new instances of managed objects can be created at runtime.

ECA policies are called obligation policies in Ponder2 and are specified with the language PonderTalk. PonderTalk has a high-level syntax that is based on the syntax of Smalltalk and is used to configure and control the Ponder2 system. Basically, everything in Ponder2 can be realized with PonderTalk, i.e. define and load managed objects, specify policies, or throw events that trigger policies. In order to realize a policy system in Ponder2, the respective PonderTalk code has to be implemented.

Example Scenario. Now, an example scenario is presented where ECA policies are used to manage the behavior of a communication system. Further sections will refer to this scenario when presenting examples.

The signal quality of wireless connections is subject to frequent fluctuations due to position changes of sender and receiver or to changing weather conditions. A possibility to react to those fluctuations is adjusting transmission power. A good tradeoff between transmission power and signal quality is desired. Too high transmission power causes additional expenditures whereas signal quality suffers from too less transmission power.

In that scenario signal strength is managed autonomously by a policy system using ECA policies. A *Transmitter* adjusts transmission power with the actions *increase_power* and *decrease_power*, both of them expecting a value by which power should be increased and decreased. Whenever a change in signal quality is noticed, an *intensityChange* event is thrown that contains the *id* of the affected receiver and the signal quality's *oldValue* and *newValue*.

Two ECA policies *lowQuality* and *highQuality* are responsible for adjusting transmission power. They are triggered whenever an *intensityChange* events occurs and in their condition check the old and new signal quality enclosed in the event. If the transmission power falls below a value of 50, the *lowQuality* policy executes a call of `increase_power(10)` to increase transmission power by 10 at the *Transmitter*. The other way round, the *highQuality* policy executes a call of `decrease_power(10)` at the *Transmitter* if transmission power goes beyond a value of 80.

The behavior of the transmission system can new be adjusted at runtime via the policies. The accepted signal quality is specified in the conditions of the two determining policies by means of the two boundaries 50 and 80. Changing those boundaries has immediate effect on the transmission power and signal quality.

3   Modeling and Transforming ECA Policies

In this section the overall approach to graphically model ECA policies and transform those policy models into executable code is presented. Various aspects have to be considered for the approach to be effective. Figure 1 illustrates how the various aspects of the approach are related to each other.

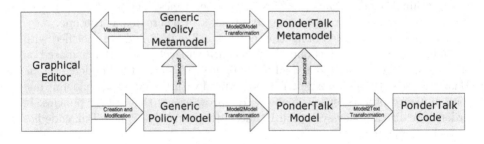

Fig. 1. From Graphical Policies to PonderTalk Code

First, a generic policy metamodel contains common concepts of ECA policies. It abstracts from special features and technical details that are specific to a certain policy language and thus allows to specify ECA policies independently of a particular language. Any ECA policy is initially represented as instance of that metamodel to offer an abstract view onto the policy from a functional point-of-view. As only common concepts are contained in the generic policy metamodel, such a generic policy model can be transformed into executable code of a concrete policy language later. Next, the concepts of the generic policy metamodel have a graphical representation so the generic policy model is visualized as a diagram. A graphical editor offers functionality to create and modify models in a comfortable way.

Once an ECA policy has been modeled as a diagram, transformation into the target language can start. The starting point for defining that transformation is the generic policy metamodel, and a metamodel for the target language, namely the PonderTalk metamodel. As no formal metamodel was available for the PonderTalk, a metamodel was created from the language documentation [12]. A model-to-model transformation is defined on the metamodels and executed on the model. It takes the generic policy model as input and generates the respective PonderTalk model as output, which is an instance of the PonderTalk metamodel. Finally, a model-to-text transformation takes the PonderTalk model as input and generates a textual representation of that policy containing the respective PonderTalk code.

The following subsections present further details about the metamodels, the graphical visualization, and the model transformations. Various aspects will be illustrated by means of the example scenario presented in section 2.2.

3.1   Generic Policy Metamodel

The generic policy metamodel comprises common concepts of well-known policy languages such as PonderTalk [11], KAoS [?], and Rei [?]. It covers the essential aspects of those languages and contains classes that are needed to define the basic functionality of an ECA policy, i.e. events, conditions, and actions, amongst others as described in the following. The generic policy metamodel is specified as Essential MOF (EMOF) model. EMOF is a subset of the Meta Object Facility (MOF) [?] that allows simple metamodels to be defined using simple concepts. EMOF provides the minimal set of elements that are required to model object-oriented systems. Figure 2 shows the generic policy metamodel as UML class diagram.

The class *Entity* represents the components of the policy system. Each *Entity* has a *name* attribute and three more technical attributes. Those attributes may contain code fragments that are specific to the target language and that need to be included into the generated code so it is executable. In case of PonderTalk, *accordingClass* e.g. specifies the name of the respective Java class implementing that *Entity* as managed object in Ponder2. This is somehow contrary to the aspect of language independency, but on the other side it is a simple possibility to to generate code that is executable without further modification.

Entities can be organized in a *Domain* hierarchy, similar to the folders of a file system. A *Domain* is a collection of *Entities* that belong together with regards to content. Events, conditions, and actions can also be contained in a *Domain*. A *Domain* is an *Entity* itself as it can also be controlled by *Policies*.

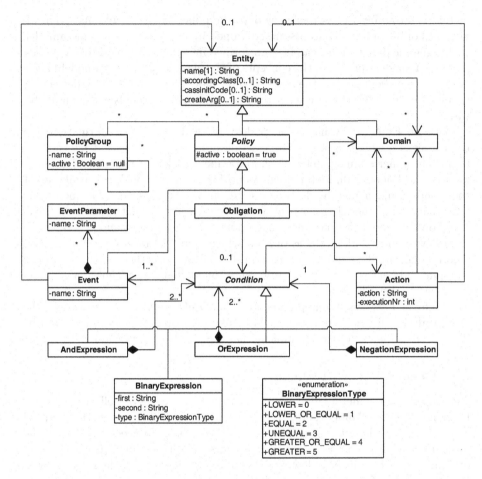

Fig. 2. Generic Policy Metamodel

The class *Policy* is the abstract superclass of all types of policy, whereas only one concrete type *Obligation* is included so far, representing ECA policies. The *active* attribute of a *Policy* describes its status and may be set to `true` or `false`. Only active *Policies* are triggered at runtime in a policy system.

Policies can be grouped in *PolicyGroups*. Groups are named and can contain other groups. In contrast to a *Domain*, a *PolicyGroup* has an administrative purpose and is used for activating and deactivating a set of *Policies* all at once using the *active* attribute of the group, which may be set to `true`, `false`, or `undefined`. The active status of a *Policy* is determined by the closest *PolicyGroup* in the group hierarchy with *active* set to `true` or `false` and where that *Policy* is contained. If at the respective level of the hierarchy some active and inactive *PolicyGroups* contain that *Policy*, the policy is regarded as being active. The *active* attribute of a *Policy* is only deciding if all *PolicyGroups* which contain this *Policy* have *active* set to `undefined`.

The classes *Event, Condition,* and *Action* represent the actual content of an *Obligation*. Each *Obligation* requires at least one *Event,* optionally has a *Condition,* and has an arbitrary number of *Actions* associated. An *Obligation* is triggered by at least one *Event* whereas at runtime the occurrence of one respective *Event* suffices to trigger that *Obligation*. An *Event* can be thrown by any *Entity,* has a *name,* and can contain a set of parameters represented by the class *EventParameter*. *EventParameters* are named and can be referred within the *Condition* that is associated with the respective *Obligation*.

A *Condition* is a boolean expression. A *BinaryExpression* is the simplest form of a *Condition* and compares two strings with each other. These strings represent the left land side (LHS) and right hand side (RHS) of the *Condition,* denoted by the attributes *first* and *second*. Those strings can contain the name of an *EventParameter,* which allows to analyze the *Event* that triggered the *Obligation*. Or, they can directly contain a simple value in the form of an enclosed string or numeric value. The comparison operator is defined by the attribute *type* and may be one of $<, \leq, =, \neq, \geq$, and $>$. An expression can additionally be negated using the class *NegationExpression,* or combined as conjunction or disjunction using the classes *AndExpression* and *OrExpression* respectively.

If the *Condition* of an obligation evaluates to true, the associated *Actions* are executed. Executing an *Action* means calling an *Operation*. The attribute *action* within the class *Action* specifies which *Operation* is called. The attribute *executionNr* must be used to denote the sequence of execution if two or more *Actions* are associated with an *Obligation*. Arbitrary numbers may be used as long as they are different from each other. They need not be consecutive, which provides some flexibility when associating multiple *Actions* to multiple *Obligations*.

3.2   PonderTalk Metamodel

As a next step, the PonderTalk metamodel is defined as the target of the model-to-model transformation. That metamodel is again specified as EMOF model; figure 3 shows it as UML class diagramm. It refers to the current version 2.840 of Ponder2 and contains only those concepts that are needed to represent an ECA policy in PonderTalk. Other functionalities of PonderTalk such as authorization policies are not adressed as they go beyond the expressiveness of the ECA policy metamodel. In the following, the PonderTalk metamodel is described with respect to its differences to the generic policy metamodel.

In PonderTalk an *Entity* is called *ManagedObject*. Apart from naming there is no difference between those two classes. The same applies to an *Obligation,* which is now called *ObligationPolicy*. The classes *Domain, Policy, Condition, BinaryExpression, NegationExpression, AndExpression, OrExpression,* and *BinaryExpressionType* do not differ from the generic policy metamodel.

PonderTalk does not know the concept of groups. Thus, a way has to be found to represent *PolicyGroups* when transforming into PonderTalk. This has an effect on the *active* attribute of a *Policy* and is described later.

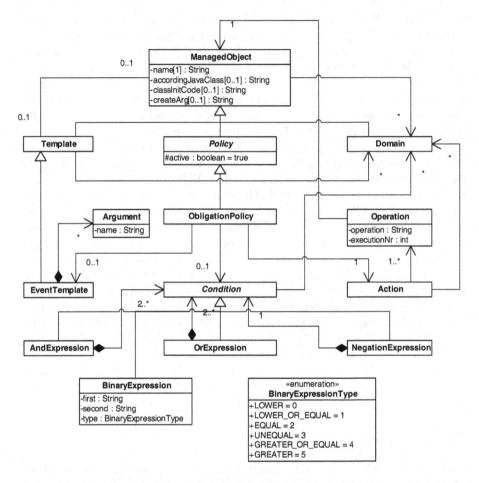

Fig. 3. PonderTalk Metamodel

On the other hand, PonderTalk introduces a new class *Template*. *Templates* are used to create new instances of *ManagedObjects*, *Policies*, or *Domains*. A *Template* itself is also a *ManagedObject*.

An *Event* in the generic policy metamodel is called *EventTemplate* in PonderTalk. An *EventTemplate* can contain an arbitrary number of named *Arguments*, which represent the respective *EventParameters*. A noticeable difference is that an *ObligationPolicy* in PonderTalk cannot be triggered by an arbitrary number of *EventTemplates*, but is triggered by at most one. This is taken into consideration by the transformation later. Additionally, an *EventTemplate* is an instance of *ManagedObject* in PonderTalk.

The condition part of an *ObligationPolicy* exactly corresponds to the generic policy metamodel, but there are important differences in the action part. An *ObligationPolicy* in PonderTalk does not execute an arbitrary number of *Actions*, but executes exactly one *Action*. An *Action* uses at least one *Operation*

to execute commands on a *ManagedObject*. The attribute *operation* is used to
specify a particular PonderTalk command.

3.3   Graphical Visualization

Now, a graphical representation of a policy is created as a diagram. For this
purpose, the classes of the generic policy metamodel that were instantiated
when modeling the policy are visualized with all necessary information. Abstract
classes in the metamodel do not have a graphical representation as no instances
of them can be created. A visualization of the classes in the PonderTalk metamodel is not required either as that metamodel is only used as intermediate step
in the transformation later and needs not be available as a diagram.

Figure 4 shows the graphical representation of the example scenario presented
in section 2.2. Additionally to the scenario description, the two policies *lowQuality* and *highQuality* are put into a policy group named *quality*.

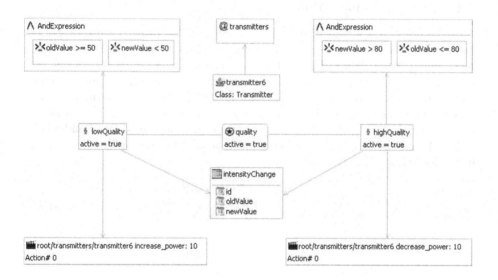

Fig. 4. Visualization of a Generic Policy Model

For visualizing the instantiated classes simple rectangular shapes were chosen
that ressemble the way classes are visualized in UML. As header of each shape,
a symbol and an identifying text are displayed to characterize it. That text
contains the *name* attribute if existant in the respective class. For an action, the
action attribute is used instead. For a binary expression, a textual representation
of its attributes is used to visualize the LHS, RHS, and the operator, and for
the other expression classes, the name of the class itself is used.

Further details of the classes are displayed in the shape body, which usually
contains the attributes with their value. Event parameters are not visualized
as rectangular shape, but they are visualized within the body of the enclosing

event, which can be seen with the event *intensityChange* and the contained event parameters *id*, *oldValue*, and *newVaule*. Conditions that are used within another condition are directly visualized within the body of the parent condition as shown with the two *AndConditions*, which both contain two binary conditions. This way of integrating event parameters and nested conditions reduces the overall number of shapes in the policy diagram.

Finally, associations between classes are displayed as directed lines as known from UML class diagrams. The direction represents the visibility of the classes as defined by the respective associations in the metamodel.

The chosen way of visualizing classes and associations omits complex shapes and technical details, so it focuses on the essential information and developers should easily get familiar with it. In section 4 a graphical editor is presented that allows to create ECA policies based on the generic policy metamodel and the graphical visualization.

3.4   Model Transformations

An ECA policy is now specified as generic policy model using the generic policy metamodel and the graphical visualization. The next step is generating a representation of that policy as PonderTalk code as an implementation for Ponder2. For this purpose, model transformations take the generic policy model as input and generate the respective PonderTalk code. The overall transformation process is divided in two steps. First, the generic policy model is transformed into a PonderTalk model. That model is in a second step transformed into PonderTalk code. The necessary transformations are summarized in figure 5 and described in the following.

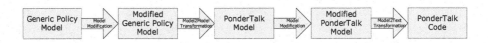

Fig. 5. Model Transformations

From a Generic Policy Model to a PonderTalk Policy Model. When transforming a generic policy model, a check is performed first whether the model fulfills the structural requirements of the metamodel with respect to the cardinalities of the associations. Furthermore, domains must not contain themselves nor contain two domains that are named equally. The same applies to policy groups. All entities of the model must have a name and names must be unique amongst obligation policies and amongst direct entities (without subtypes). Finally, any action must specify its action attribute. If all checks are passed, the model is well-formed and ready to be transformed.

Now, a model modification is executed to modify the source model. A modification does not create a new model as target, but the result of the modification is the modified source model itself. Model modifications are used in order to enrich a model with additional information that was not modeled explicitly, or to modify details of a model to simplify further transformations. In the generic policy model, an *Obligation* can be triggered by various *Events* whereas in the PonderTalk model only one *EventTemplate* is allowed per *ObligationPolicy*. For this purpose, a model modification duplicates *Obligations* with two or more associated *Events* into several *Obligations* of with each one is associated with one of the original *Events*. The rest of the *Obligation* is duplicated without changes. This structural change allows the straightforward generation of the PonderTalk model from the modified generic policy model.

Then, a model-to-model transformation takes the modified generic policy model as input and generates the respective PonderTalk model as output. For this purpose, the transformation translates the concepts of the generic policy metamodel in a way so they are expressed by the concepts of the PonderTalk metamodel. The transformation is defined on the classes of the metamodel and is executed on the instances of those classes in the model. As a result, the PonderTalk model is generated as follows.

First, all instances of *Entity* are transformed into *ManagedObjects* one after another. The attribute values of an entity are copied to the respective managed object. Transforming the *Entities* includes transforming *Domains* and *Obligations* as they are *Entities* as well. When transforming an *Entity*, any associated *Entity* (i.e. the domain of an entity) is transformed immediately, and this is a recursive process.

It is important to notice that an *Entity* can be referenced multiple times by other *Entities* and whenever one reference is processed, the transformation of that *Entity* is called straightforward. However, a caching mechanism ensures that an *Entity* is actually transformed only once and with any further transformation call to the same *Entity*, the cached result is used instead. This ensures that any model element is created only once in the target model and the model elements need not be processed in a special sequence during the transformation.

Events are transformed into *EventTemplates* and *EventParameters* into *Arguments*. In contrast to the generic policy metamodel, *EventTemplates* are subclasses of *ManagedObject* in PonderTalk, so the generated *EventTemplates* are internally marked to be *ManagedObjects* as well. Transforming the *Conditions* is performed by simply copying them as no differences exist between the two metamodels with respect to the condition part. *Actions* are transformed into *Operations*. The *Action* objects in the PonderTalk model are created newly. For each *ObligationPolicy*, one *Action* object is created and associated with that *ObligationPolicy*.

When transforming *Obligations*, the associated *PolicyGroups* are processed along the group hierarchy to determine the active status of the *Obligation* as described in section 3.1. *PolicyGroups* do not have a representation in PonderTalk and thus no more appear in the PonderTalk model. Their only purpose for the transformation was to determine the active status of policies whose *active* attribute was undefined.

From a PonderTalk Policy Model to PonderTalk Code. Executable PonderTalk code requires the standard *Domains* **root**, **policy**, and **event** to be

```
    // Create Domains
 2  root at: "transmitters" put: root/factory/domain create.

 4  // Create event intensityChange
    event := root/factory/event create: #( "id" "oldValue" "newValue" ).
 6  root/event at: "intensityChange" put: event.

 8  // Load the Transmitter class file
    root/factory at: "transmitter" put: ( root load: "Transmitter" ).
10
    // Create an instance named transmitter6 and put it in each associated domain
12  instance := root/factory/transmitter create.
    instance intensityChangeEvent: root/event/intensityChange.
14  root/transmitters at: "transmitter6" put: instance.

16  // Create policy lowQuality
    policy := root/factory/ecapolicy create.
18  policy event: root/event/intensityChange;
          condition: [ :id :oldValue :newValue | ((oldValue >= 50) & (newValue < 50)) ];
20        action: [ :id :oldValue :newValue | root/transmitters/transmitter6 increase_power: 10 ].
    root/policy at: "lowQuality" put: policy.
22  policy active: true.

24  // Create policy highQuality
    policy := root/factory/ecapolicy create.
26  policy event: root/event/intensityChange;
          condition: [ :id :oldValue :newValue | ((newValue > 80) & (oldValue <= 80)) ];
28        action: [ :id :oldValue :newValue | root/transmitters/transmitter6 decrease_power: 10 ].
    root/policy at: "highQuality" put: policy.
30  policy active: true.
```

Listing 1.1. Generated PonderTalk Code

specified. A PonderTalk model might not explicitly contain those *Domains*. For this purpose, a model modification checks whether they are modeled and if not, inserts them into the model. Furthermore, that modification adds any *Obligation-Policy* that is not contained in the **policy** *Domain* to that *Domain* and it also ensures that any *EventTemplate* is contained in the **event** domain. Finally, it adds any *Domain* that is not contained in another *Domain* to the **root** *Domain*.

Now, a model-to-text transformation takes the modified PonderTalk model as input and generates the respective PonderTalk code as output. That transformation is also called code generation as it generates code for a programming language. The transformation defines for each class of the PonderTalk metamodel a respective textual representation as PonderTalk code. When the transformation is executed on the PonderTalk model, the respective code for the enclosed classes is generated step by step. Listing 1.1 shows the resulting PonderTalk code that corresponds to the policy diagram shown in figure 4.

In PonderTalk it is important to specify the statements in the correct sequence. *Domains* must first be declared before they can be referenced by other *ManagedObjects*. For this purpose, a sorting algorithm initially creates an ordering of the *Domains* along the hierarchy and ensures that code for the **root** *Domains* is generated before proceeding with the next level in the hierarchy, etc. Now, code for all domain declarations is generated with respect to that ordering. It is also worth to be mentioned that the transformation does not generate any code for the top-level *Domains root, policy,* and *event* as Ponder2 internally creates those *Domains* at startup before any PonderTalk code is executed at all.

Now, code for *EventTemplates* is generated. In PonderTalk a factory object is used to create an *EventTemplate* together with the enclosed *Arguments*. In the PonderTalk model, any *EventTemplate* is associated with the **event** *Domain*, which also results in a respective PonderTalk statement.

As next step, *ManagedObjects* (without subtypes) are transformed into code. For any *ManagedObject*, the respective Java class specified in the *accordingClass* attribute is loaded as factory object and put into the respective factory domain. For *ManagedObjects* that are associated with a *Domain* in the PonderTalk model, an instance is created additionally and added to that *Domain*. Arguments required for instantiation are specified in the *createArg* attribute of the *ManagedObject* and are added to the statement that creates the instance. *ManagedObjects* that are not associated with any *Domain* are only loaded as factory. This is be useful if instances of *ManagedObjects* should only be created at runtime.

Finally, code for *ObligationPolicies* is generated including the referenced *EventTemplate*, *Condition*, and *Action*. First, an *ObligationPolicy* is created with the policy factory. Next, the triggering *EventTemplate* is associated with that *ObligationPolicy*. If the *ObligationPolicy* contains a *Condition*, a textual representation of that *Condition* is generated for PonderTalk. Next, the *Action* is transformed into appropriate code including the referenced *Operations* in the sequence as defined by their attribute *executionNr*. Finally, the *ObligationPolicy* is put into all associated *Domains* and the status of the *ObligationPolicy* is set according to its *active* attribute.

4   Implementation

In order to demonstrate the approach, an implementation was developed as a set of plugins for the software development platform Eclipse. The implementation is called *PolicyModeler* and can be integrated into any Eclipse 3.5 (Galileo) installation via the update site `http://policymodeler.sf.net/updates`. Alternatively, a complete Eclipse installation including the PolicyModeler is available for instant usage at `http://policymodeler.sf.net/eclipse.zip`. This section presents important aspects about the implementation.

For specifying the metamodels, the Eclipse Modeling Framework (EMF) [16] is used in the PolicyModeler. EMF stores the specified metamodels in the Ecore format, which is an implementation of EMOF. With EMF a tree-like editor is generated to create and modify a metamodel as well as instances of that metamodel. However, that editor was not used for the creation of the generic policy metamodel and the PonderTalk metamodel; instead, annotated Java interfaces were used as they are a more effective way to specify metamodels in EMF.

For the graphical representation of the policies the Eclipse Graphical Modeling Framework (GMF) [17] is used. GMF offers the generation of a graphical editor that allows to create and modify a generic policy model as a diagram. For the generation of that graphical editor some input is required. First, the generic policy metamodel is referenced as instances of that metamodel are to be visualized. Second, a graphical representation is created for each model element to define its visualization in the diagram. Third, a toolbar is defined that offers means to create model elements and associations between them. Finally, all input is combined to define which element of the toolbar is used to create which model element and how that model element is visualized in the diagram.

The model transformations are developed with the Eclipse Modeling Framework Technology (EMFT) [18] and Model To Text (M2T) [19] projects, which support the implementation of various kinds of model transformations. They offer all functionality required for the transformations used in the approach, i.e. special languages to realize checks, model modifications, model-to-model transformations, and model-to-text transformations. The projects are available as Eclipse plugins themselves and thus offer good integration with the PolicyModeler.

An ECA policy is created in the PolicyModeler by composing the desired model elements into a diagram using the toolbar. If a generic policy model already exists as Ecore file without a graphical representation, that graphical representation can be generated automatically by the PolicyModeler. The transformation into corresponding PonderTalk code can be started directly from the diagram. The resulting code can then be executed within a Ponder2 installation.

5 Conclusion

In this paper an approach to graphically model ECA policies and generate executable code for the language PonderTalk from such models was presented. It is the first the approach innovative and transfers benefits of MDE to the development of policies such as reduction of development time. The generic policy metamodel allows to model ECA policies independently from a particular language and allows to generate code in an automated way. PonderTalk is used as target language, but the approach may be extended to target other policy languages by integrating the metamodel of the respective language and setting up the necessary model transformations. The metamodel covers important features of ECA policies. A developer might require expressiveness for his policies that is not covered by the metamodel. Further policy types and concepts may be integrated by extending the generic policy metamodel so it can express more than only ECA policies. However, a tradeoff must be made here. Full code generation is only possible if the target language can represent all concepts of the generic policy metamodel in an appropriate way. This is why the metamodel is basically limited to the important concepts of ECA policies. Adressing more policy languages and types are subject to future work. The same applies to reverse engineering of PonderTalk code into a graphical model, which is currently not possible. The approach is fully implemented as Eclipse plugin [20]. A small example was shown, but the approach was also applied to a larger case study that realizes the hospital scenario from the Ponder2 tutorial [21]. That case study regards all structural details of the two metamodels and is included in the Eclipse download mentioned in section 4.

References

1. Kephart, J.O., Chess, D.M.: The Vision of Autonomic Computing. Computer 36(1), 41–50 (2003)
2. Damianou, N., Dulay, N., Lupu, E., Sloman, M.: The Ponder Policy Specification Language. In: Sloman, M., Lobo, J., Lupu, E.C. (eds.) POLICY 2001. LNCS, vol. 1995, pp. 18–38. Springer, Heidelberg (2001)

3. Wagner, G., Antoniou, G., Tabet, S., Boley, H.: The Abstract Syntax of RuleML - Towards a General Web Rule Language Framework. In: IEEE/WIC/ACM International Conference on Web Intelligence, pp. 628–631. IEEE Computer Society Press, Los Alamitos (2004)
4. Strassner, J.C.: Policy-Based Network Management: Solutions for the Next Generation. Morgan Kaufmann Publishers, San Francisco (2003)
5. Kaviani, N., Gasevic, D., Milanovic, M., Hatala, M., Mohabbati, B.: Model-Driven Engineering of a General Policy Modeling Language. In: IEEE Workshop on Policies for Distributed Systems and Networks, pp. 101–104. IEEE Computer Society, Los Alamitos (2008)
6. Distributed Management Task Force: Common Information Model (CIM) Specification. DSP0004 (June 1999)
7. Distributed Management Task Force: CIM Policy Model White Paper. DSP0108 (June 2003)
8. Strassner, J.C.: DEN-ng: Achieving Business-Driven Network Management. In: IEEE/IFIP Network Operations and Management Symposium, pp. 753–766. IEEE Computer Society, Los Alamitos (2002)
9. Bézivin, J.: On the Unification Power of Models. Software and Systems Modeling 4(2), 171–188 (2005)
10. Flater, D.W.: Impact of Model-Driven Standards. In: Annual Hawaii International Conference on System Sciences, vol. 9, pp. 3706–3714. IEEE Computer Society, Los Alamitos (2002)
11. Twidle, K., Lupu, E., Dulay, N., Sloman, M.: Ponder2 - A Policy Environment for Autonomous Pervasive Systems. In: IEEE Workshop on Policies for Distributed Systems and Networks, pp. 245–246. IEEE Computer Society, Los Alamitos (2008)
12. Imperial College London: Ponder2. (June 2009), http://ponder2.net
13. Uszok, A., Bradshaw, J.M., Jeffers, R.: KAoS: A policy and domain services framework for grid computing and semantic web services. In: Jensen, C., Poslad, S., Dimitrakos, T. (eds.) iTrust 2004. LNCS, vol. 2995, pp. 16–26. Springer, Heidelberg (2004)
14. Kagal, L., Finin, T., Joshi, A.: A Policy Language for a Pervasive Computing Environment. In: IEEE International Workshop on Policies for Distributed Systems and Networks, June 2003, pp. 63–74 (2003)
15. Object Management Group: Meta Object Facility (MOF) Core Specification (January 2006), http://www.omg.org/spec/MOF/2.0/PDF
16. The Eclipse Foundation: Eclipse Modeling Framework (EMF). (June 2009), http://www.eclipse.org/modeling/emf
17. The Eclipse Foundation: Graphical Modeling Framework (GMF) (June 2009), http://www.eclipse.org/modeling/gmf
18. The Eclipse Foundation: Eclipse Modeling Framework Technology (EMFT) (June 2009), http://www.eclipse.org/modeling/emft
19. The Eclipse Foundation: Model To Text (M2T) (June 2009), http://www.eclipse.org/modeling/m2t
20. University of Augsburg: PolicyModeler (August 2009), http://policymodeler.sf.net
21. Imperial College London: Ponder2 Tutorial (May 2009), http://www.ponder2.net/cgi-bin/moin.cgi/Ponder2Tutorial

A Rule-Based Approach to Match Structural Patterns with Business Process Models

Jens Müller

SAP Research CEC Karlsruhe
Vincenz-Prießnitz-Str. 1, 76131 Karlsruhe, Germany
jens.mueller@sap.com

Abstract. Business process models may contain certain sets of related elements of interest to modellers. External constraints on business process models may, for example, require or prohibit the presence of a specific set of related model elements or a temporal relationship between different sets. To automatically evaluate such constraints, the existence of these sets must be verified beforehand. In this paper, we present a modelling language for structural patterns using a graphical notation. These patterns are used to describe related model elements of interest. Furthermore, we introduce a technique to match pattern models with business process models using a rule-based system.

1 Introduction

The current industry standard for modelling business processes is the Business Process Modeling Notation (BPMN) [1]. When modelling business processes with BPMN, business users have to adhere to certain domain-specific constraints, which refer to the structure and semantics of business process models. For instance, a constraint could state that a sequence of model elements that have a certain meaning (e.g., tasks with certain labels or links to web service operations) must not exist within business process models. Another constraint could state that all sequences of model elements with certain labels must be followed by a specific model element (e.g., an event). According to these examples, constraints can be seen as conditions on sets of model elements that have a certain meaning. In case constraints are violated in business process models, their execution (by humans or machines) may lead to undesired results or, in the worst case, critical situations. Problems arise if constraints are not explicitly known to the modelling tool. The more constraints a business process model has to comply with, the harder it gets for modellers to adhere to them and the more difficult it becomes to judge whether they are violated. In addition, constraints might change over time, which means that existing process models have to be re-examined every time constraints are altered.

In [2], we outlined that such problems can be solved by explicitly modelling and automatically evaluating constraints. During the modelling phase, it is necessary to (1) specify structural patterns, which describe related model elements of interest, as well as (2) conditions on these patterns, whereas the evaluation

G. Governatori, J. Hall, and A. Paschke (Eds.): RuleML 2009, LNCS 5858, pp. 208–215, 2009.

phase requires (3) matching structural patterns with business process models and (4) evaluating the corresponding conditions. This work focuses on providing solutions for the first and the third aspect of this approach. The remainder of this paper is structured as follows. Chapter 2 introduces basic concepts and presents a modelling language for structural patterns. A method for matching pattern models with business process models using a rule-based system and implementation details are presented in chapter 3. Chapter 4 summarises related work. Finally, chapter 5 concludes the paper.

2 Modelling and Transforming Structural Patterns

BPMN diagrams are based on an internal model, which is typically based on a meta-model that defines modelling constructs and rules to create valid models. Thus, business process models can be considered as a set of BPMN model elements. To specify conditions on certain sets of related model elements that may be subsets of business process models, a corresponding pattern is needed to describe these sets. In [2], we presented a scenario from the aviation industry that motivates our work. In this scenario, a modeller is, for example, interested in a certain pattern of model elements that describes a sequence of two tasks, which may be contained within aircraft maintenance process models. This pattern consists of two tasks. The first task represents the removal of an aircraft part from storage, whereas the second one represents an inspection of the accompanying documents. Furthermore, the pattern describes that these two tasks match corresponding tasks within business process models that are either directly connected or separated by at most one intermediate task and an arbitrary number of intermediate events. This pattern is depicted in Fig. 1. In this case, two business process models contain a set of tasks that the structural pattern matches, although slightly different labels are used. These sets of model elements are called instances of the pattern. In order to explicitly model such patterns, we developed a modelling language that addresses the type of generality and flexibility of the example, which is called Process Pattern Modeling Language (PPML).

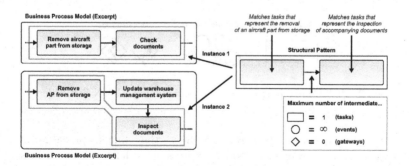

Fig. 1. Example of a structural pattern

PPML is a modelling language for explicitly specifying structural patterns. Since we assume that constraints on business process models, i.e. structural patterns and corresponding conditions, are modelled by the same users that model business processes, we developed a graphical notation that is similar to what modellers already know, which in our case is BPMN. Another reason for choosing a graphical notation similar to BPMN is that structural patterns closely relate to business process models, since they correspond to subsets of these models. The main focus of PPML is to provide generic and flexible modelling constructs that allow combining similar patterns within a single pattern model. Furthermore, PPML provides modelling constructs that allow connecting different pattern models and thus ensure modularity.

As already indicated, we use a rule-based system to check if a pattern model matches one or more subset(s) of elements within a business process model. The advantage of using a rule-based system for this purpose is that it helps to solve the combinatorial problem, since rules just describe when a match occurs instead of specifying how the matching process is performed. The main idea of our approach is to transform pattern models into single rules by translating model elements into a set of conditions. A match occurs when a rule fires during the evaluation phase, i.e. all conditions evaluate to **true**. To check certain dependencies between model elements within a condition, we developed specific algorithms that are not part of the rule language used. Another advantage of using rules is that they typically allow accessing such user-defined functions. In the following, we describe the most important PPML modelling constructs as well as the corresponding transformations.

Generic Flow Objects. PPML contains generic versions of the BPMN modelling constructs for modelling tasks, events, and gateways (cf. Fig. 2). These modelling constructs have several attributes that can be set in our modelling tool and are used to refine matching criteria. The modelling construct *GenericIntermediateEvent*, for example, has an attribute to store a list of triggers. These triggers define which types of intermediate events within BPMN models a generic intermediate event matches.

Fig. 2. PPML modelling constructs for matching BPMN flow objects

The modelling construct *GenericTask* shares the same graphical notation as BPMN tasks and is a versatile tool to express a modeller's interest in specific tasks that have a certain meaning. In BPMN, the meaning of tasks is expressed by labelling them using natural language. Within executable process models, service tasks are linked to web service operations. A more powerful way to express the meaning of tasks is to assign them to non-ambiguous domain concepts, which can be described using ontology classes. Although this method, which we call semantic tagging, poses additional overhead for modellers, business process analysis benefits from machine-readable semantics in the long term. Modellers can use generic tasks to describe their interest in tasks that are assigned to certain ontology classes.

The `GenericTask` meta-model class has three optional attributes: a list of web service operations, a regular expression, and a semantic expression (cf. Fig. 3). A semantic expression represents a boolean term that consists of references to ontology classes and the $\wedge$, $\vee$, and $\neg$ operators. The semantic expression $DomainConcept_A \vee DomainConcept_B$, for example, matches tasks within a business process model that are assigned to the ontology class $DomainConcept_A$, $DomainConcept_B$, or one of their subclasses.

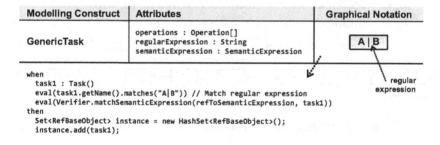

Modelling Construct	Attributes	Graphical Notation
GenericTask	operations : Operation[] regularExpression : String semanticExpression : SemanticExpression	A | B

```
when
    task1 : Task()
    eval(task1.getName().matches("A|B")) // Match regular expression
    eval(Verifier.matchSemanticExpression(refToSemanticExpression, task1))
then
    Set<RefBaseObject> instance = new HashSet<RefBaseObject>();
    instance.add(task1);
```

regular expression

Fig. 3. PPML modelling construct GenericTask

During the transformation of a pattern model into a single rule, any generic task within the model is translated into one or more condition(s), which are added to the **when** section of the rule. The generic task depicted in Fig. 3, for example, is translated into three conditions. In this example we assume that a modeller specified a regular expression as well as a semantic expression. The first of the three conditions matches any task. In case this condition matches a task, the task is bound to the variable **task1**. The second condition is satisfied if the regular expression of the generic task matches the name of **task1**. Finally, the third condition is satisfied if the semantic expression of the generic task evaluates to **true**, based on the ontology classes that **task1** is assigned to. This condition relies on a helper function (**matchSemanticExpression**), since its evaluation is more complex and cannot be performed using the built-in functionality of the rule language. The first parameter of the helper function is a reference (**refToSemanticExpression**) to the semantic expression within the pattern model. In case all conditions are satisfied, the model elements that constitute an instance of the structural pattern being searched for are then further processed within the **then** section of the rule. Like generic tasks, generic events and generic gateways are translated into rules in a similar way.

Flexible Sequence Flow. To provide flexibility when modelling sequences of flow objects, we introduced the modelling construct *FlexibleSequenceFlow* (cf. Fig. 4). Flexible sequence flows connect generic tasks, events, and gateways within pattern models and are used to match corresponding sequences of elements within business process models. However, these elements do not necessarily have to be directly connected. Instead, a modeller can specify the minimum and maximum amount of intermediate tasks, events, and gateways. In the generated rule, the helper function `matchFlexibleSequenceFlow` calculates the number of model elements between two flow objects of a process model.

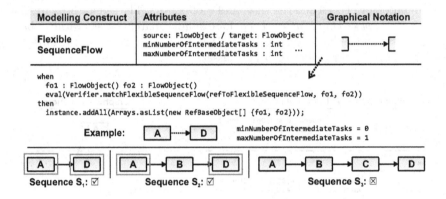

Fig. 4. PPML modelling construct FlexibleSequenceFlow

Figure 4 shows an example of a flexible sequence flow. The modeller of this structural pattern is interested in a sequence of tasks labelled A and D. Moreover, corresponding tasks within business process models with at most one intermediate task, should match the pattern as well. As depicted, sequences S_1 and S_2 within a business process model match the pattern, whereas sequence S_3 does not. PPML also provides the modelling construct *SequenceFlow*, which is syntactic sugar for a flexible sequence flow with no intermediate flow objects.

Pattern Connectors and Pattern References. The modelling constructs *IncomingPatternConnector* and *OutgoingPatternConnector* are used to mark the boundaries of a structural pattern, i.e. its inputs and outputs. In contrast to business process models, structural patterns may have several inputs and outputs, since they correspond to fragments of business process models. Pattern connectors are a prerequisite for the modelling construct *PatternReference*. Pattern references are used to combine pattern models and support modularity. Figure 5 depicts two pattern models, P_1 and P_2. P_2 contains a pattern reference that refers to P_1 as well as one of the incoming and one of the outgoing pattern connectors of P_1. Pattern references are not translated during rule transformation. Instead, the pattern reference within P_2 is resolved before transformation. The resulting pattern model is shown on the right side of Fig. 5.

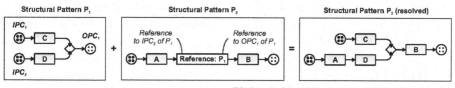

Fig. 5. Pattern connectors and pattern references

3   Rule-Based Search for Structural Patterns

Figure 6 depicts the various steps involved in searching for instances of a structural pattern that is expressed by a pattern model. Step 1 is triggered once a modeller saves a pattern model. In this step, an algorithm we developed recursively traverses the pattern model and transforms it into a corresponding rule. For each model element that the algorithm encounters, parts of this rule are generated according to the type of model element and its attributes. In step 2, the rule is inserted into the rule set of the inference engine, after the rule is generated. In step 3, the elements of the business process model that needs to be searched are inserted into the working memory of the inference engine (i.e., all flow and connecting objects). These elements thus constitute the search space. In step 4, the inference engine starts the matching process. Finally, matched rules are fired in step 5, which means that an instance of a structural pattern is found. Further checks based on the search result may be performed, after all instances of a structural pattern within a business process model are found.

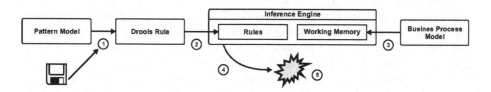

Fig. 6. Rule-based search for structural patterns

We implemented the presented concepts on top of an existing BPMN modelling tool that is based on Eclipse. Therefore, our implementation consists of a set of Eclipse plug-ins. The modelling tool uses an infrastructure that adheres to the Meta Object Facility (MOF) specification [3], which defines a standardised language for specifying technology neutral meta-models. Both BPMN models and structural patterns are internally represented as MOF models. Furthermore, we defined ontologies as MOF models using the Ontology Definition Metamodel (ODM) [4]. In order to assign tasks to domain concepts, we added a suitable MOF association between the BPMN meta-model and the ODM meta-model.

Structural patterns can be modelled with a graphical editor. To match structural patterns with business process models, the responsible plug-in uses the business logic integration platform Drools [5].

4   Related Work

Several authors proposed visual languages to query BPEL processes [6] and BPMN diagrams [7,8]. Although some PPML modelling constructs share the same semantics as the modelling constructs of these languages, PPML introduces novel modelling constructs that focus on generality (generic tasks, events, and gateways), flexibility (flexible sequence flow), and modularity (pattern connectors, pattern reference). Furthermore, the goals of these query languages are partially different. In [7], for example, the query language contains a modelling construct that returns the model elements between two specific flow objects within a business process model. In contrast, the PPML modelling construct *FlexibleSequenceFlow* checks if two flow objects within process models are connected in a certain way, but does not return potential intermediate model elements. Moreover, PPML is not just a graphical notation but a modelling language based on a standard-compliant modelling infrastructure, which offers many advantages, such as metadata management and serialisation [3].

In [9], a declarative service flow language is discussed. This language allows the visual specification of constraints to check the conformance of service flows. A comparable language for specifying quality constraints on process models is presented in [10]. Both languages focus on run-time aspects and allow, for example, to specify that the execution of a single task A must be eventually followed by the execution of a single task B. Although we will incorporate related concepts in the future, our approach focuses on identifying whole sets of model elements within the structure of process models.

There are several approaches to enrich business process models with additional semantics. In [11,8], business process diagrams are represented as ontologies. In addition, techniques to link business process ontologies with domain ontologies using an auxiliary layer are proposed. In contrast, our approach is entirely based on meta-models and links business process models with domain ontology models without an additional layer.

To the best of our knowledge, no existing approach uses a rule-based system for searching instances of structural patterns within business process models. In graph theory, there is a plethora of algorithms for solving the subgraph isomorphism problem [12]. Given two graphs G and H, these algorithms detect one or more occurrence(s) of H as a subgraph of G. However, instances of structural patterns are not necessarily graphs (cf. S_2 in Fig. 4). In [8], a method is discussed to transform visual queries into SPARQL [13] queries. However, this method requires that business processes are represented as ontologies. In contrast, our rule-based approach directly operates on process models and does not involve additional transformation steps. Moreover, SPARQL does not offer the possibility to call user-defined functions within queries, which is essential for the translation of some of our modelling constructs.

5   Conclusions

In this paper, we argued that explicitly modelling structural patterns and searching for instances of these patterns within business process models is a prerequisite for evaluating conditions on structural patterns. The purpose of this research is to provide corresponding standard-compliant solutions for BPMN models. We presented a modelling language that goes beyond related approaches in terms of generality, flexibility, and modularity. Although the language is tailored to BPMN, it could be adapted to other process modelling languages. Furthermore, we showed that rule-based systems are powerful tools to search for instances of structural patterns within process models. The key concept of our approach is to transform pattern models into rules. In addition, we introduced a method to enrich tasks with machine-readable semantics using ontology models.

In the future, we will explore ways to specify and evaluate conditions on structural patterns using model checking techniques. Such conditions could, for example, require the presence of certain temporal relationships between instances of different patterns.

References

1. OMG: Business Process Modeling Notation: Version 1.2 (2009)
2. Müller, J.: Supporting Change in Business Process Models Using Pattern-Based Constraints. In: Halpin, T., et al. (eds.) BPMDS 2009 and EMMSAD 2009. LNBIP, vol. 29, pp. 27–32. Springer, Heidelberg (2009)
3. OMG: Meta Object Facility (MOF) Core Specification: Version 2.0 (2006)
4. OMG: Ontology Definition Metamodel: Version 1.0 (2009)
5. JBoss Community: Drools, http://www.jboss.org/drools/
6. Beeri, C., Eyal, A., Kamenkovich, S., Milo, T.: Querying Business Processes with BP-QL. Information Systems 33(6), 477–507 (2008)
7. Awad, A.: BPMN-Q: A Language to Query Business Processes. In: Enterprise Modelling and Information Systems Architectures: Concepts and Applications. LNI, vol. P-119. Gesellschaft für Informatik, Bonn (2007)
8. di Francescomarino, C., Tonella, P.: Crosscutting Concern Documentation by Visual Query of Business Processes. In: Ardagna, D., et al. (eds.) RGU 1974. LNBIP, vol. 17, pp. 18–31. Springer, Heidelberg (2009)
9. van der Aalst, W.M.P., Pešić, M.: DecSerFlow: Towards a Truly Declarative Service Flow Language. In: Bravetti, M., Núñez, M., Zavattaro, G. (eds.) WS-FM 2006. LNCS, vol. 4184, pp. 1–23. Springer, Heidelberg (2006)
10. Förster, A., Engels, G., Schattkowsky, T., van der Straeten, R.: Verification of Business Process Quality Constraints Based on Visual Process Patterns. In: TASE 2007: First Joint IEEE/IFIP Symposium on Theoretical Aspects of Software Engineering, pp. 197–208. IEEE Computer Society Press, Los Alamitos (2007)
11. Thomas, O., Fellmann, M.: Semantic EPC: Enhancing Process Modeling Using Ontology Languages. In: Proceedings of the Workshop on Semantic Business Process and Product Lifecycle Management (SBPM 2007), pp. 64–75 (2007)
12. Read, R.C., Corneil, D.G.: The Graph Isomorphism Disease. Journal of Graph Theory 1(4), 339–363 (1977)
13. World Wide Web Consortium: SPARQL Query Language for RDF (2008)

Usage of the Jess Engine, Rules and Ontology to Query a Relational Database

Jaroslaw Bak, Czeslaw Jedrzejek, and Maciej Falkowski

Institute of Control and Information Engineering, Poznan University of Technology,
M. Sklodowskiej-Curie Sqr. 5, 60-965 Poznan, Poland
{Jaroslaw.Bak,Czeslaw.Jedrzejek,Maciej.Falkowski}@put.poznan.pl

Abstract. We present a prototypical implementation of a library tool, the Semantic Data Library (SDL), which integrates the Jess (Java Expert System Shell) engine, rules and ontology to query a relational database. The tool extends functionalities of previous OWL2Jess with SWRL implementations and takes full advantage of the Jess engine, by separating forward and backward reasoning. The optimization of integration of all these technologies is an advancement over previous tools. We discuss the complexity of the query algorithm. As a demonstration of capability of the SDL library, we execute queries using crime ontology which is being developed in the Polish PPBW project.

Keywords: Jess engine, rules, reasoning, ontology, relational database.

1 Introduction

Knowledge representation and processing methodologies require efficient and complete tool environments. Many toolkits exist [1] that contain reasoning engines of varying functionalities that provide a solution to knowledge management. Recently, effective rule processing and relational database support has increased in importance, because most of the data processed by modern applications is stored in relational databases. Today, however, there is need for not only data management, but also for scaleable knowledge management, as well as for processing gathered data with richer means than is offered by RDBMS (Relational Database Management System) and SQL (Structured Query Language) language. Ontologies describe and extend data, and allow for queries to gain additional knowledge. The most common query language associated with ontologies and semantic data representation is the SPARQL [2] language.

The most formal approaches consider very expressive languages, i.e., important fragments of OWL-DL in order to determine the decidability of query answering and establish its computational complexity [3].

To enable semantic access to relational data, it is necessary to express relational concepts in terms of ontology concepts, that is to define mapping between the relational schema and ontology classes and relations. Given such a mapping, one can transform relational data to RDF triples and process that copy in semantic applications. This method has obvious drawbacks, such as maintaining

G. Governatori, J. Hall, and A. Paschke (Eds.): RuleML 2009, LNCS 5858, pp. 216–230, 2009.

synchronization, and others. Another method is to create a data adapter based on query rewriting. Such adapters can rewrite SPARQL query to SQL [4] query and execute it in RDBMS. This method could be fast in data retrieval, but without a reasoner, the full potential of ontology cannot be exploited. The third method is to generate semantic data from the relational data 'on-the-fly', on demand for the requesting application, and then process that data with a reasoner. We use this method to fill a gap between the relational data representation and the semantically described data.

The main goal of this work is to facilitate development of ontology-driven applications by construction of a tool named Semantic Data Library (SDL). Its prototypical implementation enables integration to the relational database (MS SQL) [5], OWL (Web Ontology Language) [6] ontology with SWRL (Semantic Web Rule Language) [7], which extends the expressivity of OWL, and the Jess (Java Expert System Shell) [8] reasoning engine. This paper is a continuation of work presented in [9]. In comparison to the previous work this article contains query algorithm optimizations (concerns grouping SQL queries presented in Section 3.4) and adaptations to OWL2Jess [10,11] implementation. These adaptations concern triple representation of Jess facts and extension, which enables querying relational database 'on-the-fly'.

This paper is organized as follows. Section 2 presents preliminaries, related work and background of our motivation. Section 3 describes the SDL library, its architecture and the integration process which integrates relational databases, OWL ontology and SWRL rules with Jess language. Section 4 presents the query algorithm, its computational complexity and shows the hybrid reasoning process. Section 5 presents an example of SDL library use. Section 6 contains concluding remarks and presents our future work.

2 Preliminaries

In this work we present an approach that integrates relational databases, OWL ontology and SWRL rules with Jess language. Extended support of the OWL semantics and SWRL rules is provided by the OWL2Jess [12] and SWRL2Jess [13] tools. The triple representation of the OWL Meta-model is employed. Unfortunately, the transformation of OWL ontology with SWRL rules must be done in two steps. In the first step we need to transform OWL ontology to Jess language script; in the second step SWRL rules to another Jess script are transformed. Then we have to merge these files. To the best of our knowledge, the OWL2Jess tool is the most powerful of the tools mentioned above. We use the OWL Meta-model [14] in the forward chaining mode to exploit the full potential of the OWL ontology.

There are a number of tools that enable a query relational data, and results are added in terms of ontology concepts/relations. One of the tools is DataMaster [15,16], a Protégé-OWL [17] plug-in that allows importing relational database structure or content into an OWL ontology. At present it only populates the ontology with data from the relational database and save this ontology to file.

Another tool is D2RQ [18], an RDF adapter to relational data. The KAON2 [19] reasoning engine is a tool that allows reasoning with OWL ontology and SWRL rules; it enables connection to the relational database 'on-the-fly' through relational database ontology (it has to be included into OWL ontology as an import). It also supports the SPARQL query language.

None of the tools that concern OWL and SWRL transformation to Jess enables backward reasoning (only forward) or database access. Our tool enables both forward and backward reasoning, ontology and rules transformation to Jess, and relational database access, which is a significant advantage in real-life applications. The SDL library allows querying a relational database 'on-the-fly', according to the semantics specified in ontology with rules. The query mechanism currently exploits the 'is-a' relation and SWRL rules, but more sophisticated reasoning is available using OWL Meta-model from OWL2Jess.

The next Section presents an SDL overview, architecture and integration process of the Jess, OWL+SWRL ontology and relational database schema.

3 SDL Architecture and Integration Process

The SDL library is a prototypical implementation of a tool which enables to query relational databases in terms of ontology concepts and reasoning with rules. It was implemented in Java language. An answer for a query is obtained during the reasoning process in Jess engines (forward and backward chaining). The Jess facts gathered in this way can be easily processed with an OWL2Jess reasoning engine because facts are represented as triples (the same approach as described in [10]). The queries are in the form of directed graphs (for example see Section 5).

Jess is a rule-based environment for building expert systems. This engine, written in Java by Ernest Friedman-Hill at Sandia National Laboratories, has its own scripting language called Jess language which allows direct access to all Java classes and libraries (due to the *call* function). Jess language script can contain templates (data structures in Jess), rules, functions, queries and facts (data). Such script can be loaded into Jess engine by *batch* function. Due to all the Jess functionalities, an embedding of it in Java applications could be done in a simple way. Jess as a rule engine enables backward and forward reasoning, with Jess language used to query engine working memory. We use both methods of chaining: the backward method is responsible for gathering data from the relational database and the forward chaining is used to answer a given query.

In the Jess engine the forward chaining is executed using a very efficient algorithm called Rete [20]. In this mode Jess does not require any special declarations in contrast to the backward one.

The backward chaining method in Jess requires a special declaration for templates *(do-backward-chaining)*. The *do-backward-chaining* definitions are added to all *deftemplates* declarations which are used in backward chaining (for example: *(do-backward-chaining triple)*). One can define rules to match backward reactive templates. The rule compiler rewrites such rules and adds the *need-* prefix to inform the Jess engine when this rule has to be fired (when we need some fact). The *need-* prefix can be added manually during the rules creation. To fire a rule Jess needs a fact with a *need-* prefix in its working memory. Such fact can be added automatically (during reasoning) or manually (by the user), for example *(need-triple (predicate "hasSeller") (subject 3) (object ?y))*. If the rule fires and there is a way to obtain needed facts, they appear in the Jess working memory. The *need-* facts are the so called triggers (in Jess language terminology). These facts correspond to the goals in the backward reasoning method.

3.1   SDL Architecture and Functionalities

The architecture of the SDL library is shown in Figure 1. The tool consists of two main modules:

- SDL-API (Application Programming Interface) which provides all SDL functions, mainly: reasoning processes management (in backward and forward chaining), executing queries, and scripts generation in Jess language,
- SDL-GUI (Graphical User Interface) which exploits SDL-API functions for defining the mapping between relational database schema and ontology concepts/relations or between relational schema and Jess templates (data structures in Jess).

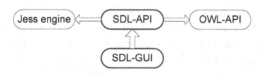

Fig. 1. The architecture of the Semantic Data Library

The SDL library utilizes some well accepted tools: Jena Ontology API to handle OWL files; Jess; SQL JDBC (Java Database Connectivity) library which enables access to a relational database; and MS SQL Server 2005 - relational database server.

The SDL-API module provides the following functionalities (they do not occur in SDL-GUI):

- executing SQL query or procedure (results are in the form of Jess facts),
- executing a Jess query, which consists of the concepts from ontology or templates defined in Jess language,
- Jess engine and hybrid reasoning management.

Due to the SDL-GUI module the library enables execution of the following functions:

- a reading ontology and a view of concepts/relations hierarchies; the view contains classes hierarchy, object an datatypes properties hierarchies,

- generating OWL ontology from relational database schema (tables as classes, columns as properties),
- a relational database schema view which contains tables, views, columns and data types,
- mapping between a relational database schema and ontology concepts/Jess templates,
- populating ontology with data from a relational database according to the specified mapping,
- transformation of OWL+SWRL ontology to Jess script in both forward and backward chaining mode (two different Jess scripts are the result).

The SDL library provides some other useful functions, but for clarity they are omitted in this paper.

3.2   Integration Process of Relational Database, OWL+SWRL Ontology and Jess Engine

The main goal of the integration process is to have one common format for data, rules and ontology to enable query relational database according to defined semantics. We have chosen the Jess language as a common format. Such integration requires mapping between relational database schema and ontology concepts/relations. Next, we need to transform the mapping and ontology with rules to the Jess language format. The result of the integration process is a Jess script; in the forward or the backward chaining mode. Generated scripts in backward mode consist of the described mapping, SWRL rules (included in the OWL ontology) and taxonomy rules. Such script is generated from a Jess script in forward mode, and after that, the mapping rules are created and added. In forward mode, the mapping does not occur. The idea of the integration process is presented in Figure 2. Previous tools including OWL2Jess are not capable of full integration (to map a relational database, a use backward chaining). Our tool for integration and transformation does not have these restrictions. However, SDL library can not exploit the full potential of the ontology, so we need to use one of the tools mentioned above (see Section 3.4).

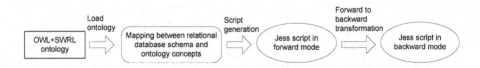

Fig. 2. The integration process executed within the SDL

In the script resulting from integration all data and metadata are in one language; enabling full advantage of the Jess engine (all functionalities available in a forward and backward chaining mode).

3.3 Mapping between Relational Database Schema and Ontology Concepts

This section presents a method for mapping between a relational database schema and ontology concepts. The SDL library also enables mapping between a relational schema and Jess templates defined in a Jess engine. The idea of the mapping is the same in both methods, so we present the first one. The mapping process is supported by SDL-GUI. We describe the main idea of this process and present some examples.

We assume that every "essential" concept or relation has its appropriate SQL query. "Essential" means that the instance of the concept/relation can not be obtained from the concepts/relations taxonomy or rules. It can be obtained only in the direct way (as the result of the SQL query or written). For example, for the hierarchy of classes $Institution \rightarrow Company \rightarrow Buyer$, the $Buyer$ class is an "essential" concept. The mapping process requires defining SQL queries for all "essential" ontology classes and properties. We assume also that the ontology which is used is properly constructed (the taxonomy is computed and classified; without inconsistencies).

We assume that every SQL query has the following form:

$$\text{SELECT [R] FROM [T] <WHERE> <C, AND, OR>}$$

where:

- R are the result columns (one or two according to class or property),
- T are the tables which are queried,
- WHERE is an optional clause to specify the constraints,
- C are the constraints in the following form: <column, comparator, value>, for example: Age>21,
- AND, OR - are the optional SQL commands.

For the example, assume that we have a table *persons* with the following columns: *id, name, age* and *gender*. The example SQL query for the concept *AdultPerson* can be defined as follows: SELECT *id* FROM *persons* WHERE *age>21*. When we want to use more constraints, we need to use an OR or AND clause. For query which obtains all adult women, we would define the following query: SELECT *id* FROM *persons* WHERE *age>21* AND *gender='Female'*.

3.4 Transformation to Jess Language

This section presents the transformation method of the ontology with rules and mapping to scripts in Jess language. The transformation process is done automatically. The SDL library generates Jess scripts in forward and backward modes. Their transformation method differs in the technical details. We describe only the backward mode processing, because it is more complicated. For clarity in this article, we do not present full URI addresses (only short names) and we use the following shortcuts: p - predicate, s - subject, o - object, and for http://www.w3.org/1999/02/22-rdf-syntax-ns#type - rdf:type.

Currently implementation has some restrictions and limitations of the transformation of ontology, rules and mapping to Jess language script:

- the taxonomy of classes and properties are transformed as rules; subsumption is the only ontology relation that we use in backward chaining mode (in forward chaining mode we use OWL Meta-model and SWRL rules); extensions are planned,
- rules can only add new facts to the working memory (in future implementation, they will be able to retract and modify the information in a relational database and Jess working memory),
- the SWRL rule can be extended by SWRL built-ins [21] only with the following built-ins: swrlb:equal ('='), swrlb:notEqual ('≠'), swrlb:greaterThan ('>'), swrlb:greatherThanOrEqual ('≥'), swrlb:lessThan ('<') and swrlb:lessThanOrEqual ('≤').

The generation of the Jess script in backward mode is done in the following way:

- the template *triple* is created: *(deftemplate triple (slot p)(slot s)(slot o))* and information that *triple* is backward-reactive is added: *(do-backward-chaining triple)*,
- SWRL rules are directly transformed to Jess; for example the rule (*?x* and *?y* are the companies names and *?InV* is a number of issued invoice): *issuedVATIn(?x, ?InV), receivedVATIn(?y, ?InV) → TransactionBetween(?x, ?y)* is transformed into the following rule:

      (defrule Def-TransactionBetween
          (need-triple (p "TransactionBetween")(s ?x)(o ?y))
          (triple (p "issuedVATIn") (s ?x) (o ?InV))
          (triple (p "receivedVATIn") (s ?y) (o ?InV))
       => (assert (triple (p "TransactionBetween")(s ?x)(o ?y))))

- for the taxonomy of concepts/relations the appropriate rules are created; for example, for the hierarchy *Document→ VATInvoice* the following rule is created:

      (defrule HierarchyDocument
          (need-triple (p "rdf:type")(s ?x)(o "Document"))
          (triple (p "rdf:type")(s ?x)(o "VATInvoice"))
       => (assert (triple (p "rdf:type")(s ?x)(o "Document"))))

- defined mappings are transformed as rules with SQL queries in their heads.

In our approach every mapping is transformed into one Jess rule. It means that we need exactly as many rules as defined mappings. The transformation is done according to the specified template:

Rule name: "Def-" + name of the mapping concept/relation
Body: ?r←(need-triple (p "name of concept") (s ?x) (o ?y)
Head: (call of the *runQueriesFromJess* function with its parameters)
    (retract ?r)

The *need-* facts which are triggers to fire rule are deleted from Jess working memory *(retract ?r)*. For this reason duplicates of firing the same rule do not occur. The example rule for property *MoneyTransferTo* between ID of the money transfer and the receiver's company name is shown below:

```
(defrule Def-MoneyTransferTo
  ?r<-(need-Triple (p "MoneyTransferTo")(s ?x)(o ?y))
  =>
  (bind ?query (str-cat "SELECT id, receiver FROM transfers;"))
  (?*access* runQueriesFromJess
    "Def-MoneyTransferTo"
    ?query
    "s;id;o;receiver;p;MoneyTransferTo;"
    (str-cat ?x ";" ?y ";")
    "triple" ?*conn* (engine))
  (retract ?r))
```

Function *runQueriesFromJess* comes from the JessDBAccess class, which allows accessing a relational database. It has the following parameters:

- name of the rule,
- SQL query defined for mapping,
- names of columns used to obtain results,
- variables values (if determined),
- name of the template used to add Jess fact (e.g. triple),
- connection to the relational database (?*conn*),
- instance of Jess engine where facts should be added.

During the execution of the rules, the SQL queries responsible for gathering instances of the same concept/relation are grouped. If a new concept/relation occurs, one aggregate SQL query is executed and results are added to the Jess engine as triples (or other templates, because SDL supports more mapping possibilities - see Section 3.1). We also developed the data types mapping method, which enables transformation of relational data types to the Jess data types. Figure 3 shows a grouping algorithm of SQL queries. The grouping algorithm

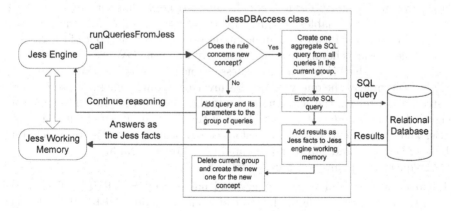

Fig. 3. The grouping algorithm

is enabled due to the very simple form of the SQL queries presented in Section 3.3. The algorithm goes beyond the scope of this article, so details are omitted.

The transformation to forward chaining script can be done in two ways: with our method, or with the OWL2Jess [10]. In the first case, only the SWRL and taxonomy rules are transformed; in the second case the rules, classes and properties are transformed. Both transformations can be done automatically by SDL-GUI. In forward mode we do not transform the mappings. The result is the Jess script which contains template *triple*, rules and facts (if individuals occurred in ontology). When we use OWL2Jess transformation, the more sophisticated reasoning is supported after data from the relational database as triples is gathered (during the execution of the Jess query).

4   Query Algorithm and Hybrid Reasoning Process

This querying method uses a hybrid reasoning process to answer a given query. This means that we use two Jess engines: one for the forward chaining and for the backward one. The queries are constructed in Jess language in terms of ontology concepts. A mapping between the relational database schema and ontology classes and properties is used to achieve the semantics of the data. Data itself is stored in a relational database. The ontology and the mapping rules transformed into Jess language format provide the additional semantic layer to the relational database. Such an approach allows for querying a relational database and reasoning using Jess, rules and ontology.

The reasoning process is fully executed by the Jess engine and managed by the SDL library. We need to use two Jess engines, because backward chaining mode is very inefficient during queries execution. The reason for this inefficiency is that the Jess engine creates trigger facts (with *need-* prefix) during execution of a query and then calculates rules activations (but it does not fire any of the rules). This procedure does not occur in the forward chaining mode, so the answering process is much faster.

The backward chaining engine is responsible only for gathering data from the relational database. Data is added (asserted in Jess terminology) as triples into the engine working memory. The forward chaining engine can answer a query with all constraints put on variables in a given query ($=, !=, <, >$ etc.). During the execution of a query the forward chaining engine does not reason (none of the rules is fired). The forward chaining engine is run only when the OWL2Jess transformation script (or other) is loaded. Every Jess engine has its own working memory.

The beginning of the querying process involves loading a backward script generated (or written) in the Jess language into the backward engine. In the forward engine the template *triple* is created. Then the user can query about the properties and classes defined in the transformed ontology. A query is constructed in the Jess language and can be represented as a directed graph.

The query algorithm is defined as follows:

1. Create a rule from a given query and name it QUERYRULE. The query is the body of the rule, and the head is empty. Add QUERYRULE to the forward chaining engine.

2. In backward chaining engine, for all concepts/relations occurring in the query, do:

   a) Add need-X fact/facts to the engine (where X is the current concept/ relation) with bounded variables (if it exists).

   b) Run the engine - the reasoning process begins and during it the instances of the X concept/relation are obtained from a relational database.

   c) If the group of queries in JessDBAccess class is not empty, force the aggregation and execution of the SQL query. Results are added as triples to Jess working memory.

   d) Copy results to the forward chaining engine, remembering variables bindings. If there is no result, the engine stops.

   e) Clean the working memory of the backward chaining engine.

3. In the forward chaining engine, get activations of the QUERYRULE. These activations contain facts that are results for a given query.

4. If there is a need for more sophisticated reasoning, load the OWL2Jess transformation script into the forward chaining engine and run it. The results are presented to the user as new triples.

Steps 1, 3 and 4 are executed in the forward reasoning engine, and step 2 in the backward reasoning engine. The Jess engine allows querying its working memory using a special function called *runQueryStar*. We decided to get facts from rule activation because it is the most efficient way to obtain Jess query results (according to the Jess implementation [22]).

For better understanding of the presented method, in Figure 4 we illustrate an example with the following query: 'Find companies that received invoices issued by company *Comp1* on product ID=10'.

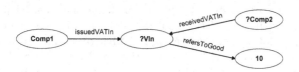

Fig. 4. An example of a query involving three triples

The query is written in Jess language:

```
(defquery ExampleQuery
    (triple (p "issuedVATIn") (s "Comp1") (o ?VIn))
    (triple (p "receivedVATIn") (s ?Comp2) (o ?VIn))
    (triple (p "refersToGood") (s ?VIn) (o 10)))
```

Our method is used to execute the query. In the first step, the QUERYRULE is created and added to the forward engine, so the querying process goes to the second step. The second step is executed three times because three relations in the query occurred. In this step the *(need-triple (p "issuedVATIn") (s "Comp1") (o ?VIn))* is asserted, and then the backward chaining engine is run. All results are copied to the forward chaining engine and bindings of the *?VIn* variable are remembered. Then the working memory of the backward chaining engine is cleared. The second step is executed again, but now the *(need-triple (p "receivedVATIn") (p ?Comp2) (o ?VIn))* facts are asserted with bindings of the *?VIn* variable (for example *(need-triple (p "receivedVATIn") (p ?Comp2) (o "8/2008")))*. When results are copied to the forward chaining engine the second step is executed again and the *(need-triple(p "refersToGood") (s ?VIn) (o 10))* are asserted with bindings of the *?VIn* variable.

After reasoning in the backward engine, the querying process goes to the third step. The example query is executed in the forward chaining engine and the query results are obtained. In the fourth step (if required) the OWL2Jess transformation script (or forward chaining script with SWRL and taxonomy rules) is loaded and additional information about gathered data is produced.

It is hard to define the computational complexity of our method, because the Rete algorithm which is used to reason is too complex to be described in general. That is because performance depends on declared rules and the data that is processed by them. However, the computational complexity of the querying method is between $O(RF^P)$ and $O(RFP)$, where R is the number of rules, F is the number of facts in the working memory, and P is the number of patterns per rule body. One can see that this complexity is the same as in the Jess reasoning engine. The computational complexity of the executed SQL queries should be added. Due to the simple form of queries and the Rete algorithm, the computational complexity of SQL queries can be skipped.

5   Example Use of the SDL Library

This section presents an example use of the SDL library. This demonstration is done using the crime ontology, the so-called minimal model [23,24], which has been developed in the PPBW [25] project. The ontology contains classes, object properties, datatype properties and SWRL rules. The ontology describes invoices, money transfers and tax information, and relations of decisive people to companies. We use data related to a real case of a criminal investigation of a fuel crime. Only some fragments of ontology and related queries are shown here. The case database contains information about 6500 money transfers which were done between 400 companies. The database schema contains one table named moneyTransfers with the following columns:

- ID - identifier of the money transfer in the table,
- senderAccount - number of sender account,
- sender - name of the sender,
- date - date of the money transfer,
- amount - amount of the money transfer,
- receiver - name of the receiver of the money,
- title - title of the money transfer,
- bank - bank from which the sender transferred money.

The processing procedure is as follows. First, we load the crime ontology into SDL-GUI. Next, we connect it to the relational database, and define a mapping between ontology concepts/relations and relational database schema. The following step in SDL-GUI is a generation of a Jess script in the backward chaining mode. The script is loaded into Jess engines using SDL-API. Finally, the analytical queries are executed with our query algorithm and hybrid reasoning process. These queries are shown as directed graphs in Figures 5, 6 and 7. The first query searches for a chain of different companies which transfer money. The chain's length is arbitrarily set at 6. The name of the first company in the chain is *TRAWLOLLEX*. Dates of transactions and the amounts of money transfers must occur in an increasing order. Existence of such a chain is a red flag for money laundering. The second query asks about a money transfer chain between companies, where *TRAWLOLLEX* is the first and the last company. Dates must occur in increasing order. Amounts must occur in decreasing order (due to provisions). Such a chain suggests that we probably found a VAT carousel fraud. The third query asks about money interchange between two such companies, where transferred amounts are exactly the same. Transaction dates must occur in an increasing order. Results for such a query suggest that a fictional flow of goods was found. We created and executed appropriate SQL queries. The performance results are shown in Table 1. The results show that our method is less efficient than SQL queries, but delivers an easier way for query creation. The most significant difference between SDL and SQL performance appears in the third query, because then the query contains only variables. In such a case, data concerning all concepts/relations occurring in the query need to be loaded as triples from a relational database. The efficiency of our method is satisfactory, and we are convinced it can be improved.

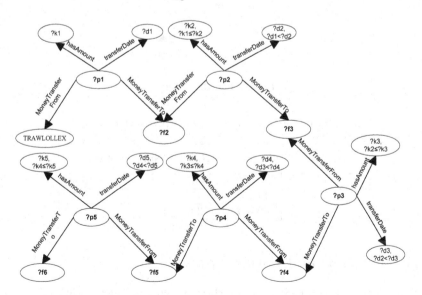

Fig. 5. The first query

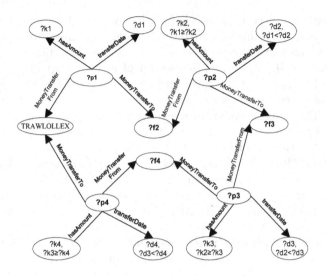

Fig. 6. The second query

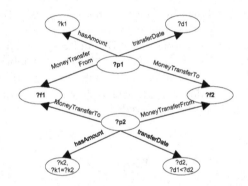

Fig. 7. The third query

Table 1. Performance comparison between SDL and SQL

Query / Tool	SDL [ms]	SQL [ms]	Number of results
1	33 359	3 075	15 464
2	14 156	473	711
3	49 257	295	230

6   Conclusions and Future Work

In summary, we presented an approach that extends functionality of the Jess reasoning engine and OWL2Jess transformational tool. The SDL library enables to query a relational database in terms of ontology concepts/relations and allows the integration of ontology, rules and database expressed in one format acceptable by the Jess engine. We believe that our query method, which enables complex queries to be created in a simple way, has advantages over SQL querying (creation of appropriate queries in SQL is more difficult). The SDL can also be used in expert systems, which require many rules and lots of data from relational databases. In this approach an answer is always up-to-date, because a query is executed on the current state of the relational database. This is a very useful feature, because we do not have to prepare data to begin an execution of the query (in contrast to the forward chaining method).

We are going to make SDL available online, under an Open Source license [26].

The presented approach has some significant limitations. Currently, the head of every rule contains only the *assert* command (it adds facts). We would like to be able to handle the *modify* and *retract* commands in the heads of the rules. We have to read data from a database and then we can test it (in the Jess terminology this means comparing values of variables). So we have to load some excess information. It would be better if we could test data during the load process and exclude data not fulfilling constraints.

We are working on extending our approach to use an OWL2Jess transformation in the backward chaining mode. In this case the transformation from OWL to Jess requires modifications which are needed to generate the Jess backward script (i.e. *do-backward-chaining* declaration etc.). It is possible that such an approach will reduce the efficiency of our method, but would enable more sophisticated reasoning about semantics (more than 'is-a' relation) in the backward chaining mode.

The SDL-GUI extensions are planned to support a query relational database using graphical interface to create a query in terms of ontology concepts/relations as a directed graph (as presented in Section 5).

A more formal approach than in our system, the ontology-based data access (OBDA), for forward reasoning, has been presented [26]. In this approach the DL-$Lite_A$ ontology is used to access a relational database and answer queries. The mapping between ontology concepts and relational database schema is done with the use of SQL queries (similar to our approach).

This work has been supported by the Polish Ministry of Science and Higher Education, Polish Technological Security grant 0014/R/2/T00/06/02 and by 45-083/08/DS grant.

References

1. Papataxiarhis, V., Tsetsos, V., Karali, I., Stamatopoulos, P., Hadjiefthymiades, S.: Developing Rule-Based Applications for the Web: Methodologies and Tools. In: Handbook of Research on Emerging Rule-Based Languages and Technologies: Open Solutions and Approaches, May 2009, ch. XVI. Information Science Reference (2009)
2. SPARQL Query Language for RDF. In: Seaborne, A., Prud'hommeaux, E. (eds.) W3C Recommendation, January 15 (2008)

3. Calvanese, D., De Giacomo, G., Lenzerini, M.: Conjunctive query containment and answering under description logics constraints. ACM Transactions on Computational Logic 9(3) (2008)
4. Falkowski, M., Jedrzejek, C.: An efficient SQL-based querying method to RDF schemata. In: KKNTPD Conference, Poznan, pp. 162–173 (2007)
5. Microsoft SQL Server, http://www.microsoft.com/sql/default.mspx
6. OWL Web Ontology Language Reference. In: Dean, M., Schreiber, G. (eds.) W3C Recommendation, February 10 (2004),
   http://www.w3.org/TR/2004/REC-owl-ref-20040210/
7. SWRL - Semantic Web Rule Language, http://www.w3.org/Submission/SWRL/
8. Jess (Java Expert System Shell), http://jessrules.com/
9. Bak, J., Jedrzejek, C.: Querying relational databases using ontology, rules and Jess reasoning engine. Studies in Automation and Inf. Technology, vol. 33, pp. 25–44 (2008)
10. Mei, J., Paslaru Bontas, E., Lin, Z.: OWL2Jess: A Transformational Implementation of the OWL Semantics. In: Chen, G., Pan, Y., Guo, M., Lu, J. (eds.) ISPA-WS 2005. LNCS, vol. 3759, pp. 599–608. Springer, Heidelberg (2005)
11. OWL2Jess, http://www.ag-nbi.de/research/owltrans/
12. Mei, J., Paslaru Bontas, E.: Reasoning Paradigms for OWL Ontologies, FU Berlin, Fachbereich Informatik, Technical Reports B 04-12 (2004)
13. Mei, J., Paslaru Bontas, E.: Reasoning Paradigms for SWRL-enabled Ontologies, Protégé With Rules, Workshop, Madrid (2005)
14. OWL Meta-model, http://www.ag-nbi.de/research/owltrans/owlmt.clp
15. O'Connor, M.J., Shankar, R.D., Tu, S., Nyulas, C.I., Das, A.K.: Developing a Web-Based Application using OWL and SWRL. In: Conference Proceedings, AAAI Spring Symposium, Stanford, CA, USA (2008)
16. O'Connor, M.J., Tu, S.W., Das, A.K., Musen, M.A.: Querying the Semantic Web with SWRL. In: Paschke, A., Biletskiy, Y. (eds.) RuleML 2007. LNCS, vol. 4824, pp. 155–159. Springer, Heidelberg (2007)
17. Protégé (ed.): http://protege.stanford.edu/
18. D2RQ, http://www4.wiwiss.fu-berlin.de/bizer/D2RQ/
19. KAON2, http://kaon2.semanticweb.org/
20. Charles, F.: Rete: A Fast Algorithm for the Many Pattern/Many Object Pattern Match Problem. Artificial Intelligence 19, 17–37 (1982)
21. SWRL Built-ins, http://www.w3.org/Submission/2004/SUBM-SWRL-20040521/
22. Friedman-Hill, E.: Jess in Action. Manning Publications Co., (2003)
23. Jedrzejek, C., Bak, J., Falkowski, M.: Graph Mining for Detection of a Large Class of Financial Crimes. In: 17th International Conference on Conceptual Structures, Moscow, Russia, July 26-31 (2009)
24. Jedrzejek, C., Cybulka, J., Bak, J.: Application Ontology-based Crime Model for a Selected Economy Crime. In: CMS 2009, Cracow (2009) (to be published)
25. PPBW, the Polish Platform for Homeland Security,
   http://www.ppbw.pl/en/index.html
26. Open Source, http://www.opensource.org/licenses
27. Poggi, A., Lembo, D., Calvanese, D., De Giacomo, G., Lenzerini, M., Rosati, R.: Linking Data to Ontologies. J. on Data Semantics 10, 133–173 (2008)

An XML-Based Manipulation and Query Language for Rule-Based Information

Essam Mansour and Hagen Höpfner

International University in Germany
Campus 3, D-76646 Bruchsal, Germany
essam.mansour@ieee.org, hoepfner@acm.org

Abstract. Rules are utilized to assist in the monitoring process that is required in activities, such as disease management and customer relationship management. These rules are specified according to the application best practices. Most of research efforts emphasize on the specification and execution of these rules. Few research efforts focus on managing these rules as one object that has a management life-cycle. This paper presents our manipulation and query language that is developed to facilitate the maintenance of this object during its life-cycle and to query the information contained in this object. This language is based on an XML-based model. Furthermore, we evaluate the model and language using a prototype system applied to a clinical case study.

1 Introduction

Several applications of information systems utilize monitoring processes to support their activities. Examples include health care applications (i.e., disease and medical-record management) and financial applications (i.e., customer relationship and portfolio management). These information systems are "standardized" by best practices, which refer to the best way to perform specified activities [13]. Information extracted from the best practices is specified in form of rules as a pre-step for monitoring the changes of interest in these applications.

We developed a framework [6,9] for managing best practices. In our framework, the best practices are modeled as a skeletal plan, which contains sets of rules defined using the user terminologies. As object-oriented model, the skeletal plan is similar to a class, from which several objects could be instantiated. The instance of the skeletal plan is called an entity-specific (ES) plan, in which the rules of the skeletal plan are mapped into low-level rules such as SQL triggers. For short, we refer to the skeletal and entity-specific plans as *rule-based information*. The entity-specific plan go through a life-cycle, in which the plan is created from a particular skeletal plan, activated, deactivated, terminated or completed. The skeletal and entity-specific plans are modeled using AIMSL (Advanced Information Management Specification Language) [7,11] and DRDoc (Dynamic Rule-Based Document) [10], respectively.

The execution history of the *rule-based information* represents several information scenes. The ability of manipulating and querying these scenes enhances the reporting and decision-support capabilities in organizations. This ability facilitates the information analysis and mining to discover and understand information trends.

G. Governatori, J. Hall, and A. Paschke (Eds.): RuleML 2009, LNCS 5858, pp. 231–245, 2009.

There is a need to move the complexity of manipulating and querying the *rule-based information* and its execution history from user/application code to a high level declarative language. We developed a language called AIMQL (Advanced Information Management Query Language) to facilitate the management of the *rule-based information* as a first-class object. The main functional requirements of AIMQL are to assist in: 1) Manipulating the skeletal plan and ES plan. The changes are made to AIMSL specification might be required to be propagated to the corresponding ES plan; and 2) Retrieving this information. This includes the ability to replay the ES plan or a specific part of it within a specific time period.

This paper presents the manipulation support of the AIMQL language and discusses the evaluation of a proof-of-concept system, which implements AIMQL using XML technologies and database utilities. A clinical case study is used in our experiments.

The reminder of this paper is organized as follows: Section 2 discusses the related work. Section 3 presents the management life-cycle of and examples for the *rule-based information*. Section 4 outlines the manipulation and query requirements. Section 5 presents our manipulation support for the *rule-based information*. Section 6 highlights our proof-of-concept system and discusses experimental results. Section 7 concludes the paper and gives an outlook on future research.

2   Related Work

In the area of active XML, an event-driven mechanism based on the Event-Condition-Action (ECA) rule paradigm [14] is incorporated into XML to provide an advanced active behavior. We have classified the languages, which have been produced in this area, into three categories: The languages in the *first category* play the same role as the high level SQL trigger standard, such as Active XQuery [4], or an Event-Condition-Action language for XML [3]. These languages support the reactive applications at the level of rules and triggers. The languages in the *second category*, such as AXML [1], Active XML Schemas [16], and XChange [2], utilize the event-driven mechanism to support a specific reactive application, such as the Web content management. The languages of the *third category* use XML only to standardize the rule-based information as individual rules, such as ARML [5].

The RuleML languages aim at providing a standard rule language that is interoperability platform [17,18]. The RuleML language has been utilized to support semantic Web and business applications, such as in [12,15]. However, the RuleML languages formalize the *rule-based information* as individual rules, not as a unified distinct entity that could be instantiated. Languages proposed in both areas, active XML and RuleML, overlook the need to manipulate the *rule-based information* and keep its evolution history as a pre-step to analyze and mine the execution history of such information.

The maintenance support management presented in this paper is part of the AIM language, which has been developed by Mansour [6,7] for specifying, instantiating and maintaining *rule-based information*. The AIM language is based on XML and ECA rule paradigm, and has been implemented using DBS utilities to support the advanced management required for the *rule-based information*. Our maintenance support provides a high-level method for manipulating the *rule-based information*.

3   Modeling the Rule-Based Information

In our framework, the *rule-based information* is either a skeletal plan or an entity-specific (ES) plan. A skeletal plan is static in the sense that it does not have a state transition. An ES plan is dynamic in the sense that it has state transitions. An example is the generic specification of the test ordering protocol developed for diabetic patients, from which several patient plans are generated to suit particular patients. Patient plans change over the time. This section focuses on modeling the maintenance of the ES plan.

3.1   The Management Life-Cycle

The state transitions of the ES plan, as shown in Figure 1, are predefined and context-sensitive. The context-sensitive means that the ES plan's state is affected by changes in the application information, such as increasing the patient temperature. These state transitions are applied to the ES plan and its *knowledge action* component, which represents sets of modularized ECA rules.

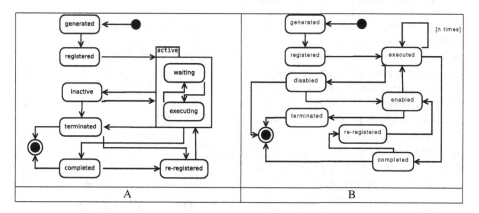

Fig. 1. The life-cycle of A) an entity-specific (ES) plan and b) an ES plan rule

When the ES plan is generated from the skeletal plan, the ES plan and its sets of rules go from the initial state into *generated* state (see Figure 1.A). In *generated* state, the ES plan is not yet a subject to execution, it should be firstly authorized to be then registered. The ES plan is authorized by an domain expert, who is in charge of the ES plan. Once it is authorized, the ES plan and its sets of rules go into the *registered* state. In the registered state, all rules of the ES plan are installed in the system. In this state, no rule has fired yet.

On the first occurrence of an event of interest to one or more of the ES plan's rules, the ES plan goes into the *active* state, and one or more rules are fired and go into the *executed* state. As shown in Figure 1.A the *active* state includes two sub-states: *waiting* and *executing*. In the *waiting* state, all ES plan's rules are waiting for events that are of interest to them. In the *executing* state, at least one rule is being executed. Once the rule execution completes, the ES plan returns to the *waiting* state. Between the *waiting* and *executing* states of the ES plan, the rules are considered to be executed, as shown in Figure 1.B . The *executed* state is a state for the rules. On this state, a rule is being executed and after the execution the rule is waiting until the next event occurrence of interest. The ES plan might be transited from *active* state to *inactive*, *terminated*, or *completed* states, as shown in Figure 1.A.

The *inactive* state means that all the ES plan rules become disabled. The ES plan might be transited from *inactive* state to *active* state. That means enabling the rules of the ES plan. The *terminated* state means that all the ES plan rules were removed from the system, but are not removed from the ES plan itself. The *completed* state of a rule means that the execution of the rule was successful and the rule will not be subject to any further execution. After all the enabled rules in the ES plan were completed the ES plan goes into the *completed* state. The *completed* state of the ES plan could be determined by a domain user, who is in charge of the ES plan. After the ES plan went to the *completed* state, all the ES plan rules are removed from the system. An ES plan could be re-registered, after it had been terminated or completed.

3.2   Instantiation and Execution History

This sub-section discusses the ES plan instantiation and execution history using an example. We assume that there is a skeletal plan for diabetic patients, which consists of two rules MAP_1 and MAP_2 grouped at the same schedule. The rule MAP_1 is to be fired two hours after patient admission to order an Albumin/Creatinine Ratio (ACR) test for the patient. The rule MAP_2 is to be fired once the result of the ACR test is received. If the result is greater than 25 then the ACR test is repeated every two days after patient admission, as an action. This

```
-<protocol id="PRO124">
    <name>microalbuminuria protocol (MAP) </name>
    <categoryID>CID124</categoryID>
    +<header>
    -<Schedules>
        -<schedule id="SIDMAP">
            <name>Basic MAS</name>
            +<header>
            -<scheduleRules>
                +<rule id="MAP1">
                +<rule id="MAP2">
            </scheduleRules>
        </schedule>
    </Schedules>
</protocol>
```

Fig. 2. AIMSL specification

action is formalized as a new rule (MAP_3). Figure 2 illustrates the AIMSL specification of the skeletal plan, which has the ID *PRO124*, and belongs to the category, whose ID is *CID124*.

For each diabetic patient, a patient plan is generated from the generic plan shown in Figure 2 by customizing the rules to a particular patient, i.e. using the admission time of the patient. It is assumed that the patient plans were registered at time point 1, and the result of the ACR test is received three hours after patient admission. The result was 33 and is greater than 25.

According to this scenario, the patient plan was maintained as shown in Figure 3 that illustrates part of the patient plan four hours after the patient admission. This patient plan is modeled using DRDoc [10]. The rule MAP_1 and MAP_2 were generated at time point zero and registered at time point 1. The *generated* status is a system-defined status that happens at the generation time of an entity-specific plan. The rule MAP_1 was fired two hours after patient admission. Therefore, the status *registered* of rule MAP_1

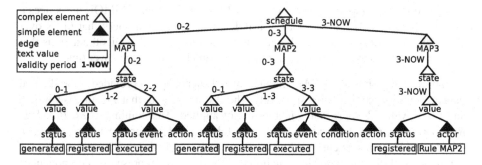

Fig. 3. A part of a patient plan modeled using DRDoc

was valid from 1 to 2. The status *executed* was added with validity period 2 to 2. The actual evaluation of the *event* and *action* of MAP_1 were recorded. The rule MAP_2 was fired three hours after patient admission. Therefore, the status *registered* of rule MAP_2 was valid from 1 to 3. The status *executed* was added with validity period 3 to 3. The evaluation of the *event, condition*, and *action* of MAP_2 were recorded. The rule MAP_3 was added at time point 3 and it is recorded that MAP_2 caused such modification, as represented by the *actor* element shown in Figure 3.

4   The Manipulation and Query Requirements

This section presents main requirements and functionalities of the Advanced Information Management Query Language (AIMQL). The *rule-based information* and the execution history are represented and stored as XML-based document, such as the AIMSL specification or DRDoc documents that are presented in Section 3. AIMSL and DRDoc are compatible with the XML model and could be queried using XQuery. However, maintaining these documents demands special manipulation and query operators.

The general functional requirements, which should be provided, are: 1) **Declarativity**, the AIMQL language should be independent of any particular platform or query evaluation strategy; 2) **Temporal Support**, the AIMQL language should be able to record the history of executing the ES plan and to query it; and 3) **XQuery-based**, the AIMSL specification and DRDoc are represented as XML document. Therefore, AIMQL should be based on XQuery. Several extensions to XQuery are required in order to achieve the AIMQL requirements as the following:

 – **Manipulation Operations:** AIMQL introduces seven manipulation operations (expressions). These expressions include add, remove, modify, activate, deactivate, terminate and fire. The AIMQL manipulation operations are distinguished in the sense that they do not only potentially modify the AIMSL specification or ES plan, but also propagate the modification to the corresponding ES plan documents and modify the corresponding triggers created in the system. Furthermore, the manipulation expressions log the changes occurring to ES plan documents; and
 – **Query Support:** AIMQL provides support to query AIMSL specification and ES plan document, as the domain information, plus special query capabilities, replay

function and temporal query support for ES plan document. AIMQL introduces a
new functionality called replay. AIMQL replay query is a query that plays over
again the history of the rule-based information to show in details the actions that
cause changes on the rule-based information and how it evolved over time.

Tables 1 and 2 show the AIM manipulation and query support provided to the skeletal
plan and entity-specific plan, respectively. The √ value denotes that a feature is applied,
and the × value denotes that a feature is not applied. The columns (*Cat, Pro, Sch,
Rule, Trm, Eve, Con, Act,* and *Ent*) shown in Tables 1 and 2 refer to category, protocol,
schedule, rule, terms, event, condition, action, and domain entity. These are the main
components of the rule-based information [6]. Each column represents a component of
either the skeletal or entity-specific plan.

Table 1. AIMQL function applicability for the skeletal plan

Category	Function	Cat	Pro	Sch	Rule	Trm	Eve	Con	Act
					Skeletal Plan				
Manipulation	Add	√	√	√	√	√	√	√	√
	Remove	√	√	√	√	√	√	√	√
	Modify	√	√	√	√	√	√	√	√
	Activate	×	×	×	×	×	×	×	×
	Deactivate	×	×	×	×	×	×	×	×
	Terminate	×	×	×	×	×	×	×	×
	Fire	×	×	×	×	×	×	×	×
Query	Normal	√	√	√	√	√	√	√	√
	Replay	×	×	×	×	×	×	×	×

Table 2. AIMQL functions applicability for the entity-specific plan

Category	Function	Ent	Pro	Plan	Sch	Rule	Trm	Eve	Con	Act
Manipulation	Add	×	×	×	√	√	√	√	√	√
	Remove	×	×	√	√	√	√	√	√	√
	Modify	×	×	×	√	√	√	√	√	√
	Activate	×	×	√	√	√	×	×	×	×
	Deactivate	×	×	√	√	√	×	×	×	×
	Terminate	×	×	√	√	√	×	×	×	×
	Fire	×	×	×	×	√	×	×	×	×
Query	Normal	√	√	√	√	√	√	√	√	√
	Replay	×	×	√	√	√	×	×	×	×

For the skeletal plan, the *add, remove* and *modify* operations are applied to all
skeletal plan components. However, the *activate, deactivate, terminate* and *fire* operations are used to facilitate the execution of the entity-specific plan. Therefore, these
operations are not used with the skeletal plan components, but used with the *plan,
schedule* and *rule* components of the entity-specific plan. The *fire* operation is used

only with the *rule* component. The entity-specific plan is generated for a specific domain entity from a specific protocol (skeletal plan). The domain entity and protocol of the entity-specific plan are not changeable. Therefore, the *add, remove* and *modify* operations are not applied to the domain entity nor the protocol components. Moreover, the *add* and *modify* operations are not applied to the *plan* component.

This research work focuses on the execution history of the entity-specific plan. Consequentially, the AIMQL replay query is provided to the entity-specific plan, specially the components (*plan, schedule* and *rule*) that are called re-playable components. The other components of the entity-specific plan could be replayed as a part of the replayable components. The reader is refered to [10] for more details concerning our replay support.

5   The High-Level Manipulation Operations

The manipulation operations are applied to the skeletal plan, entity-specific plan and its corresponding triggers created as an implementation for the execution process of this plan. The changes made to the skeletal plan or the entity-specific plan might need to be propagated to the corresponding plan or triggers, respectively. The manipulation operations could be issued in the action component associated with the AIMSL *rule* element. The supported manipulation operations are:

- **Add** a skeletal plan (protocol), entity-specific plan, or one of their components.
- **Remove** a protocol, entity-specific plan, or one of their components.
- **Modify** a protocol, entity-specific plan, or one of their components.
- **Activate** an entity-specific plan, schedule, or rule components.
- **Deactivate** an entity-specific plan, schedule, or rule components.
- **Terminate** an entity-specific plan, schedule, or rule components.
- **Fire** a rule component.

5.1   Add

The *add* operation is an manipulation operation that add copies of one or more protocol specification or ES plan components into a designated position with respect to a target component. Figure 4.A shows the XML schema of the *add* operation as follows:

- The `AddedExpr` represents one of the protocol (skeletal plan) or ES plan components.
- The `as` value could be one of this values (Category, Protocol, Schedule, Terms, Event, Condition, Action, or domain entity), or the values (schedule rule, protocol rule or global rule).
- the `AddedTargetExpr` represents a targeted component in a specific protocol or ES plan.
- If `into` is specified without *Before* or *After*, `AddedExpr` becomes a child of the `AddedTargetExpr`. Else, `AddedExpr` becomes a child of the parent of `AddedTargetExpr`.

A	B

`<xsd:complexType name="addDT">`
  `<xsd:sequence>`
    `<xsd:element name="addedExpr" />`
    `<xsd:element name="as" />`
    `<xsd:element name="into">`
      `<xsd:complexType>`
        `<xsd:sequence>`
          `<xsd:element name="posBA"/>`
          `<xsd:element name="AddedTargetExpr"/>`
        `</xsd:sequence>`
      `</xsd:complexType>`
    `</xsd:element>`
    `<xsd:element name="propagation" minOccurs="0"/>`
  `</xsd:sequence>`
`</xsd:complexType>`
	`<add">`
	  `<addedExpr>`
	    `+<rule id="rul123">`
	  `</addedExpr>`
	  `<as>scheduleRule</as>`
	  `<into>`
	    `<AddedTargetExpr>`
	    `protocol[id="pro123"]//schedule[id="sch123"]`
	    `</AddedTargetExpr>`
	  `</into>`
	`</add">`

Fig. 4. A: the structure of the add operation. B: an example for add operation.

- the `propagation` values are (Yes or No), and the default value is No. The `Added Expr` will not be propagated to the corresponding ES plans, if the value is No. If the value is Yes, the `AddedExpr` will be propagated to all the corresponding plans.

The semantics of an *add* expression are as follows:

- `AddedExpr` must be a valid AIMSL component for the protocol or ES plan; otherwise a static error is raised. The result of this step is either an error or a sequence of components to be added.
- `AddedTargetExpr` must refer to a valid AIMSL component; otherwise a static error is raised.
- The result of the *add* expression must be a valid AIMSL component for a protocol or ES plan; otherwise a dynamic error is raised.
- If the *add* expression is applied for a plan, the validity period associated with the `AddedTargetExpr` and its children should be changed to reflect the new changes that have been made by the *add* expression.

Figure 4.B shows an example for an *add* operation that adds a rule as a schedule rule under the schedule, whose id is *sch123* and its parent is a protocol, whose id is *pro123*. This rule will not be propagated as the default value of the propagation is No.

5.2 Remove

A *remove* expression removes at least one of AIMSL components from a protocol or ES plan. Figure 5.A shows the syntax of a *remove* expression as follows:

- The `RemovedTargetExpr` refers to one of the protocol or ES plan components.
- the `propagation` values are (Yes or No), and the default value is No.
  The `RemovedTargetExpr` will not be propagated to the corresponding ES plans, if the value is No. Otherwise, it will be propagated to all corresponding plans.

< xsd:complexType name="removeDT"> < xsd:sequence> < xsd:element name="RemoveedTargetExpr"/> < xsd:element name="propagation" minOccurs="0"/> </xsd:sequence> < /xsd:complexType>	< remove"> < RemoveedTargetExpr> protocol[id="pro123"]//schedule[id="sch123"]//rule[id="rul123"] < /RemoveedTargetExpr> < propagation> Yes < /propagation> < /remove">
A	B

Fig. 5. A: the structure of the remove operation. B: an example for a remove operation.

The semantics of a *remove* expression are as follows:

- The `RemovedTargetExpr` must refer to a valid AIMSL component; otherwise a static error is raised.
- After removing the `RemovedTargetExpr`, the parent of the removed component must be a valid AIMSL component or null, otherwise a dynamic error is raised.
- If the *remove* expression is applied for an ES plan component, the `RemovedTarget Expr` is logically removed. That means the component is not deleted, but it is marked as a deleted component. Also, the validity period associated with the parent of `RemovedTargetExpr` should be changed to reflect the new changes that have been made by the *remove* expression.

Figure 5.B shows an example for a *remove* operation that removes a rule, whose id is *rul123* and its schedule id is *sch123*. This schedule is under a protocol, whose id is *pro123*. This *remove* operation will be propagated as the propagation value is Yes.

5.3 Modify

A *modify* operation might modify a component as a whole or only the values. Figure 6.A shows the syntax of the *modify* operation as follows:

< xsd:complexType name="modifyDT">
   < xsd:sequence>
     < xsd:element name="value-of" minOccurs="0"/>
     < xsd:element name="ModifyTargetExpr"/>
     < xsd:element name="with"/>
     < xsd:element name="propagation" minOccurs="0"/>
   < /xsd:sequence>
 < /xsd:complexType>
	< modify>
	   < ModifyTargetExpr>
	   protocol[id="pro123"]//rule[id="rul123"]//event[id="EID123"]
	   < /ModifyTargetExpr>
	   < with>
	     +¡event id="EID127"¿
	   < /with>
	   < /modify>

A	B

Fig. 6. A: the structure of the modify operation. B: an example for a modify operation.

- The `value-of` element determines whether the *modify* operation updates a value or a component.
- The `ModifyTargetExpr` element represents a targeted component in a specific protocol or ES plan.
- The `with` element represents a protocol or ES plan components or a valid value for a protocol or ES plan components.
- the `propagation` values are (Yes or No), and the default value is No. The *modify* operation will not be propagated to the corresponding ES plans, if the value is No. If the value is Yes, it will be propagated to all corresponding plans, if applicable.

Modify Component. If the `value-of` element is not specified, the *modify* operation modifies one valid AIMSL component with a new valid AIMSL component. The semantics of this form of the *modify* operation are as follows:

– The `ModifyTarggetExpr` must refer to a valid AIMSL component; otherwise a static error is raised. The `ModifyTarggetExpr` is evaluated. The result of this step is either an error or a sequence of component to be modified.
– The `with` element must be a valid AIMSL component; otherwise a static error is raised.
– The result of the modify expression must be a valid AIMSL component.
– If the *modify* operation is applied for a plan, instead of modifying the component targeted by `ModifyTarggetExpr`, a copy of this component will be modified by the *with* element and added as a sibling to the `ModifyTarggetExpr`. Also, the validity period associated with the `ModifyTarggetExpr` should be changed to reflect the new changes that have been made by the *modify* operation.

Modify the Value of a Component. If the `value-of` is specified, the *modify* operation modifies only the value of a valid AIMSL component. The semantics of this form of the *modify* operation are as follows:

– The `ModifyTarggetExpr` must refer to a valid AIMSL component that does not contain another component; otherwise a static error is raised.
– The `ModifyTarggetExpr` is evaluated. The result of this step is either an error or a sequence of components to be modified.
– The `with` element must be a valid value for the *ModifyTarggetExpr* according to AIMSL Schema; otherwise a static error is raised.
– The result of the modify expression must be a valid AIMSL component.

Figure 6.B shows a *modify* operation that replaces the event, whose id is *EID123*. This event is under a rule, whose id is *rul123*, and the rule's parent is the protocol, whose id is *pro123*.

5.4   Activate

The *activate* operation activates a *plan, schedule* or *rule* component in a specific ES plan. This means these components will be ready for the execution process. Figure 7.A shows the syntax of the *activate* operation. Figure 7.B shows an example for activating a rule, whose id is *rul123*, in an ES plan, whose *proid* is *pro123*. The semantics of *activate* are :

– The `ActTargetExpr` element must refer to a valid re-playable component (*plan, schedule* or *rule*), or to a component containing at least one of these components, such as the *scheduleRules* component.
– As a result to the *activate* operation, the state of the activated component will be transited to the *active* state, and the corresponding triggers will be activated.

A	B
<xsd:complexType name="activateDT"> <xsd:sequence> <xsd:element name="ActTargetExpr"/> </xsd:sequence> </xsd:complexType>	<activate> <ActTargetExpr> plan[proid="pro123"]//rule[id="rul123"] </ActTargetExpr> </activate">

Fig. 7. A: the structure of the activate operation. B: an example for an activiate operation

5.5 Deactivate

The *deactivate* operation deactivates a *plan, schedule* or *rule* component in a specific ES plan. This means these components will be off. Figure 8.A shows the syntax of the *deactivate* operation. Figure 8.B shows an example for deactivating a rule, whose id is *rul123*, in a plan, whose *proid* is *pro123*. The semantics of Deactivate are :

- The DeactTargetExpr element must refer to a valid re-playable component (*plan, schedule* or *rule*), or to a component containing at least one of these components, such as the *scheduleRules* component.
- As a result to the *deactivate* operation, the state of the deactivated component will be transited to the *inactive* state, and the corresponding triggers will be deactivated in the system.

A	B
<xsd:complexType name="deactivateDT"> <xsd:sequence> <xsd:element name="DeacTargetExpr"/> </xsd:sequence> </xsd:complexType>	<deactivate> <DeacTargetExpr> plan[proid="pro123"]//rule[id="rul123"] </DeacTargetExpr> </deactivate>

Fig. 8. A: the structure of the deactivate operation. B: an example for a dectiviate operation

5.6 Terminate

The *terminate* operation halts a *plan, schedule* or *rule* component in a specific ES plan. This means these components will be not in use anymore. Figure 9.A shows the syntax of the *terminate* operation. Figure 9.B shows an example for terminating a rule, whose id is *rul123*, in a plan, whose *proid* is *pro123*. The termination semantics are:

- The TermTargetExpr element must refer to a valid re-playable component (*plan, schedule* or *rule*), or to a component containing at least one of these components, such as the *scheduleRules* component.
- As a result to the *terminate* operation, the state of the terminated component will be transited to the *terminated* state, and the corresponding triggers will be deleted from the system.

<xsd:complexType name="terminateDT"> <xsd:sequence> <xsd:element name="TermTargetExpr"/> </xsd:sequence> </xsd:complexType>	<terminate> <TermTargetExpr> plan[proid="pro123"]//rule[id="rul123"] </TermTargetExpr> </terminate>
A	B

Fig. 9. A: the structure of the terminate operation. B: an example for a terminate operation

5.7 Fire

The *fire* operation is applying only to the *rule* component in a specific plan. This means the rule's action will be carried out if the rule condition is evaluated to true. Figure 10.A shows the syntax of the *fire* operation. Figure 10.B shows an example for firing a rule, whose id is *rul123*, in a plan, whose *proid* is *pro123*. The semantics of *fire* are:

- The FireTargetExpr element must refer to a valid rule component in an ES plan.
- As a result to the *fire* operation, the corresponding triggers will be activated regardless their event.

<xsd:complexType name="fireDT"> <xsd:sequence> <xsd:element name="FireTargetExpr"/> </xsd:sequence> </xsd:complexType>	<fire> <FireTargetExpr> plan[proid="pro123"]//rule[id="rul123"] </FireTargetExpr> </fire>
A	B

Fig. 10. A: the structure of the fire operation. B: an example for a fire operation.

6   Evaluation

This section highlights the proof-of-concept system, which implements our maintenance support, and the evaluation of the maintenance efficiency.

6.1   A Prototype System

We have utilized DB2 express-C 9.5 and Sun Java 1.6, to develop a proof-of-concept system, called *AIMS* [8], for managing the rule-based information. AIMS maps the AIMQL queries and operations into XQuery scripts that are to be executed by DB2.

The conceptual architecture of *AIMS* is illustrated in Figure 11. The main components of *AIMS* are the *Complex Information Manager, Rule Manager, Information Manager*, and *Communication Manager*. The *Complex Information Manager* supports the management of the rule-based information at a high level. The domain users and information providers, such as patient information systems, deal with *Complex Information Manager* through the *Communication Manager*. The *Information Manager* extends the XML support provided by DBSs to provide temporal support and utilizes the DBS to validate and store the DRDoc document.

The *Rule Manager* extends the triggering mechanism of the DBS to support the advanced features of the *rule-based information*, and recording the execution history. One of the main roles of *Rule Manager* is to map the AIMQL operations into XQuery scripts. The translator knows the structure (elements and attributes) of the AIMSL and DRDOC documents. The translator generates the equivalent XQuery script, which is to be executed using the XQuery engine provided within DB2.

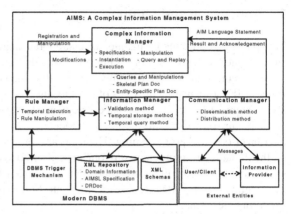

Fig. 11. The AIMS conceptual architecture

6.2 Experimental Results

Our evaluation to the maintenance efficiency has been tested on Debian 4, a Linux system, and an Intel Pentium III processor machine, whose configuration is one Gigabyte RAM and 40 Gigabyte hard disk. The DRDOC document is a temporal XML document that records all the changes produced by updating the DRDOC document. Most of these changes add a new state to an element of the DRDOC document. For example, executing the rule MAP_3 every two days adds a new *executed* state under the *rule* element. These changes might be also adding a new rule, such as MAP_2 that adds a new rule, MAP_3. Consequentially, the storage management of the DRDOC document is of critical importance and the main factor of the AIMS storage management performance.

This experiment compares the size of the DRDOC document with the number of updates that take place in them. The growing in the plan size is almost linear to the number of updates, as shown in Figure 12. This linearity assists in estimating the DR-DOC document size after N number of updates, such that most of the updates are changes on the rule state. This change is represented using a *value* element of almost fixed size, see Figure 3. In conclusion, the AIMS storage management is stable to the number of updates.

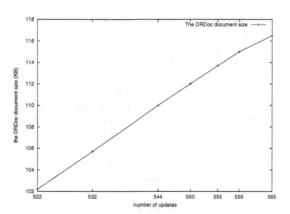

Fig. 12. The maintenance efficiency

7 Conclusion and Outlook

This paper has presented a maintenance support management for information formalized as rules to assist in monitoring processes that are required in activities, such as disease management, and customer relationship management. We have developed a framework for managing the rules of monitoring processes as one object (plan) that has a management life-cycle. This paper focussed on the modeling mechanism for these rules. In our framework, these rules are modeled similary to the class-object concept in the object-oriented programming. The rules are to be specified at a generic level based on the domain terminologies using our AIMSL model. From the AIMSL specification, several entity-specific plans could be generated. Our DRDOC model maps the generic rules defined in the AIMSL specification into SQL triggers, and keeps the history of executing these rules.

The paper also has presented the manipulation support of AIMQL language. AIMQL provides manipulation operations, such as activate, deactivate, terminate and fir, to support the management life-cycle of rules modeled using our framework. The AIMQL is distinguished by the ability to specifying declaratively the manipulated information. Furthermore, the AIMQL is able to manage the *rule-based information* as first class object. Moreover, the paper highlighted our proof-of-concept system (AIMS) that implements our maintenance support by mapping the AIMQL operations into XQuery update scripts that are to be executed by the DBS. Our evaluation of the maintenance efficiency shows positive results.

Currently we are doing additional experiments with different workloads and query sets. There is a need to develop a method that provides automatic discovery of information from the execution history of the *rule-based information*. This discovered information can assist in auditing, analyzing and improving already enacted *rule-based information*.

References

1. Abiteboul, S., Benjelloun, O., Manolescu, I., Milo, T., Weber, R.: Active XML: A Data-Centric Perspective on Web Services. In: Levene, M., Poulovassilis, A. (eds.) Web Dynamics - Adapting to Change in Content, Size, Topology and Use, pp. 275–300. Springer, Heidelberg (2004)
2. Bailey, J., Bry, F., Eckert, M., Pătrânjan, P.-L.: Flavours of XChange, a Rule-Based Reactive Language for the (Semantic) Web. In: Adi, A., Stoutenburg, S., Tabet, S. (eds.) RuleML 2005. LNCS, vol. 3791, pp. 187–192. Springer, Heidelberg (2005)
3. Bailey, J., Poulovassilis, A., Wood, P.T.: An Event-Condition-Action Language for XML. In: Proceedings of the 11th international conference on World Wide Web, pp. 486–495. ACM, New York (2002)
4. Bonifati, A., Braga, D., Campi, A., Ceri, S.: Active XQuery. In: Proceedings of the 18th International Conference on Data Engineering, p. 403. IEEE Computer Society, Washington (2002)
5. Cho, E., Park, I., Hyun, S.J., Kim, M.: ARML: an active rule mark-up language for heterogeneous active information systems. In: Schroeder, M., Wagner, G. (eds.) Proceedings of the International Workshop on Rule Markup Languages for Business Rules on the Semantic Web, CEUR Workshop Proceedings, vol. 60. CEUR-WS.org. (2002)

6. Mansour, E.: A Generic Approach and Framework for Managing Complex Information. PhD thesis, Dublin Institute of Technology, DIT (2008), http://arrow.dit.ie/sciendoc/51/
7. Mansour, E., Dube, K., Wu, B.: AIM: An XML-Based ECA Rule Language for Supporting a Framework for Managing Complex Information. In: Paschke, A., Biletskiy, Y. (eds.) RuleML 2007. LNCS, vol. 4824, pp. 232–241. Springer, Heidelberg (2007)
8. Mansour, E., Dube, K., Wu, B.: Managing complex information in reactive applications using an active temporal XML database approach. In: Cardoso, J., Cordeiro, J., Filipe, J. (eds.) ICEIS 2007 - Proceedings of the Ninth International Conference on Enterprise Information Systems, pp. 520–523 (2007)
9. Mansour, E., Höpfner, H.: A rule-based approach and framework for managing best practices – An XML-Based Management Using Puer Database System Utilities. In: Filipe, J., Cordeiro, J. (eds.) ICEIS 2009. LNBIP, vol. 24. Springer, Heidelberg (2009)
10. Mansour, E., Höpfner, H.: Replay the Execution History of Rule-Based Information. In: DBKDA 2009: Proceedings of the 2009 First International Conference on Advances in Databases, Knowledge, and Data Applications, pp. 28–35. IEEE Computer Society, Washington (2009)
11. Mansour, E., Wu, B., Dube, K., Li, J.X.: An Event-Driven Approach to Computerizing Clinical Guidelines Using XML. In: Proceedings of the IEEE Services Computing Workshops, pp. 13–20. IEEE Computer Society, Washington (2006)
12. Nagl, C., Rosenberg, F., Dustdar, S.: VIDRE–A Distributed Service-Oriented Business Rule Engine based on RuleML. In: Proceedings of the 10th IEEE International Enterprise Distributed Object Computing Conference, pp. 35–44. IEEE Computer Society, Washington (2006)
13. O'Leary, D.E.: Empirical analysis of the evolution of a taxonomy for best practices. Decision Support Systems 43(4), 1650–1663 (2007)
14. Paton, N.W. (ed.): Active Rules in Database Systems. Springer, New York (1999)
15. Pontelli, E., Son, T.C., Baral, C.: A Framework for Composition and Inter-operation of Rules in the Semantic Web. In: Proceedings of the Second International Conference on Rules and Rule Markup Languages for the Semantic Web, pp. 39–50. IEEE Computer Society Press, Los Alamitos (2006)
16. Schrefl, M., Bernauer, M.: Active XML Schemas. In: Arisawa, H., Kambayashi, Y., Kumar, V., Mayr, H.C., Hunt, I. (eds.) ER Workshops 2001. LNCS, vol. 2465, pp. 363–376. Springer, Heidelberg (2002)
17. Wagner, G., Antoniou, G., Tabet, S., Boley, H.: The Abstract Syntax of RuleML – Towards a General Web Rule Language Framework. In: Proceedings of the 2004 IEEE/WIC/ACM International Conference on Web Intelligence (WI 2004), pp. 628–631. IEEE Computer Society, Washington (2004)
18. Wagner, G., Giurca, A., Lukichev, S.: A General Markup Framework for Integrity and Derivation Rules. In: Bry, F., Fages, F., Marchiori, M., Ohlbach, H.-J. (eds.) Principles and Practices of Semantic Web Reasoning, number 05371 in Dagstuhl Seminar Proceedings, Schloss Dagstuhl, Germany. Internationales Begegnungs- und Forschungszentrum für Informatik, IBFI (2005)

Exploration of SWRL Rule Bases through Visualization, Paraphrasing, and Categorization of Rules

Saeed Hassanpour, Martin J. O'Connor, and Amar K. Das

Stanford Center for Biomedical Informatics Research,
MSOB X215, 251 Campus Drive, Stanford, California, USA 94305
{saeedhp,martin.oconnor,amar.das}@stanford.edu

Abstract. Rule bases are increasingly being used as repositories of knowledge content on the Semantic Web. As the size and complexity of these rule bases increases, developers and end users need methods of rule abstraction to facilitate rule management. In this paper, we describe a rule abstraction method for Semantic Web Rule Language (SWRL) rules that is based on lexical analysis and a set of heuristics. Our method results in a tree data structure that we exploit in creating techniques to visualize, paraphrase, and categorize SWRL rules. We evaluate our approach by applying it to several biomedical ontologies that contain SWRL rules, and show how the results reveal rule patterns within the rule base. We have implemented our method as a plug-in tool for Protégé-OWL, the most widely used ontology modeling software for the Semantic Web. Our tool can allow users to rapidly explore content and patterns in SWRL rule bases, enabling their acquisition and management.

Keywords: Rule Management, Rule Abstraction, Rule Patterns, Rule Visualization, Rule Paraphrasing, Rule Categorization, Knowledge Representation, OWL, SWRL.

1 Introduction

Rules are increasingly being used to represent knowledge in ontology-based systems on the Semantic Web. As the size of such rule bases increases, users face a perennial problem in understanding and managing the scope and complexity of the specified knowledge. To support rapid exploration of rule bases and meet the scalability goals of the Semantic Web, automated techniques are needed to provide simplified interpretations of rules as well as high-level abstractions of their computational structures. In particular, rule paraphrasing and rule visualization can help non-specialists understand the meaning of logically complex rules. Abstraction of common patterns in rule bases can also enable automatic or semi-automatic categorization of rules into related groups for knowledge management. Such categorized patterns could ultimately form the basis of rule elicitation tools that guide non-specialists entering new rules.

We are addressing the need for such rule management solutions in our development of tools for the Semantic Web Rule Language (SWRL) [35]. In prior work, we developed SWRLTab [37], a plug-in for editing SWRL rule bases within Protégé-OWL [38].

G. Governatori, J. Hall, and A. Paschke (Eds.): RuleML 2009, LNCS 5858, pp. 246–261, 2009.
© Springer-Verlag Berlin Heidelberg 2009

Protégé-OWL is freely available, open-source knowledge management software that is widely used to specify OWL ontologies for Semantic Web applications. In this paper, we describe a novel approach for exploration of SWRL rule bases through three related techniques: (1) rule visualization, (2) rule paraphrasing, and (3) rule categorization. These three techniques are based on a method of syntactic analysis of SWRL rules. We use the data structure output of this analysis to graphically present the structure of a rule for rule visualization. We use the structural information, along with general heuristics, to paraphrase SWRL rules into simplified, readable English statements. We also apply a pattern recognition algorithm to the structural information to automatically categorize rules into groups that share a common syntactic representation. We show how these techniques can be used to support exploration and analysis of SWRL rule bases, allowing users to more easily comprehend the knowledge they contain. We evaluate the use of our approach by applying these techniques to several biomedical ontologies that contain SWRL rule bases. Finally, we discuss the development of a Protégé-OWL plug-in, called Axiomé that provides these three management techniques for users and developers of SWRL rule bases.

2 Background

OWL [34] is the standard ontology language of the Semantic Web and is rapidly becoming one of the dominant ontology languages in the development of knowledge bases. OWL provides a powerful language for building ontologies that specify high-level descriptions of Web content. These ontologies are created by constructing hierarchies of classes describing concepts in a domain and relating the classes to each other using properties. OWL also provides a powerful set of axioms for precisely defining how to interpret concepts in an ontology and to infer information from these concepts.

The Semantic Web Rule Language (SWRL) [35] is an extension to the OWL language to provide even more expressivity. The SWRL language allows users to write Horn-like rules that can be expressed in terms of OWL concepts and that can reason about OWL individuals. SWRL thus provides deductive reasoning capabilities that can infer new knowledge from an existing OWL ontology. For example, a SWRL rule expressing that a person with a male sibling has a brother can be defined using the concepts of 'person', 'male', 'sibling' and 'brother' in OWL. Intuitively, the concept of person and male can be captured using an OWL class called `Person` with a subclass `Male`; the sibling and brother relationships can be expressed using OWL properties `hasSibling` and `hasBrother`, which are attached to `Person`. The rule in SWRL would be[1]:

```
Person(?x) ^ hasSibling(?x,?y) ^ Male(?y) → hasBrother(?x,?y)
```

Eecuting this rule would have the effect of setting the `hasBrother` property of x to y. Similarly, a rule that asserts that all persons who own a car should be classified as drivers can be written as follows:

```
Person(?p) ^ hasCar(?p, true) → Driver(?p)
```

---

[1] The SWRL Submission [35] does not detail a standard syntax for language presentation; the examples shown in this paper reflect the presentation syntax adopted by the Protégé-OWL SWRL Editor.

This rule would be based on an OWL ontology that has the property `hasCar` and the class `Driver`. Executing this rule would have the effect of classifying all car-owner individuals of type `Person` to also be members of the class `Driver`.

One of SWRL's most powerful features is its ability to support user-defined methods or *built-ins* [37]. A number of core built-ins for common mathematical and string operations are defined in the SWRL W3C Submission. For example, the built-in `greaterThan` can be used to determine if one number is greater than another. A sample SWRL rule using this built-in to help classify as adults any person who has an age greater than 17 can then be written as:

```
Person(?p)^ hasAge(?p,?age) ^ swrlb:greaterThan(?age,17) → Adult(?p)
```

When executed, this rule would classify individuals of class `Person` with a `hasAge` property value greater than 17 as members of the class `Adult`.

SWRL rules can also establish relationships between entities in an ontology. For example, the following rule from the California Driver Handbook [36] provides California's driving regulations about minor visitors:

An individual under the age of 18 as a potential driver of a vehicle with a weight of less than 26,000 lbs if they possess an out-of-state driver's license and are visiting the state for less than 10 days.

can be written in SWRL as:

```
Person(?p) ^ has_Driver_License(?p,?d) ^ issued_in_State_of(?d,?s) ^
swrlb:notEqual(?s,"CA") ^ has_Age(?p,?g) ^ swrlb:lessThan(?g,18) ^
number_of_Visiting_Days_in_CA(?p,?x) ^ swrlb:lessThan(?x,10) ^ Car(?c)
^ has_Weight_in_lbs(?c,?w) ^ swrlb:lessThan(?w,26000) →
can_Drive(?p,?c)
```

As mentioned, all classes and properties referred to in this rule must preexist in an OWL ontology.

Table 1. SWRL atom types and example atoms from the Californian Driver Handbook rule

SWRL Atom Type	Example Atom
Class atom	`Person(?p), Car(?c)`
Individual property atom	`has_Driver_License(?p,?d)` `issued_in_State_of(?d,?s)` `can_Drive(?p,?c)`
Same/Different atom	`sameAs(?x, ?y)` `differentFrom(?x, ?y)`
Datavalued property atom	`has_Age(?p,?g)` `number_of_Visiting_Days_in_CA(?p,?x)` `has_Weight_in_lbs(?c,?w)`
Built-in atom	`swrlb:notEqual(?s,"CA")` `swrlb:lessThan(?g,18)`
Data range atom	`xsd:double(?x)`

As can be seen from these examples, SWRL rules have a simple Horn-like rule structure. A rule is composed of a body and a head, each of which contain conjunctions of atoms. SWRL does not support disjunction. There are six main types of SWRL atoms defined in the W3C Submission for SWRL. Table 1 lists these atom types and provides example atoms based on the previous rule from the California Driver Handbook.

3 Related Work

Rule management is a very active application area in the business rules domain. These systems are used to define, execute, monitor and maintain the rules used by operational systems [30]. There are a wide variety of commercial rule management tools, which are used to help business organizations standardize and enhance the visibility and consistency of their rule bases. These tools typically provide business user-friendly rule formats, multiple data models for rules implementation, rule testing and refinement, high level rules management interfaces and editors, in addition to other capabilities such as rules versioning, access control, and justification capabilities [31, 32].

Rules are increasingly being used for knowledge management in combination with ontologies [1-3]. As these rule bases grow larger, standard business rule management solutions are being investigated to deal with the resulting complexity. The intimate interactions between rules and the underlying ontology formalisms often require novel solutions [4, 5], however. In particular, the formal underpinnings of the technologies can sometimes be exploited to automatically infer information that may not be possible with the more loosely coupled interactions that are typical between business rules and underlying data.

Some of the traditional approaches used with expert systems can be utilized for certain management tasks. For example, a substantial amount of work has been done in automatic extraction of rules from data [6-11, 33]. Comparatively little work has been done in mining rule bases themselves to assist user comprehension. Rule argumentation techniques [12-15] do typically examine the relationships between rules in a rule base. However, these techniques do not focus on making the rule bases themselves easier to understand. Instead, the goal is to explain the reasoning steps that have been operationalized by the rules. Similarly, descriptive user-friendly text has been used in expert systems to explain the behavior of systems [16, 18], but, again, these textual descriptions have not primarily aimed to explain rule bases and their structure. Other work on rule visualization has mostly focused on showing the connections between rules themselves or and ontology entities or connections between rules and their supporting data, not the structure of the rules bases [19, 20]. UML-based visualization techniques have been used in the business rules domain [43, 44] but these methods are typically designed to provide very detailed views of rule interactions and are not designed for high level rule base exploration.

Principled methods to examine structural patterns in rule bases may significantly aid user comprehension. These approaches can help users to rapidly explore and understand large unfamiliar rule bases. They can also be used to help users understand their own rule bases and spot non-obvious knowledge patterns, which can ultimately help them better structure both the rule bases and the associated ontologies.

4 Methods

In Section 4.1, we discuss a rule abstraction method that parses a SWRL rule and provides as output a tree data structure to represent the rule. We then describe how we use this data structure to visualize, paraphrase and categorize SWRL rules, respectively, in Sections 4.2, 4.3 and 4.4. In Section 4.5, we present a plug-in for Protégé-OWL that supports these three techniques.

4.1 Rule Abstraction

As a first step in our rule management approach, we apply a rule abstraction method to analyze the syntactic structure of a SWRL rule. This method scans the atoms in the body and head of each rule using lexical analysis, reorders the atoms using a set of heuristics, and maintains them in a tree data structure.

Table 2. SWRL atom types and their corresponding rule abstraction priority

SWRL Atom Type	Priority
Class atom	1
Individual property atom	2
Same/Different atom	3
Data-valued property atom	4
Built-in atom	5
Data range atom	6

We give each of the six main types of SWRL atom (shown in Table 2) an ordinal ranking from 1 to 6 that indicates an intuitive sense of the semantic importance of each atom type. Class atoms (e.g., `Person(?p)`) are given the highest priority since they typically refer to the entities of primary interest in a rule. This ranking is followed by object property atoms (e.g., `Can_Drive(?x, ?y)`), which capture relationships between these entities. Same as and different from atoms indicate relationships of similarity or difference between entities and are given a lower priority because their use is typically complementary to the use of object property atoms. Data valued atoms (e.g., `has_Age(?x, ?y)`) specify the values of properties of particular entities, so are given less priority than inter-entity relationships. Built-in and data range properties operate on these data values so are hence given a lower ordering.

We then reorder the atoms in the head and body of each rule using these priorities. Figure 1 shows the resulting representation for the body of our sample California Driver Handbook rule.

After performing this atom reordering, we build a tree data structure that reflects the information captured by the variable chains in the rule together with the priority information associated with each atom. These trees are generated by a depth-first

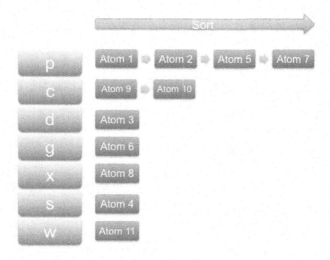

Fig. 1. Example tree data structure that uses a set of priority heuristics to reorder atoms for the sample rule from the California Driver Handbook. The left-hand column (orange boxes) contains variables used in the first position of a SWRL predicate. The atom number (blue boxes) represents the original ordering provided in the SWRL rule.

search of each variable chain in a rule. Once a variable is chosen as a root of a particular tree, atoms that contain that variable as their first argument are created as nodes of the tree at the same level. Any variables that appear as the second arguments of atoms are used to recursively expand the tree to the next level. Loops are avoided by keeping track of atom use.

If several variables share atoms with the same priority we break the tie by giving a higher priority to the variable with a longer list of atoms that start with that variable. If there still is a tie, we use the original ordering of atoms that the rule writer used in creating the rule to determine the first variable to expand. For trees with multiple disconnected roots we chose a new root from atoms not contained in earlier trees and begin the process again. We continue this process until we have scanned all the atoms in the atom list.

4.2 Rule Visualization

Our canonical representation of a SWRL rule can be used to provide a visual representation of the rule. Figure 2 show the visual representation of this data structure for the California Driver Handbook sample rule. This representation allows complex rules that have many classes and properties to be shown as an easily understood nested diagram. In Section 4.5, we show how this graphical representation is used to visualize and browse individual rules in the Axiomé rule management Protégé-OWL plug-in.

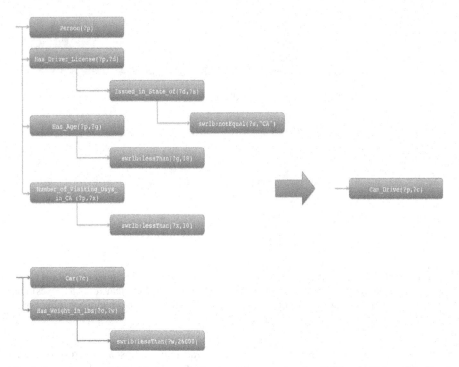

Fig. 2. Data structure showing a reorganization of atoms for the California Driver Handbook sample rule based on the described priority heuristics and variable chains

4.3 Rule Paraphrasing

We also use the tree data structure created by the rule abstraction method to generate paraphrases of SWRL rules, which are more understandable than the syntactic form. We have developed a textual template for each type of atom to generate these paraphrases. The templates use the first atom argument as the subject and the second argument (if any) as the object of sub phrases. An atom's predicate is used in an appropriate form in the template to convey the semantics. In general, we use the name of the underlying OWL classes and properties when generating paraphrases, but we can also support the use of OWL annotations to provide these names. We use heuristics for special cases such as property predicate names or annotation text starting with "has", articles before letters, and predicates beginning with silent 'h'. The ordering of paraphrased atoms is based according to their position in the rule abstraction tree and the paraphrased atoms are connected with appropriate conjunctions. Indentation is used to indicate atom depth. The same margins are used for phrases that are in the same tree level. Each successively deeper level has a larger margin.

Built-in atoms require more elaborate processing because SWRL built-ins can have a variable number of arguments and it is generally not possible to automatically paraphrase the built-in operation by simply using its name. So in the case of built-ins we have defined an annotation ontology that can be used to associated text with a built-in that can be used directly in paraphrases. We have defined annotations for a set of the standard built-ins defined in the core SWRL built-in ontology [37]. We specially process some standard mathematical operators such as less than and equal and generate condensed paraphrases that omit the mention of some variables to produce more concise text.

For sameAs atoms we make the equality between the variables in the rule explicit when paraphrasing. While we are scanning the atom list to build the tree data structure we note variable pairs that are described to be the same as each other with the sameAs atom. After building the tree we then merge the entries that have been noted to be the same. We also scan the pair list to discover the pairs that are not mentioned explicitly to be the same in the rule but can be inferred to be the same based on the transitivity of the sameAs property. These new discovered pairs are also added to the sameAs pairs list. In paraphrasing the rule we then use only one variable name for the equivalent variables.

Our paraphrasing approach can produce concise and easy-to-read English forms of SWRL rules. The following, for example, is the text that is generated for our sample rule earlier California Driver Handbook (see Section 2):

```
IF
        "p" IS A Person
        AND "p" HAS Driver License "d"
                WHERE "d" Issued in State of "s" WHERE "s" IS NOT EQUAL-
                TO "CA"
        AND "p" HAS Age LESS THAN 18
        AND "p" HAS VALUE "x" FOR Number of Visiting Days in CA WHERE-
        "x" IS LESS THAN 10

AND IF
        "c" IS A Car
        AND "c" HAS Weight in lbs "w" WHERE "w" IS LESS THAN 26000

THEN
        "p" Can Drive "c"
```

Our rule management tool, Axiomé, can generate English paraphrases of rules as a part of the Protégé-OWL plug-in.

4.4 Rule Categorization

We can use the tree data structure to categorize SWRL rules based on the patterns of atoms used. To undertake this rule management technique, we first establish a rule signature for each SWRL rule to capture the structure of the atoms in our abstracted representation. The rule signature is based on a regular expressions language that is composed of an alphabet Σ, and a set of quantifiers Q, such that:

$$\Sigma = \{1, 2, 3, 4, 5, 6\}$$
$$Q = \{-, \hat{\ }, (\ ), \#, +\}$$

Literals in the alphabet Σ represent each of the six main atom types in Table 1. The quantifiers Q are used in the following ways: (1) '-' separates the atoms in the body from the atoms in the head; (2) '^' separates different trees; (3) parenthesis pairs are placed around direct descendants of a node; (4) a '#' is used to expansion of an atom in the data structure and is placed before the next level's atoms; and (5) A '+' is used to show repeated use of the same atoms. Table 3 summarizes the role of each quantifier.

Table 3. Signature quantifiers in rule signature regular expression language

Rule Quantifier	Role
-	Body-Head separator
^	Tree separator
( )	Direct descendents of a node
#	Node expansion
+	Repetition

Consider, for example, rule from family history ontology, which defines a paternal aunt relationship:

```
has_natural_father(?a,?b) ^ has_natural_sister(?b,?c) →
has_paternal_aunt(?a,?c)
```

Each atom in this rule is an individual property atom. Using the rule abstraction method from section 4.1 we can generate a tree structure for the rule, which can be paraphrased as:

```
IF
        "a" HAS Natural Father "b"
                WHERE "b" HAS Natural Sister "c"

THEN
        "a" HAS Paternal Aunt "c"
```

The rule signature is represented:

```
(2#(2))-(2)
```

Using the notations of our regular expressions language, we can define the signature of the example rule California Driver Handbook as:

```
(12#(2#(5))4+#(5)#(5))^(14#(5))-(2)
```

We then use these signatures to group rules into categories. In the Axiomé rule management tool, we support invocation of this categorization technique and graphically show the resulting categories in a tree table.

4.5   Rule Management Tool

We have implemented the three rule-management techniques in a tool called Axiomé, Axiomé is developed as a Protégé-OWL plug-in with functional areas for each of these techniques. These are available as sub-tabs within the plug-in: (1) a Rule Visualization tab to visualize individual rules; (2) a Rule Paraphrasing tab that displays an English-like text explanation for each rule; and (3) a Rule Categorization tab to automatically categorize rules in a rule base. A Rule Browser component is permanently displayed to show a tree-table representation of the SWRL rules in an ontology. This tree-table enables users to explore the rule base and lunch any of three sub-tabs for the rule being explored.

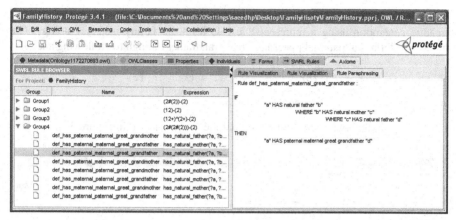

Fig. 3. An screenshot from Axiomé plug-in with three tabs and the SWRL rule browser. Figure shows the paraphrasing tab for one of the rules in the family history ontology.

5 Results

To evaluate the usefulness and efficacy of our visualization, paraphrasing and categorization techniques, we applied our method to four OWL ontologies containing SWRL rules bases. Each of these ontologies was developed as part of a biomedical application and designed by a knowledge engineer or domain expert who was not one of the authors.

The first set of rules that we analyzed is part of an ontology for family medical history [39]. This rule base is composed of 146 rules, which define possible relations between people in a family. We applied our method on the rule base to generate and visualize the data structures and paraphrase them. Our categorization method found four types of rule signatures and thus divides the rule base into four groups. The number of members in each group and their signatures are shown in Table 4. Because the rule base contains general knowledge about family relatedness, we were able to verify the integrity and clarity of these results directly.

Table 4. Rule categorization for family history ontology

Rule signature	Number of instances in the rule base	Examples
(2#(2))-(2)	110	Two link relations, e.g., Uncle and Aunt
(12)-(2)	22	One link relations, e.g., Son and Daughter
(2#(2#(2)))-(2)	8	Three link relations, e.g., Great Grandfather/Grandmother
(12+)^(2+)-(2)	6	Natural/half relations, e.g., Half Brother/Sister
Total Number of rules in the rule base:	*146*	-

Given space limitations, we provide one example in this paper for paraphrasing and signature generation using a representative rule from the family history ontology. Our sample rule defines a 'paternal maternal' great grandfather through the following ancestry:

```
has_natural_father(?a,?b) ^ has_natural_mother(?b,?c) ^
has_natural_father(?c,?d) →
has_paternal_maternal_great_grandfather(?a,?d)
```

The rule is graphically structured by the Axiomé tool as shown in Figure 3.

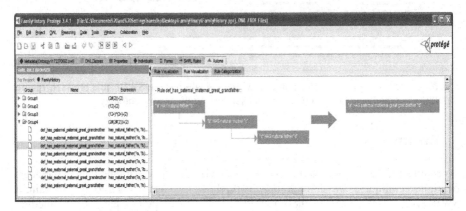

Fig. 4. Axiomé tool as a Protege-OWL plug-in tab, showing a visualization of a simple rule from the family history ontology

The text generated by our paraphrasing method for this rule is:

```
IF
        "a" HAS natural father "b"
                WHERE "b" HAS natural mother "c"
                        WHERE "c" HAS natural father "d"

THEN
        "a" HAS paternal maternal great grandfather "d"
```

And the rule signature is:

```
(2#(2#(2)))-(2)
```

The second rule base we evaluated was developed as part of an ontology of disease phenotypes, or genetically relevant clinical characteristics, for the neurodevelopmental disorder of autism. The ontology and rules will support concept-based querying of the National Database of Autism Research (NDAR), a public resource funded by the National Institutes of Health for archiving, sharing, and analyzing data collected in autism research [40]. NDAR uses the ontology as an information model representing research and clinical data about study subjects and as a domain ontology that defines terms and relationships in autism. The SWRL rules define how each phenotype is to be derived from a set of clinical findings. The terms, relationships, and abstractions for building the autism ontology are gathered by a literature search of the PubMed database [41].

We applied our categorization technique to the SWRL rules in the autism ontology to find rule signatures. The 14 rules in the current rule base are divided into five groups; one of the groups contains 6 rules with a common structure. The signature and the numbers in each group are shown in Table 5. To check the validity of our results, we asked the developer of the autism ontology to review them. The developer confirmed that our graphical representation and English paraphrases of the rules are semantically equivalent to those in the rule base, and that our categories include the two major types of patterns he used to develop the rule base.

Table 5. Rule categorization for autism ontology

Rule Signature	Number of instances in the rule base
(14+)^(5+)-(12+4+)	6
(14#(5))^(4+)-(12+4+)	3
(14+#(5))^(4)-(12+4+)	2
(14+)^(5)^(5+)-(12+4+)	2
(14+#(5))-(12+4+)	1
Total Number of rules in the rule base:	*14*

The third rule base to which we applied our method was part of a heart disease ontology developed at Stanford Medical School in collaboration with the European Union HEARTFAID project. The resulting THINHeart ontology contains 70 SWRLrules, each of which classifies heart conditions based on presumed cause. A domain expert encoded each of these 70 definitions using a single template. When we applied our categorization technique to the rule base, we found that all 70 rules matched a single rule signature, shown in Table 6.

Table 6. Rule categorization for THINHeart ontology

Rule Signature	Number of instances in the rule base
(14)-(2)	70
Total Number of rules in the rule base:	*70*

We applied our method to a fourth biomedical ontology that contained 63 rules to assess a patient's response to cancer treatment over time [42]. Our categorization method divided the rules into 41 groups; 37 of these groups contain less that 3 rules. Table 7 shows the rule signatures for the groups that had 3 rules or more. We confirmed with the ontology developers that the SWRL rules were intentionally written to fit into a set of distinct rule templates by analyzing and merging rules with the same structures during the authoring process.

Table 7. Rule categorization for cancer response assessment ontology

Rule Signature	Number of instances in the rule base
`(12#(2#(5))4#(5))^(12)^(12#(2+4))-(12)`	7
`(1+2#(2#(5)))^(12)^(12#(2+))^(5)^(5+)-(12+)^(2)`	4
`(1+2#(2#(5)))^(12)^(12#(2))-(12)`	4
`(12#(12#(2#(5))))^(12#(2))-(12)`	3
8 categories with 2 members	2
29 categories with 1 member	1
Total Number of rules in the rule base:	*63*

6 Discussion

Research on rule representation and rule management has been an active area of work in expert systems, active database systems, association rule mining, and business systems. Rule bases also play an increasingly important role in encoding declarative knowledge within ontology-based systems on the Semantic Web. In this paper, we present work we are undertaking to enable analysis and management of rule bases as part of providing SWRL tool support. We propose a rule abstraction approach that uses a tree data structure to represent a SWRL rule. We have shown that this simple data structure can enable three techniques for visualization, paraphrasing, and categorization of rules. This analytic approach is similar to prior work on static code analysis and formal methods verification [21-25]. Related methods from this field, such as model checking, data-flow analysis and abstract interpretation, could also be applied to perform rule base analysis and integrity checking of the rule base and ontology.

In addition to creating a plug-in tool within Protégé-OWL to make our method available to developers of SWRL rule bases, we applied the method to four existing biomedical ontologies with SWRL rule bases. We have checked the visualization and paraphrasing output of our method for each ontology, and ensured that the outputs accurately represent the rules and have face validity. We are planning a more extensive user evaluation of the Axiomé tool. Our initial application of the method to the four available ontologies was revealing about the nature of the categorizations that were created. In the family history and autism ontologies, the method revealed multiple patterns of rule signatures that we could verify as being valid ourselves or with the developers of the ontology.

The discovery of such 'hidden' rule signatures may enable rule elicitation. The most common rule elicitation methods are performed by knowledge engineers [26, 27]. Another common approach is using domain experts to provide predefined templates and categories for rules or to ask fixed questions to make the rule elicitation easier and semi-automated [28, 29]. This approach may be limited by the skill of the domain expert who provides the templates and the questions. Our rule categorization approach can be used during ontology and rule base development as an alternative method to create the design of templates. As the development of a rule base occurs, common groups of rules may be found based on their signature. The signature can then be used to create a template, which domain experts can employ to specify new rules. A rule elicitation interface using the rule signatures could also interactively suggest to users how many and what types of atoms a rule might have based on what the user has partially specified for the rule. Such suggestions may speed rule base development and increase the quality of the content.

The other two ontologies that we analyzed in this paper contained rule bases for which the developers had used one or more rule templates, although they were not explicitly represented or constrained by a user interface. We observed that our categorization method accurately identified rule templates. As a result, our categorization method could be used for post hoc analysis of a rule base to ensure that the rules do match known templates. Our rule management approach could be extended to support the design of templates that are based on ontology class restrictions rather than syntactic structure alone. For example, in the case of the cancer response assessment ontology, we can divide the classes and properties of the ontology into several independent sub-ontologies, and categorize the rules based on the sub-ontologies that they cover. Such an approach was of particular interest to the developers of the cancer response assessment ontology who are seeking further approaches to reducing the number of templates needed for rule elicitation by domain experts.

Finally, we believe that creating a simple data structure to represent rules provides an opportunity to perform machine learning on rule bases and to discover frequently occurring or higher level patterns among the rule signatures. We are thus planning to use powerful and sophisticated classifiers such as support vector machines, genetic algorithms and artificial neural networks in our future work.

Acknowledgments. The authors would like to thank Richard Waldinger for his comments on this work and manuscript, and Jane Peace, David Kao, and Mia Levy for sharing their ontologies for our analysis. This research was supported in part by NIH grant 1R01LM009607.

References

1. Maedche, A., Motik, B., Stojanovic, L., Studer, R., Volz, R.: Ontologies for Enterprise Knowledge Management. IEEE Intelligent Systems 18(2), 26–33 (2003)
2. Maedche, A., Staab, S.: Ontology learning for the Semantic Web. IEEE Intelligent systems 16(2) (2001)
3. Wang, X.H., Zhang, D.Q., Gu, T., Pung, H.K.: Ontology Based Context Modeling and Reasoning using OWL. In: Proceedings of the Second IEEE Annual Conference on Pervasive Computing and Communications Workshops, PERCOMW, vol. 18. IEEE Computer Society, Washington (2004)
4. Ostrowski, D.A.: Rule Definition for Managing Ontology Development. Advances in Rule Interchange and Applications, 174–181 (2007)
5. Dou, D., McDermott, D., Qi, P.: Ontology translation on the Semantic Web. Journal on Data Semantics (JoDS) II, 35–57 (2005)
6. Maedche, A., Staab, S.: Discovering conceptual relations from text. In: ECAI 2000, Proceedings of the 14th European Conference on Artificial Intelligence. IOS Press, Amsterdam (2000)

7. Berendt, B., Hotho, A., Stumme, G.: Towards Semantic Web Mining. In: Horrocks, I., Hendler, J. (eds.) ISWC 2002. LNCS, vol. 2342, pp. 264–278. Springer, Heidelberg (2002)
8. Liu, B., Hsu, W., Ma, Y.: Integrating classification and association rule mining. In: Appeared in KDD 1998, New York (1998)
9. Vaidya, J., Clifton, C.: Privacy preserving association rule mining in vertically partitioned data. In: Proceedings of the Eighth ACM SIGKDD international Conference on Knowledge Discovery and Data Mining, Edmonton, Alberta, Canada (2002)
10. Hipp, J., Güntzer, U., Nakhaeizadeh, G.: Algorithms for association rule mining — a general survey and comparison. SIGKDD Explor. Newsl. 2, 1 (2000)
11. Bayardo, R.J., Agrawal, R., Gunopulos, D.: Constraint-Based Rule Mining in Large, Dense Databases. Data Min. Knowl. Discov. 4, 2–3 (2000)
12. Rahwan, I., Amgoud, L.: An argumentation based approach for practical reasoning. In: Proceedings of the Fifth international Joint Conference on Autonomous Agents and Multiagent Systems, Hakodate, Japan (2006)
13. García, A.J., Simari, G.R.: Defeasible logic programming: an argumentative approach. Theory Pract. Log. Program. 4(2), 95–138 (2004)
14. Bench-Capon, T.J.M., Dunne, P.E.: Argumentation in artificial intelligence. Artificial Intelligence 171(10-15), 619 (2007)
15. Chesñevar, C., McGinnis, J., Modgil, S., Rahwan, I., Reed, C., Simari, G., South, M., Vreeswijk, G., Willmott, S.: Towards an argument interchange format. Knowl. Eng. Rev. 21(4), 293–316 (2006)
16. Core, M.G., Lane, H.C., van Lent, M., Gomboc, D., Solomon, S., Rosenberg, M.: Building explainable artificial intelligence systems. In: Proceedings of the 18th Conference on Innovative Applications of Artificial Intelligence (IAAI 2006), Boston, MA (2006)
17. Johnson, W.L.: Agents that explain their own actions. In: Proc. of the Fourth Conference on Computer Generated Forces and Behavioral Representation, Orlando, FL (1994)
18. Van Lent, M., Fisher, W., Mancuso, M.: An explainable artificial intelligence system for small-unit tactical behavior. In: Proceedings of the 16th Conference on Innovative Applications of Artificial Intelligence (IAAI 2004), San Jose, CA, pp. 900–907 (2004)
19. Wong, P.C., Whitney, P., Thomas, J.: Visualizing Association Rules for Text Mining. In: Proceedings of the 1999 IEEE Symposium on information Visualization. INFOVIS, p. 120. IEEE Computer Society, Washington (1999)
20. Blanchard, J., Guillet, F., Briand, H.: Exploratory Visualization for Association Rule Rummaging. In: KDD 2003 Workshop on Multimedia Data Mining (MDM 2003) (2003)
21. Pfleeger, S.L., Hatton, L.: Investigating the Influence of Formal Methods. Computer 30(2), 33–43 (1997)
22. Tilley, T., Cole, R., Becker, P., Eklund, P.: A survey of formal concept analysis support for software engineering activities. In: Ganter, B., Stumme, G., Wille, R. (eds.) Formal Concept Analysis. LNCS (LNAI), vol. 3626, pp. 250–271. Springer, Heidelberg (2005)
23. Blanchet, B., Cousot, P., Cousot, R., Feret, J., Mauborgne, L., Miné, A., Monniaux, D., Rival, X.: A static analyzer for large safety-critical software. In: Proceedings of the ACM SIGPLAN 2003 Conference on Programming Language Design and Implementation. PLDI 2003, San Diego, California, USA, pp. 196–207. ACM, New York (2003)
24. Bush, W.R., Pincus, J.D., Sielaff, D.J.: A static analyzer for finding dynamic programming errors. Softw. Pract. Exper. 30(7), 775–802 (2000)
25. Dean, J., Grove, D., Chambers, C.: Optimization of Object-Oriented Programs Using Static Class Hierarchy Analysis. In: Olthoff, W. (ed.) ECOOP 1995. LNCS, vol. 952, pp. 77–101. Springer, Heidelberg (1995)

26. Leite, J.C., Leonardi, M.C.: Business Rules as Organizational Policies. In: Proceedings of the 9th international Workshop on Software Specification and Design. International Workshop on Software Specifications & Design, p. 68. IEEE Computer Society, Washington (1998)
27. Wright, G., Ayton, P.: Eliciting and modelling expert knowledge. Decis. Support Syst. 3(1), 13–26 (1987)
28. Mechitov, A.I., Moshkovich, H.M., Olson, D.L.: Problems of decision rule elicitation in a classification task. Decis. Support Syst. 12(2), 115–126 (1994)
29. Larichev, A.I., Moshkovich, H.M.: Decision support system "CLASS" for R&D planning. In: Proceedings of the First International Conference on Expert Planning Systems, Brighton, England, pp. 227–232 (1990)
30. Business Rule Management Systems, http://en.wikipedia.org/wiki/BRMS
31. SAP NetWeaver,
    https://www.sdn.sap.com/irj/sdn/nw-rules-management
32. ILOG, http://www.ilog.com/products/businessrules/
33. Park, S., Lee, J.K.: Rule identification using ontology while acquiring rules from Web pages. Int. J. Hum. Comput. Stud. 65(7), 659–673 (2007)
34. McGuinness, D.L., van Harmelen, F. (eds.): OWL Web Ontology Language Overview. W3C Recommendation (February 10, 2004), http://www.w3.org/TR/2004/REC-owl-features-20040210/
35. SWRL Submission, http://www.w3.org/Submission/SWRL/
36. California Driver Handbook, http://www.dmv.ca.gov/pubs/dl600.pdf
37. O'Connor, M.J., Musen, M.A., Das, A.: Using the Semantic Web Rule Language in the Development of Ontology-Driven Applications. In: Handbook of Research on Emerging Rule-Based Languages and Technologies: Open Solutions and Approaches, ch. XXII. IGI Publishing (2009)
38. Knublauch, H., Fergerson, R.W., Noy, N.F., Musen, M.A.: The Protégé OWL Plugin: An open development environment for semantic web applications. In: Proceedings of the Third International Semantic Web Conference, Hiroshima, Japan, pp. 229–243 (2004)
39. Peace, J., Brennan, P.F.: Instance testing of the family history ontology. In: Proceedings of the American Medical Informatics Association (AMIA) Annual Symposium, Washington, DC, p. 1088 (2008)
40. Young, L., Tu, S.W., Tennakoon, L., Vismer, D., Astakhov, V., Gupta, A., Grethe, J.S., Martone, M.E., Das, A.K., McAuliffe, M.J.: Ontology-Driven Data Integration for Autism Research. In: Proceedings of the 22nd IEEE International Symposium on Computer-Based Medical Systems, IEEE CBMS (2009)
41. Tu, S., Tennakoon, L., O'Connor, M., Shankar, R., Das, A.: Using an integrated ontology and information model for querying and reasoning about phenotypes: the case of autism. In: Proceedings of the American Medical Informatics Association (AMIA) Annual Symposium, Washington, DC, pp. 727–731 (2008)
42. Levy, M.A., Rubin, D.L.: Tool support to enable evaluation of the clinical response to treatment. In: Proceedings of the American Medical Informatics Association (AMIA) Annual Symposium, Washington, DC, pp. 399–403 (2008)
43. Kulakowski, K., Nalepa, G.J.: Using UML state diagrams for visual modeling of business rules. In: International Multiconference on Computer Science and Information Technology, 2008. MCSIT 2008, October 20–22, pp. 189–194 (2008)
44. Lukichev, S.: Visual Modeling and Verbalization of Rules, KnowledgeWeb PhD Symposium (2006)

TomML: A Rule Language for Structured Data

Horatiu Cirstea, Pierre-Etienne Moreau, and Antoine Reilles

Université Nancy 2 & INRIA & LORIA
BP 239, F-54506 Vandoeuvre-lès-Nancy, France
`first.last@loria.fr`

Abstract. We present the TOM language that extends JAVA with the purpose of providing high level constructs inspired by the rewriting community. TOM bridges thus the gap between a general purpose language and high level specifications based on rewriting. This approach was motivated by the promotion of rule based techniques and their integration in large scale applications. Powerful matching capabilities along with a rich strategy language are among TOM's strong features that make it easy to use and competitive with respect to other rule based languages. TOM is thus a natural choice for querying and transforming structured data and in particular XML documents [1]. We present here its main XML oriented features and illustrate its use on several examples.

1 Introduction

Pattern matching is a widely spread concept both in the computer science community and in everyday life. Whenever we search for something, we build a so-called pattern which is a structured object that specifies the features we are interested in. Mathematics makes a full use of patterns, some quite elaborated, and this is similar in logic and computer science.

The complexity of the matching process obviously depends on the complexity of the objects it targets. Matching a shape in a picture is significantly more difficult than recognizing a word in a text. There is also a compromise between the complexity of the data and the facility to manipulate it. The eXtensible Markup Language (XML) is a specification language that proved to be an excellent option for describing data since it allows the description of complex structures in a computer friendly format well suited for matching and transformation. In particular, XML data is organized in a tree structure having a single root element at the top and this is exactly the kind of objects manipulated in classical pattern matching algorithms and corresponding tools.

Of course, matching is generally not a goal by itself: if matching is done, this is because we want to perform an action. Typically, if we have detected that zero is added to some number, we would like to simplify the expression, for example by replacing $7 + 0$ by 7. Similarly, XML documents are often transformed into HTML documents for a better presentation in web pages. This transformation can be naturally described using rewriting that consists intuitively in matching an object and (partially) replacing it by another one.

G. Governatori, J. Hall, and A. Paschke (Eds.): RuleML 2009, LNCS 5858, pp. 262–271, 2009.

The rewriting concept appears from the very theoretical settings to the very practical implementations. Several rewriting languages, like ASF+SDF [2], ELAN [3] and MAUDE [4], have been developed on top of efficient pattern-matching algorithms and use sophisticated techniques for optimizing the (strategic) rewriting computations. Nevertheless, the libraries and facilities (*e.g.* input/output, threads, interfaces, etc.) are relatively limited compared to largely used languages like JAVA, for example. The language TOM (`tom.loria.fr`) implements the concept of *Formal Island* [6] that consists in making a specific technology available on top of an existing language. In the case of TOM [5] this technology is the strategic term rewriting and the existing language is JAVA (although the connection is also possible with potentially any other language).

The TOM programs have a clear semantics based on term rewriting and thus can be subject to formal analysis using the tools available in the domain and, on the other hand, a great potential for cross-platform integration and usability. We recover the portability and re-usability features of XML and TOM looks thus a natural choice when one wants to query and transform XML documents.

We present here an extension of TOM, called TOMML, that makes available all the TOM features into a syntax adapted to XML document manipulation. In fact, this extension can be easily adapted to accommodate any other structured data provided that a tree representation can be obtained out of it.

We briefly introduce TOM in the next section. In Section 3 we illustrate via several examples the main pattern-matching features of TOMML and in Section 4 we show how strategies can be used to give more expressive power to the formalism. We finish with a brief comparison with similar tools and some concluding remarks.

2   Tom in a Nutshell

As we have already said, TOM is an extension of JAVA which adds support for algebraic data-types and pattern matching. One of the most important constructs of the language is `%match`, a pattern matching construct which is parametrized by a list of objects, and contains a list of rules. The left-hand sides of the rules are pattern matching conditions (built upon JAVA class names and variables), and the right-hand sides are JAVA statements. Like standard `switch/case` construct, patterns are evaluated from top to bottom, firing each action (*i.e.* right-hand side) whose corresponding left-hand side *matches* the objects given as arguments.

For instance, assume that we have a hierarchy of classes composed of `Account` from which inherit `CCAccount` (credit card account) and `SAccount` (savings account), each with a field `owner` of type `Owner`. Given two objects `s1` and `s2`, the following code prints the owner's name if it is the same for the two accounts and if these accounts are of type `CCAccount`, respectively `SAccount`, and just prints the text `"CCAccount"` if both accounts are of type `CCAccount`:

```
%match(s1,s2) {
  CCAccount(Owner(name)),SAccount(Owner(name)) -> { print(name); }
  CCAccount(_),CCAccount(_) -> { print("CCAccount"); }
}
```

In the above example, `name` is a variable. Notice the use of the *non-linearity* (*i.e.* the presence of the same variable at least twice in the pattern) to denote concisely that the same value is expected. The "`_`" is an anonymous variable that stands for anything. The equivalent JAVA code would be:

```
if (s1 instanceof CCAccount) {
  if (s2 instanceof SAccount) {
    Owner o1=((CCAccount)s1).getOwner();
    Owner o2=((SAccount)s2).getOwner();
    if (o1 != null && o2 != null) {
      if ((o1.getName()).equals(o2.getName())) {
        print(o1.getName());
      }
    }
  } else { if (s2 instanceof CCAccount) { print("CCAccount"); } }
}
```

Besides matching simple objects, TOM can also match lists of objects. For instance, given a list of accounts (`List<Account> list`), the following code prints all the names of the credit card accounts' owners:

```
%match(list) {
  AccountList(X*,CCAccount(Owner(name)),Y*) -> { print(name); }
}
```

`AccountList` is a variadic list operator, the variables suffixed by * are instantiated with lists (possibly empty), and can be used in the action part: here `X*` is instantiated with the beginning of the list up to the matched object, whereas `Y*` contains the tail. The action is executed for each pattern that matches the subject (assigning different values to variables). Patterns can be non-linear: `AccountList(X*,X*)` denotes a list composed of two identical sublists, whereas `AccountList(X*,x,Y*,x,Z*)` denotes a list containing two occurrences of one of its elements. Another feature of TOM patterns that is worth mentioning is the possibility to embed negative conditions using the complement symbol "`!`" [7]. For instance, a so-called *anti-pattern* of the form `!AccountList(X*,CCAccount(_),Y*)` denotes a list of accounts that does not contain a credit card account. Similarly, `!AccountList(X*,x,Y*,x,Z*)` stands for a list containing only distinct elements, and `AccountList(X*,x,Y*,!x,Z*)` for one that has at least two distinct elements. There is no restriction on patterns, including complex nested list operators combined with negations. This allows the expression of different algorithms in a very concise and safe manner.

Since its first version in 2001, TOM itself has been written using TOM. The system is composed of a compiler and a library which offers support for predefined data-types such as integers, strings, collections, and many other JAVA datastructures. The compiler is organized, in a pure functional style, as a pipeline of program transformations (type inference, simplification, compilation, optimization, generation). Each phase transforms a JAVA+TOM abstract syntax tree using rewrite rules and strategies. At the end a pure JAVA program is obtained.

The complete environment is integrated into Eclipse (`www.eclipse.org`) providing a simple and efficient user interface to develop, compile, and debug rule based applications. It has been used in an industrial context to implement several large and complex applications, among them a query optimizer for Xquery and a platform for transforming and analysing timed automata using XML manipulation. On several classical benchmarks TOM is competitive with state of the art rule-based implementations and functional languages.

3 TomML, an Extension for XML Manipulation

One of the main objectives of TOM is to be as generic as possible. The implementation of the handled data-structures is not hard-wired in the system but becomes a parameter of the compiler. For that, we have introduced the notion of *formal anchor*, also called *mapping*, which describes how a concrete data-structure (*i.e.* the trees which are transformed) can be seen as an algebraic term. This idea, related to P. WADLER's views, allows TOM to rewrite any kind of data structure, and in particular XML trees, as long as a *formal anchor* is provided.

There are two possible approaches for using TOM. Either we start from an existing application that is improved with new functionalities implemented using the rewriting features of TOM, and in this case the data-structure used by the application is already defined: we just have to define a *mapping* from this data-structure to TOM.

Alternatively, we can abstract on the data-structure by using the data-structure generator integrated in TOM. Given a signature, TOM generates a set of JAVA classes that provide static typing. A subtle hash-consing technique is used to offer maximal sharing [8]: there cannot be two identical terms in memory. Therefore, the equality tests are performed in constant time, which is very efficient in particular when non-linear rewrite rules are considered.

In this section, we show how this general approach can be tailored to transform XML documents in both an expressive and a theoretically grounded way. Although the extension we present here is completely integrated in TOM we use the name TOMML to refer to the syntax features and standard libraries specific to XML document manipulations.

For the rest of the paper we consider the following XML document:

```
<Bank name="BNP">
 <Branch name="Etoile">               <Branch name="Lafayette">
  <CCAccount id="12">                   <CCAccount id="23">
   <Owner gender="M">Bob</Owner>         <Owner gender="M">John</Owner>
    <Balance>10000</Balance>             <Balance>10000</Balance>
  </CCAccount>                          </CCAccount>
   <SAccount rate="4">                  <CCAccount id="6">
    <Owner gender="M">Bob</Owner>        <Owner gender="M">Bob</Owner>
    <Owner gender="F">Alice</Owner>      <Balance>6000</Balance>
    <Balance>100000</Balance>          </CCAccount>
   </SAccount>                         </Branch>
 </Branch>                            </Bank>
```

A *bank* consists of several *branches*, each of them containing different types of *accounts*. Whatever the representation the XML document is (DOM for instance), it can be *seen* as a tree built out of (XML) nodes. For the scope of this paper we consider the following meta-model:

```
TNode = ElementNode(Name:String, AttrList:TNodeList, ChildList:TNodeList)
      | AttributeNode(Name:String, Specified:String, Value:String)
      | TextNode(Data:String)
      | CommentNode(Data:String)
      | CDATASectionNode(Data:String)
      | ...
TNodeList = concTNode(TNode*) // denotes a list of TNode
```

An `ElementNode` has a name and two children: a list of attributes, and a list of nodes. A *formal anchor* is materialized by a file which describes the Tom view of the corresponding XML document implementation[1] and, for the purpose of this paper, we have considered a correspondence between the algebraic sort `TNode` and the `Node` class from the `w3c.dom` package. This mapping is available in the standard TomML libraries and specifies, for example, how the name of an `ElementNode` of the Tom model can be retrieved (using the method `getNodeName` of the DOM class for instance). Once we have defined this mapping, the abstract notation can be used to match an XML document and print the name of all the branches of a bank:

```
%match(xmlDocument) {
  ElementNode("Bank",_,    //_ means that any list of attributes is accepted
   concTNode(_*,
    ElementNode("Branch",concTNode(_*,AttributeNode("name",_,bname),_*),_),
    _*)) -> { System.out.println("branch name: " + bname); }
}
```

We consider in this section that `xmlDocument` corresponds to the Tom encoding of the XML document given above and, in this case, the application of this code prints the strings "branch name: Etoile" and "branch name: Lafayette" and corresponds intuitively to the following XSLT template:

```
<xsl:template match="Bank/Branch">
    branch name: <xsl:value-of select="@name"/>
</xsl:template>
```

The interest of this approach is that the semantics of the match construct is theoretically well grounded and based on associative-matching with neutral element. The above pattern is rather complex partly because of the highly decorated XML syntax but Tom provides an alternative and much simpler XML tailored syntax. For example, the above match construct can be written:

```
%match(xmlDocument) {
  <Bank><Branch name=bname></Branch></Bank> -> {
   System.out.println("branch name: " + bname);
  }
}
```

Note that XML nodes can be directly used and that the extension variables (identified by "_*" previously) not used in the right-hand side are left implicit in the left-hand side. The semantics of these match constructs is exactly the one provided by Tom, and thus can deal with nested patterns, non-linear variables,

---

[1] See `http://tom.loria.fr` for more details.

anti-patterns, *etc.*. For example, in order to print the *name* of all clients who own two credit-card accounts in two different branches of the bank, *i.e.* to match a template of the form:

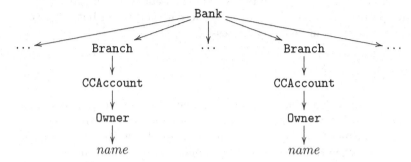

a simple non-linear pattern can be used in TOMML:

```
void ownerInTwoBranches(TNode xmlDocument) {
 %match(xmlDocument) {
  <Bank>
   <Branch name=bname1><CCAccount><Owner>name</Owner></CCAccount></Branch>
   <Branch name=bname2><CCAccount><Owner>name</Owner></CCAccount></Branch>
  </Bank> -> {
   System.out.println(name + " in " + bname1 + " and " + bname2);
  }
 }
}
```

As expected, this method prints the string "John in Etoile and Lafayette". A similar behavior is obtained for the following XSLT code

```
<xsl:template match="Bank/Branch/CCAccount">
    <xsl:for-each select="preceding::CCAccount[Owner=current()/Owner]">
        <xsl:value-of select="Owner" /> in
        <xsl:value-of select="../@name" /> and
        <xsl:value-of
            select="following::CCAccount[Owner=current()/Owner]/../@name"/>
    </xsl:for-each >
</xsl:template>
```

which is clearly less intuitive than the corresponding TOM code and becomes even more elaborated when negative conditions like the ones below are needed.

The anti-patterns are convenient if we want to specify concisely definitions of relatively complex patterns implying negative conditions. The branches whose clients are all mutually different can be printed using the following method:

```
void branchWithNoMultipleOwner(TNode xmlDocument) {
 %match(xmlDocument) {
  <Bank>
   branch@!<Branch> <_><Owner>o</Owner></_>
                    <_><Owner>o</Owner></_> </Branch>
  </Bank> -> { printXMLFromTNode(branch); }
 }
}
```

The variable `branch` can be seen as alias for the whole term matched by the pattern following the "`@`" operator; this is of course just syntactic sugar allowing for concise definitions of the consequent actions. The function `printXMLFromTNode` available in the standard ToMML libraries prints a `TNode` using an XML syntax.

The above pattern corresponds to an "all-different" constraint but other constraints like "all-equal" can be easily expressed.

4   Strategies

When programming using functions, pattern matching constructs, and more generally the notion of *transformation rule*, it is common to introduce extra functions that *control* their application. In the case of rewriting, this control describes how and when the rules should be applied. This control can be defined in the right-hand side of the rules but this is usually a bad practice since it makes the rules more complex, specific to a given application, and thus not reusable.

Rewriting based languages provide more abstract ways to express the control of rule applications, either by using reflexivity as in MAUDE, or the notion of strategy for ELAN, Stratego [9], or ASF+SDF. Strategies such as *bottom-up, top-down* or *leftmost-innermost* are higher-order features that describe how rewrite rules should be applied. This compares to some extent to the "`//`" operator of XPath which corresponds to a *depth-search*. TOM offers a flexible and expressive strategy language where high-level strategies are defined by combining low-level primitives. Among these latter primitives, we consider the *sequence* (denoted `;`), the *choice* (denoted `<+`), and two generic congruence operators called `All` and `One`. A rewrite rule is also an elementary strategy that can be applied on a term (*i.e.* a tree).

As for the `%match` construct, a user defined strategy is defined by a pattern and an action but, additionally, it also specifies a default behaviour for the case when the pattern does not match. This default behaviour can be either the `Identity` meaning that no action is performed or `Fail` in which case an exception is raised when the pattern does not match. For example, when applying the strategy

```
%strategy printOwner() extends Identity() {
  visit TNode {
    <Owner gender="M">#TEXT(name)</Owner> -> {
      System.out.println(name);
    }
  }
}
```

to the term `<Owner gender="M">Bob</Owner>` the string "Bob" is printed, while applying it to `<CCAcount><Owner gender="M"> Bob </Owner></CCAccount>` leads to no action since the pattern does not match at the root position and the default behavior is the `Identity`.

The strategy can be fired on the variable `xmlDocument` by the JAVA statement `printOwner().visit(xmlDocument)`. It is important to understand that with such a statement the strategy is only applied on the root node of the corresponding document and that there is no automatic recursive application that would search for a convenient sub-tree. In TOM, the control is explicit and should be specified by an appropriate strategy built using the available strategy primitives. Nevertheless, such higher-level strategies can be easily defined and all the classical strategies are already available in the TOM standard library. For instance, the *top-down* strategy can be recursively defined by TopDown(s) $\triangleq$ s;All(TopDown(s)) where s1;s2 means that, first, s1 is applied, and then s2 is applied on the result of s1. It fails if s1 or s2 fails. The All(s) combinator applies s to all the immediate children of a given node. TopDown(s) corresponds thus to the application of s followed by a recursive application of TopDown(s) to *all* the immediate children. This strategy fails if the application of s fails. Note that the application of All(s) to a constant (*i.e.* a leaf of the tree) does not fail but it simply does nothing.

The execution of TopDown(printOwner()).visit(xmlDocument) prints the names of all the account owners in a bank (independently of the branch). Using other combinators such as <+[2] and One[3], it is easy to define other general purpose strategies such as BottomUp, Innermost, *etc.* More complex tasks can be accomplished by strategies using elaborated non-linear patterns involving (explicit) list matching. For example, if we want to update the initial document and give a 15% bonus to all account owners that have opened a savings account in the same branch then, the following strategy can be used:

```
%strategy bonus() extends Identity() {
    visit TNode {
        <Branch> (A1*,
                  <CCAccount>
                    (X1*,owner,X2*,<Balance>#TEXT(bal)</Balance>,X3*)
                  </CCAccount>,
                  A2*,
                  sa@<SAccount>owner</SAccount>,
                  A3*)
        </Branch> -> {
          TNode newbal =
            xml(<Balance>#TEXT(Double.parseDouble(bal)*1.15)</Balance>;
          return xml(<Branch> A1*
                    <CCAccount>X1* owner X2* newbal X3*</CCAccount>
                    A2* sa A3* </Branch>);
}  }  }
```

In this example we have used explicit lists of the form (X1*,...,X2*,...,X3*) to retrieve the context information (*i.e.* the other XML nodes) needed in order to build the XML tree in the right-hand side of the rule. The TOM construct xml(...) can be used to build a tree (a DOM object in our case) using an XML notation. Once again, the notation sa@... indicates that the matched node is stored in the variable sa and thus, this variable can be used in the right-hand side of the strategy. Due to lack of space, this has not be exemplified, but note that the right-hand side of a rule is an arbitrary list of JAVA and TOM statements

---

[2] s1<+s2 tries to apply s1; if it succeeds, the result is returned, otherwise s2 is applied.

[3] One(s) searches for an immediate children where s can be applied.

and therefore, recursive function calls as well as nested calls to strategies can be freely used.

5   Comparison with Similar Tools

There exist numerous languages aiming at manipulating XML documents. We briefly present some of them and emphasize the main differences with respect to the TOM approach.

XPath is a language providing a concise and efficient syntax for selecting parts of an XML document and querying XML documents. It can be used to describe the search of a particular node in the document in breadth and arbitrary depth. All XPath queries can be encoded within a TOM strategy. However, it is not possible with XPath to have full control over the way the document is explored, as it is with TOM strategies.

XSLT [10,11] is a transformation language for XML, aiming at describing transformations from one XML dialect to another. It uses XPath to select part of the original document and query it, and offers only functional features, thus one can only loop over the results of an XPath query. The result of the application of an XSLT template on a document may only be another document. Contrary to TOM, it is not possible to execute arbitrary actions when examining the initial documents.

The OCamlDuce system [12] is a modified version of the OCaml functional language which integrates XDuce features, such as XML expressions, regular expression types and patterns, iterators. OCamlDuce fully integrates XML manipulations in the OCaml language, providing static type inference for XML expressions, by the mean of regular expression types. This provides a static insurance that a program will produce values of a given XML type. The integration of XML manipulation in TOM cannot provide such guarantee. On the other side, a main advantage of TOM is to be fully integrated in a JAVA environment.

The JAVA standard library provides a DOM implementation, that enables manipulation of XML documents through the DOM API. Additionally, the package javax.xml.xpath does provide an XPath implementation, that enables the JAVA programmer to evaluate XPath expressions over DOM documents. However, this approach is purely interpreted, and does not provide any guarantee on the transformation.

6   Conclusion

We have presented TOM, an extension of JAVA which adds support for algebraic data-types, pattern matching and strategic rewriting, focusing essentially on the XML related features of the language. The powerful pattern-matching construct of TOM allows one to express relatively complex matching conditions using concise and natural patterns. The strategies add more expressive power by providing a simple method for the traversing structured data. We should point out that all the rules as well as the corresponding guiding strategies should be explicitly given and thus no ambiguity concerning their application is possible.

Besides its strong expressive power and its solid semantics, the approach guarantees the portability of the applications that can be executed on top of any JAVA environment. We have mainly shown examples for querying XML documents but, as we have seen in the last example of Section 4, the xml(...) construct can be used to modify and build XML documents.

The applications developed in TOM are independent of the data structure implementation given that a mapping between the respective internal implementation and the TOM representation is given. The TOMML standard libraries already provide this kind of mapping for DOM classes but new mappings can be easily integrated.

References

1. Bray, T., Paoli, J., Sperberg-McQueen, C.M., Eve Maler, F.Y., Cowan, J.: Extensible markup language (XML) 1.1. Technical report, W3C, 2nd edn. (2006), http://www.w3.org/TR/2006/REC-xml11-20060816/
2. Brand, M., Deursen, A., Heering, J., Jong, H., Jonge, M., Kuipers, T., Klint, P., Moonen, L., Olivier, P., Scheerder, J., Vinju, J., Visser, E., Visser, J.: The ASF+SDF Meta-Environment: a Component-Based Language Development Environment. In: Wilhelm, R. (ed.) CC 2001. LNCS, vol. 2027, pp. 365–370. Springer, Heidelberg (2001)
3. Kirchner, H., Moreau, P.E.: Promoting rewriting to a programming language: A compiler for non-deterministic rewrite programs in associative-commutative theories. Journal of Functional Programming 11(2), 207–251 (2001)
4. Clavel, M., Durán, F., Eker, S., Lincoln, P., Martí-Oliet, N., Meseguer, J., Talcott, C.: The maude 2.0 system. In: Nieuwenhuis, R. (ed.) RTA 2003. LNCS, vol. 2706, pp. 76–87. Springer, Heidelberg (2003)
5. Moreau, P.E., Ringeissen, C., Vittek, M.: A Pattern Matching Compiler for Multiple Target Languages. In: Hedin, G. (ed.) CC 2003. LNCS, vol. 2622, pp. 61–76. Springer, Heidelberg (2003)
6. Balland, E., Kirchner, C., Moreau, P.E.: Formal islands. In: Johnson, M., Vene, V. (eds.) AMAST 2006. LNCS, vol. 4019, pp. 51–65. Springer, Heidelberg (2006)
7. Kirchner, C., Kopetz, R., Moreau, P.E.: Anti-pattern matching. In: De Nicola, R. (ed.) ESOP 2007. LNCS, vol. 4421, pp. 110–124. Springer, Heidelberg (2007)
8. van den Brand, M.G.J., de Jong, H.A., Klint, P., Olivier, P.: Efficient annotated terms. Software-Practice and Experience 30, 259–291 (2000)
9. Visser, E., Benaissa, Z.e.A., Tolmach, A.: Building program optimizers with rewriting strategies. In: Proceedings of the 3rd ACM SIGPLAN International Conference on Functional Programming, pp. 13–26. ACM Press, New York (1998)
10. Kay, M.: XSL transformations (XSLT) version 2.0. Technical report, W3C (2007), http://www.w3.org/TR/2006/REC-xml11-20060816/
11. Kay, M.: XSLT 2.0 and XPath 2.0 Programmer's Reference (Programmer to Programmer). Wrox Press Ltd., Birmingham (2008)
12. Frisch, A.: Ocaml + xduce. In: Castagna, G., Raghavachari, M. (eds.) PLAN-X, BRICS, Department of Computer Science, pp. 36–48. University of Aarhus (2006)

Geospatial-Enabled RuleML in a Study
on Querying Respiratory Disease Information

Sheng Gao[1], Harold Boley[2], Darka Mioc[1,3], Francois Anton[4], and Xiaolun Yi[5]

[1] GGE, University of New Brunswick, Fredericton, NB, Canada
[2] Institute for Information Technology, NRC, Fredericton, NB, Canada
[3] National Space Institute, Technical University of Denmark, Denmark
[4] Department of Informatics and Mathematical Modelling, Technical University of Denmark
[5] Service New Brunswick, Fredericton, NB, Canada

Abstract. A spatial component for health data can support spatial analysis and visualization in the investigation of health phenomena. Therefore, the utilization of spatial information in a Semantic Web environment will enhance the ability to query and to represent health data. In this paper, a semantic health data query and representation framework is proposed through the formalization of spatial information. We include the geometric representation in RuleML deduction, and apply ontologies and rules for querying and representing health information. Corresponding geospatial built-ins were implemented as an extension to OO jDREW. Case studies were carried out using geospatial-enabled RuleML queries for respiratory disease information. The paper thus demonstrates the use of RuleML for geospatial-semantic querying and representing of health information.

1   Introduction

Geospatial location provides a solution to link multiple sources in the same area. The spatial component of health data can show the geographical distribution of disease outbreaks, hospitals, air quality, and census. Basic geometric information of location is recorded in spatial data collections, using spatial reference and coordinate arrays. Utilizing spatial information allows the spatial analysis and visualization of health data. For example, with the geometric information of the Georges L. Dumont Hospital in Moncton and the New Brunswick Route 15, the neighboring spatial relationship between them can be deduced. The Semantic Web aims to improve machine understanding of Web-based information and its effective management. By employing Semantic Web (e.g., Web rule) techniques, part of the meaning of the information can be captured by machines, thus enabling more precise information queries and interoperation. To enhance the ability to query health information, its spatial component can also be represented and deduced by rules.

The Semantic Web environment, in which data are given well-defined meaning, can facilitate health data query and knowledge discovery. Similar to the non-spatial attributes of data, the spatial attributes can also be represented in the Semantic Web. The use of spatial information in Semantic Web can support dynamic spatial relationship discovery for health data, and furthermore, new concepts and new instances can

G. Governatori, J. Hall, and A. Paschke (Eds.): RuleML 2009, LNCS 5858, pp. 272–281, 2009.

be generated. For example, from the locations of infectious disease outbreaks, we can determine the sensitive areas that are within a certain distance from the disease outbreak locations. Because of the advantages in supporting the representation of a spatial component, we endeavor to include spatial information in the Semantic Web environment to enhance the ability to query and to represent health data. This paper builds on and extends the eHealthGeo results in Gao et al., [1], and includes the geometric representation in RuleML to enhance information reasoning and inference.

2 Semantic Web and Geospatial Semantics

Semantics-level interoperability among heterogeneous information sources and systems can be achieved by the Semantic Web. According to Sheth and Ramakrishnan [2], three kinds of important applications of the Semantic Web are (1) semantic integration, (2) semantic search and contextual browsing, and (3) semantic analytics and knowledge discovery. Ontologies, as shared specifications of conceptualizations [3], constitute an important notion in the Semantic Web. Many XML-based languages, such as RDF(S) and OWL, have been developed for the representation of ontologies. Description Logic (DL) is usually used to represent ontologies. When concepts are defined using ontologies, three types of relation can be distinguished: taxonomic, functional, and partonomic [4]. With the meaning and relations of concepts defined by ontologies, semantic data classification, integration, and deduction can be implemented. One limitation of DL is that it is impossible to represent relationships between a composite property and another (possibly composite) property in the ontology representation; however, the use of rules can establish more complex relationships between properties [5]. Rules encode machine-interpretable conditional knowledge ("if ... then ...") for automatic reasoning [6]. Rules can describe concepts by using the relation of instances through different property paths. Many different kinds of approaches in combining ontologies and rules have been surveyed (see [7]). RuleML [8] is the de facto open-language standard for Web rules.

Spatial relations can exist between two spatial objects (concepts or instances), and exploring them can advance information query and discovery. Three types of major spatial relations between spatial objects are topological, direction, and metrical relations [9]. Topological relations formalize the notion of neighborhood; directional relations require the existence of a vector space; and metric relations are measuring distances. Topological relations are invariant under continuous translations while directional and metric relations may change during these translations. A well-known method by which to formalize topological relationship between spatial objects in two-dimensional space is the Nine Intersection Model (9IM), developed by Egenhofer, that considers boundaries, interiors, and complements intersection of two spatial objects [10]. The further improved model, the Dimensionally Extended Nine Intersection Model (DE-9IM), considers the 9IM of two spatial objects with the dimensions of -1 (no intersection), 0, 1, or 2 [11, 12]. The commonly known topological predicates described by the DE-9IM include overlaps, touches, within, contains, crosses, intersects, equals, and disjoint.

With possible spatial relations existing in the data, several studies have been done on the capture of geospatial semantics for facilitating data integration, query, and discovery. Kieler [13] discussed the feasibility of identifying semantic relations between different ontologies by exploring the geometric characteristics of the instances. To represent spatial relations, the explicit storage or dynamic computation of spatial relations is possible. Explicating all the possible spatial relations between every two spatial objects is usually not necessary. While the weakness of dynamic computation is that it is time-consuming, the weakness of explicit storage requires significant storage space and involves reliability issues because of the imprecise nature of relations [14]. Klien and Lutz [15] illustrated the definition of geospatial concepts based on spatial relations and automatic annotation of geospatial data using a reference dataset. The annotation process uses DL in reasoning and focuses on the concept level. Smart et al. [16] distinguished multi-representations, implicit spatial relations, and spatial integrity of geospatial data, claiming that rule expression for geo-ontologies needs to consider spatial reasoning rules and spatial integrity rules. Kammersell and Dean [17] proposed GeoSWRL, which is a set of geospatial SWRL built-ins. GeoSWRL allows users to include spatial relation operators in queries; however, spatial data representation and processing abilities are not fully integrated in the GeoSWRL system.

In addition, spatial operations can generate new spatial objects from existing spatial objects, such as spatial intersection and spatial union. Because rules are able to describe relations through complex property paths, it would be feasible to represent spatial operations and spatial relations of geospatial objects as rules in knowledge deduction. Cartographic principles can also be applied as rules in the deduction. In this paper, we not only enable geometric representation support for RuleML reasoning, but also apply ontologies and rules in health information reasoning, query, and representation. The respiratory disease information queries are used as examples.

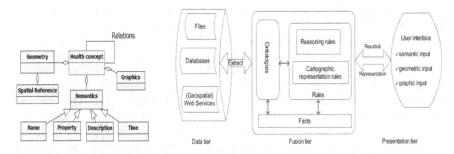

Fig. 1. Metamodel of health concepts   **Fig. 2.** Health data query and representation framework

3   Framework for Health Information Query and Representation

Health concepts can be described with semantic, geometric, and (carto)graphic components, as shown in Figure 1. The semantic component deals with the definition of the concepts. The geometric component provides shapes to locate the concepts. The graphic component solves the issues of how to represent these concepts through maps. For example, in the case of a hospital, the semantic component can describe its name and attributes, the geometric component can describe its polygon shape, and the graphic component can describe its map style. Moreover, relations, including non-spatial and spatial relations, exist between health concepts.

3.1 Framework

Semantic health data queries need to find data with corresponding semantic and geometric attributes. Cartographic representation of the query results allows users to visualize health information. Figure 2 describes the framework for the semantic health information query and representation, including a data tier, a fusion tier, and a presentation tier. (1) Data tier. The health data are obtained from various organizations through files, databases, or (Geospatial) Web Services. Following the ontology implementation, data are extracted to the knowledge base as facts. (2) Fusion tier. The fusion tier contains ontologies, facts, and rules. It queries and fuses semantic, spatial, and cartographic information for representing health data homogeneously. The ontologies are the representation of health concepts and their relationships in the semantic, geometric, and graphic dimensions. Facts are generated from various health data and existing knowledge about health. Rules, supported by ontologies and facts, deduce health information and present the information to users. Two types of rules are considered: reasoning rules and cartographic representation rules. (3) Presentation tier. The user interface allows the input of semantic, geometric, and graphic criteria to find health information.

3.2 Ontologies and Rules in Health Data Fusion

A. Ontologies
Ontologies can be utilized to connect various concepts (e.g., subconcepts and superconcepts). Depending on the requirements, different application ontologies exist in health applications. To facilitate health data exchange and query, a global ontology can improve interoperability. Three types of ontologies are important in querying and representing health data: health domain ontologies, geometric ontologies, and cartographic ontologies.

Health domain ontologies are used for the definition of health information models, concepts, and terminologies. Many standards exist in this field, such as Health Level 7 (HL7), SNOMED-CT, and International Classification of Diseases (ICD-9). Geometric ontologies should be able to describe basic geometry types, such as point and polygon. The European Petroleum Survey Group (EPSG, http://www.epsg.org/) coordinate system codes are widely used in the exchange of geospatial data over the Internet. Cartographic ontologies deal with the styles in representing information. For instance, the symbol of hospitals can be represented as point graphics to show the location of hospitals. With the existence of health domain ontologies, geometric ontologies, and cartographic ontologies, the application ontology definition can easily link to these semantic, geometric, and cartographic elements.

B. Rules
Based on the ontologies and facts, rules can define and deduce new information. Besides non-spatial attribute rules, spatial rules can also be applied in this framework. Although the definition of geometric ontologies follows the same methodology for non-spatial ontologies, the inference of geometric relations is different. The utilization of geometries can incorporate the spatial analysis and cartographic representation abilities in rules. Two types of rules are distinguished: reasoning rules and cartographic representation rules.

Reasoning rules cover semantic matching, spatial relation operators, spatial operations, and cartographic comparison of data. (1) Semantic matching rules deal with the domain knowledge for understanding health data. For instance, the manifestation of several symptoms could determine that a patient may have caught a disease. (2) Spatial relation rules are used to determine the topological, directional, and metric relations between geospatial components. For instance, rules can be used to evaluate the direction and distance from the location of an emergency to hospitals. (3) Spatial operation rules can generate new concepts and instances from existing health data. For example, spatial union can combine data from neighboring regions to assist in the comparison of disease outbreaks. (4) Cartographic comparison rules are able to fuse different cartographic representations into a homogeneous form.

Cartographic representation rules focus on the distribution of information to users more efficiently and effectively. Map scale is of great significance in the geometric representation of a concept. For example, a hospital will be shown as a polygon in large scale representation and as a point in small scale representation. Cartographic rules include concept-based rules, attribute-based rules, scale-based rules, priority-based rules, and cartographic generalization rules. (1) Concept-based rules determine graphic styles based on the health concept semantics. For example, standard symbols exist in representing the concepts in national or provincial cartographic design. (2) Attribute-based rules classify health concepts based on their attributes. For example, pie charts can show the age distribution of people in each health region. (3) Scale-based rules are essential in determining what information is represented based on scales. A concept can be stored with multi-representation in the data, and scale can be used to select the optimal representation. (4) Priority-based rules emphasize high priority information. (5) Cartographic generalization (simplification, exaggeration, and displacement) rules allow the dynamic generalization of spatial information.

4 Design and Implementation

4.1 Geospatial Support for RuleML Deduction

OO jDREW is an open source RuleML engine which is used in this study because it supports RuleML's Naf Hornlog sublanguage and backward/forward reasoning [18]. RuleML's POSL presentation syntax is employed in the following. To use spatial information in the reasoning process, the representation of spatial information in the RuleML engine is needed. Therefore, a geometric ontology is designed to support basic geometry types: point, linestring, polygon, multipoint, multilinestring, multipolygon, and multimix, as shown in Figure 3. A polygon can have an out boundary and many inner holes (inner boundaries). Multipoint, multilinestring, and mulitpolygon can have one or more points, linestrings, and polygons respectively. Multimix contains collections of points, linestrings, and polygons. Figure 4 lists examples of how to represent each geometry type. Coordinate reference systems are specified with EPSG codes, and coordinates are recorded in the order of (x1,y1,x2,y2,...). With the specification of geometries, the spatial operation (union, buffer, convexhull, difference, distance, intersection) and spatial relation operators (touches, contains, within, crosses, equals, overlaps, intersects, covers, coveredby, disjoint, iswithindistance) can be incorporated into rules.

Based on this design, a geometry type was added and a parser implemented for parsing geometries in OO jDREW. For the geospatial operations and spatial relation operations, the JTS Topology Suite is used in this study. The JTS is an open source Java API for two-dimensional spatial predicates and functions, using the DE-9IM model [19]. Several geospatial built-ins, such as the gpred_intersects, gpred_within and gfunc_intersection built-ins, were created using the JTS library. The gpred_intersects built-in checks whether two geometries intersect or not; the gpred_within built-in checks whether a geometry is inside another geometry; the gfunc_intersection built-in computes the intersection of two geometries.

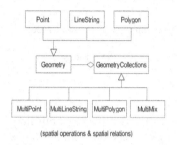

(spatial operations & spatial relations)

Point:	geo[EPSG4326, point[0,0,0.0]]:Geometry
LineString:	geo[EPSG4326, linestring[0.0,0,0,1.0,1.0]]:Geometry
Polygon:	geo[EPSG4326, polygon[outboundary[-1.0,-1.0,-1.0,1.0,1.0,1.0,1.0, -1.0,-1.0,-1.0], innerboundary[-0.5,-0.5,-0.5,0.5,0.5,0.5,0.5,-0.5,-0.5,-0.5]]]:Geometry
MultiPoint:	geo[EPSG4326, multipoint[point[0.0,0.0],point[1.0,1.0]]]: Geometry
Multi-LineString:	geo[EPSG4326, multilinestring[linestring[0.0,0.0,1.0,1.0], linestring[2.0,2.0,5.0,5.0,3.5,4.6]]]:Geometry
Multi-Polygon:	geo[EPSG4326,multipolygon[polygon[outboundary[0.0,0.0,0.0,1.0,1.0,1.0,1.0,0.0,0.0,0.0]], polygon[outboundary[-1.0,-1.0,-1.0,1.0, 1.0,-1.0,-1.0,-1.0]]]]:Geometry
MultiMix:	geo[EPSG4326, multimix[point[0.0,0.0], polygon[outboundary[-1,-1,-1,1, 1,-1,-1,-1]]]]:Geometry

Fig. 3. Geometry type designed for RuleML

Fig. 4. Examples of geometry representation

4.2 Data Sources and Ontology Definition

The health data used in this study were collected from different organizations, such as New Brunswick Lung Association, Service New Brunswick, Statistic Canada census, and Statistic Canada community health survey. Respiratory disease data are used as examples in this study. Following the disease taxonomy of respiratory diseases in the International Classification of Diseases (ICD-9), we created an ontology for respiratory diseases. Respiratory disease data are from hospital patient incidents, which record the time, postcode, disease diagnosis category, age, and gender. Different data could be collected in various spatial boundaries. Taking this study as an example, the disease rate data from the Statistics Canada community health survey were collected at *Health region* and the income data from Statistics Canada census were collected at *Census division*. From these data, the application ontologies of our case study were generated. We also created entities, such as *Health event, Hospital, Health region, Census division, Postcode, Disease rate*, and *Income*.

Health event can describe a variety of cases, such as patient incidents, health training services, etc. The following properties (POSL: "->") associated with health events are shown here: the involved participants' age and gender, the admit date, the disease category diagnosis, and the postcode. Example with a variable (POSL: "?"):

health_event (disease->?:Influenza_with_pneumonia; age->88:Integer; gender->Female; postcode->E1C; admitdate->date[2000:Integer,1:Integer,1:Integer]).

Hospital introduces general information about hospitals, with attributes: name, address, city, province, telephone, and geometry. Example:

hospital (name->Dr_Everett_Chalmers_Hospital; address->700_Priestman_St; province->NB; city->Fredericton; telephone->5064525400; totalbeds->384:Integer; geometry->...).

Health region and *Census division* are two kinds of administrative boundaries. They have name, area, perimeter, and geometry attributes. Example:

health_region (name->Health_region_1; area->10455463176.5:Real;
              perimeter-> 844278.079968:Real; geometry->geo[EPSG4326,
              multipolygon[polygon[outboundary[...]],...]]:Geometry).

Postcode shows the central location of the three-digital postcodes. Example:

pcode3 (name->E1A;
       geometry->geo[EPSG4326,point[-64.7078903603,46.0967513316]]:Geometry).

Disease rate and *Income* show the value associated with the geometry name, statistic method, and year. Example:

disease_rate (disease->?:Asthma; geometryname->Health_region_1; statistics->average;
             year->2003:Integer; rate->0.104:Real).
income (geometryname->Saint_John_County; statistics->average; year->2003:Integer;
       incomevalue->32748.56028:Real).

4.3 Scenarios

Case 1. With the collected health events, it is possible to find disease cases fulfilling semantic and geometric requirements. Since disease cases include outbreak locations using postcodes, geospatial semantic query of diseases can discover whether the location of a postcode is inside any spatial boundary. The following disease_locator rule queries a patient's age, gender, and postcode within a certain health region, disease category, age type, and period:

disease_locator (healthregionname->?name; disease->?disease:Respiratory_diseases;
                startdate->?startdate; enddate->?enddate; agetype->?agetype;
                age->?age:Integer; gender->?gender; postcode->?postcode) :-
health_event (disease->?disease:Respiratory_diseases; age->?age:Integer;gender->?gender;
             postcode->?postcode; admitdate->?date),
  age (agetype->?agetype; age->?age:Integer),
  earlier (?date, ?enddate), later (?date, ?startdate),
  health_region (name->?name; geometry->?hrgeometry:Geometry!?),
  pcode3 (name->?postcode; geometry->?pcgeometry:Geometry!?),
  gpred_within (?pcgeometry:Geometry, ?hrgeometry:Geometry).

The *disease_locator* rule conjoins several subqueries for the semantic query of disease cases. The *earlier* and *later* queries search disease cases in which the admit date is between the start date and end date. The *age* query is used to determine to which age group a certain age belongs. The *gpred_within* built-in query is used to locate postcodes in health regions.

Case 2. Since data collected from different organizations may use different kinds of spatial boundaries, the ability to integrate those data is useful. New concepts and instances will be generated in the integration process. The below disease_income_correlator rule figures out the intersection between disease rate and income. For example, a user would like to know those spatial areas where the asthma disease rate is higher than 0.1 and the average income is above $30,000 in 2008.

disease_income_correlator (disease->?disease:Respiratory_diseases; year->?year:Integer;
    minincome->?minincome:Real;minrate->?minrate:Real; geometry->?geometry:Geometry):-
    disease_rate(disease->?disease:Respiratory_diseases; year->?year:Integer;

```
                    geometryname->?dgeometryname; rate->?rate:Real!?),
income (geometryname->?igeometryname;
          year->?year:Integer;incomevalue->?incomevalue:Real!?),
health_region (name->?dgeometryname;geometry->?hrgeometry:Geometry!?),
census_division (name->?igeometryname;geometry->?cdgeometry:Geometry!?),
greaterThan (?rate:Real,?minrate:Real), greaterThan (?incomevalue:Real,?minincome:Real),
gfunc_intersection (?geometry:Geometry,?hrgeometry:Geometry,?cdgeometry:Geometry).
```

Case 3. To provide better representation of the information to users in the query process, it is beneficial to allow users to define queries with semantic, geometric, and graphic requirements. For example, a user wants to get the asthma rate in 2008 (semantic) in a spatial boundary geometry1 (geometric) with a graduated color ramp1 (graphic). Firstly, the user can define geometric and graphic requirements. The graphics here use graduated color with two categories. One category ranges from 0.0 to 0.2 in green; the other category ranges from 0.2 to 1 in red:

```
geometries (geometryname->geometry1;
          geometry->geo[EPSG4326, polygon[outboundary[...]]]:Geometry).
graduated_colors (name->ramp1; startvalue->0.0:Real; endvalue->0.2:Real; color->0x00FF00).
graduated_colors (name->ramp1; startvalue->0.2:Real; endvalue->1:Real; color->0xFF0000).
```

Then, the user can use the disease_rate_finder rule to query disease rates. This rule deduces the graphics for disease rate instances within specified geospatial boundaries.

```
disease_rate_finder (disease->?disease:Respiratory_diseases; geometryname->?geometryname;
          rampname->?rampname; year->?year:Integer;
          geometryname->?healthregionname; color->?color):-
    disease_rate  (disease->?disease:Respiratory_diseases;  geometryname->?healthregionname;
          year->?year:Integer; rate->?rate:Real!?),
    health_region (name->?healthregionname; geometry->?hrgeometry:Geometry!?),
    geometries (geometryname->?geometryname; geometry->?geometry:Geometry),
    graduated_colors (name->? rampname;startvalue->?startvalue:Real;
          endvalue->?endvalue:Real; color->?color),
    greaterThanOrEqual (?rate:Real,?startvalue:Real), lessThan (?rate:Real,?endvalue:Real),
    gpred_intersects (?geometry:Geometry,?hrgeometry:Geometry).
```

Case 4. Depending on the scale of representation, the cartographic information represented to users could be different. For example, between the scale of 1:1,000 to 1:1, hospitals are shown as polygons. Between the scale of 1:1,000,000 and 1:1,000, hospitals are shown as points. With the scale smaller than 1:1,000,000, hospitals disappear. In this case, we can add a minimum scale and maximum scale in the hospital entity for the cartographic representation purpose. The following sample fact shows one geometric representation for the multi-representation of a hospital:

```
hospital (name->Dr_Everett_Chalmers_Hospital;address->700_Priestman_St;
          city->Fredericton;province->NB; telephone->5064525400; totalbeds->384:Integer;
          minscale->0.001:Real; maxscale->1:Real; geometry->geo[EPSG4326,
          polygon[outboundary[66.65654990041024, 45.93896756130009,...]]]:Geometry).
```

With a scale input by users, this rule finds the optimal representation of hospitals:

```
hospital_locator (name->?name; scale->?scale:Real; geometry->?geometry:Geometry;
          totalbeds->?totalbeds:Integer):-
    hospital (name->?name; geometry->?geomery:Geometry; minscale->?minscale:Real;
          maxscale->?maxscale:Real; totalbeds->?totalbeds:Integer!?),
    lessThan (?scale:Real,?maxscale:Real), greaterThanOrEqual (?scale:Real,?minscale:Real).
```

Complex queries can then be supported by combining the available predicates exemplified in the above cases. For example, users can define the spatial area of interest (using customized geometries of Case 3). Then they may like to know where high disease rate and low income values exist within the area of interest (using the disease_income_correlator of Case 2). After that, users can get the information about hospitals within the previously determined high disease rate and low income areas in a certain map scale (using the hospital_locator of Case 4 and gpred_within of Case 1). All these steps can be chained into complex rules to formalize user queries.

5 Discussion and Conclusions

Our health data query and representation framework provides a solution for health experts to express knowledge as ontologies and rules (regarding semantic, geometric, and graphic dimensions) in health information integration and representation. The use of rule techniques enables health experts to exchange reasoning and representation rules on the Web. Much research has been done on semantic health information integration and query using non-geospatial information in the reasoning. However, fewer investigations utilize geometric information for dynamic spatial reasoning in this process. This research builds an integrated system that supports geospatial-enabled semantic health information retrieval. A basic geometric ontology is designed for the spatial component representation. Spatial operations and spatial relations are expressed in RuleML for knowledge representation and deduction. Basic geometries, spatial operations, and spatial relation operators for RuleML are enabled through the extension of the OO jDREW engine. This implementation thus facilitates semantic health data integration and query with the use of both non-spatial and spatial operations and relations. Complex queries and reasoning processes can be implemented to allow the use of semantic, geometric, and graphic dimensions.

The current system implementation uses the interface of OO jDREW in the query process. More customized user interfaces in the presentation tier will be implemented to facilitate health information query and representation. Moreover, as dynamic spatial reasoning and computation has demanding time and memory requirements, the balance between caching computed results and dynamic spatial computation need to be optimized for efficient health information querying. In addition, the ontology designed in this study is based on the data collected. Its implementation supports the transformation of various health data to facts in the knowledge base. To further improve data integration and query, upper-level or domain-level ontologies need to be investigated. Various health and geospatial standards can be taken into consideration, such as the HL7 ontology and Open Geospatial Consortium (OGC) standards.

With the rapid growth of health data, the semantic query of health information becomes increasingly important for health practitioners in understanding health phenomena. The support of spatial operations and spatial relation operators by rule systems is useful for health data integration, query, and representation. In this study, an integrated semantic system has been built to support geospatial-enabled query and reasoning of health information. With the use of RuleML, we have enabled geometry types, spatial operation rules, and spatial relation rules for health information query. The case scenarios in this study demonstrate the benefits of including a geospatial component in semantic health data query, permitting the fusion of various kinds of data in the semantic, geometric, and graphic dimensions. This research fosters the use of ontologies and rules in representing these dimensions of public health information. It facilitates the deduction of information collected by different health organizations. Our future work will be devoted to the exploration of ontologies and rules for further semantic integration, query, and representation of health information.

References

1. Gao, S., Mioc, D., Boley, H., Anton, F., Yi, X.: A RuleML Study on Integrating Geographical and Health Information. In: Bassiliades, N., Governatori, G., Paschke, A. (eds.) RuleML 2008. LNCS, vol. 5321, pp. 174–181. Springer, Heidelberg (2008)
2. Sheth, A.P., Ramakrishnan, C.: Semantic (Web) Technology In Action. IEEE Data Engineering Bulletin. 26(4), 40–48 (2003)
3. Gruber, T.R.: A Translation Approach to Portable Ontology Specifications. International Knowledge Acquisition 5(2), 199–220 (1993)
4. Luscher, P., Burghardt, D., Weibel, R.: Ontology-Driven Enrichment of Spatial Databases. In: 10th ICA Workshop on Generalisation and Multiple Representation, Moscow (2007)
5. Antoniou, G., Damasio, C.V., Grosof, B., Horrocks, I., Kifer, M., Maluszynski, J., Patel-Schneider, P.F.: Combining Rules and Ontologies: A survey (2005)
6. Boley, H.: Are Your Rules Online? Four Web Rule Essentials. In: Paschke, A., Biletskiy, Y. (eds.) RuleML 2007. LNCS, vol. 4824, pp. 7–24. Springer, Heidelberg (2007)
7. Bruijn, J.D.: ONTORULE: ONTOlogies meet business RULEs, State-of-the-art survey of issues, http://ontorule-project.eu/deliverables-and-resources?func=fileinfo&id=1
8. The Rule Markup Initiative, http://www.ruleml.org/
9. Rashid, A., Shariff, B.M., Egenhofer, M.J., Mark, D.M.: Natural-Language Spatial Relations between Linear and Areal Objects. Int. J. Geogr. Inf. Sci. 12, 215–245 (1998)
10. Egenhofer, M.J.: Reasoning about Binary Topological Relations. In: Günther, O., Schek, H.-J. (eds.) SSD 1991. LNCS, vol. 525, pp. 143–160. Springer, Heidelberg (1991)
11. Clementini, E., Felice, P.: A Comparison of Methods for Representing Topological Relationships. Inf. Sci. 80, 1–34 (1994)
12. Clementini, E., Felice, P.: A Model for Representing Topological Relationships between Complex Geometric Features in Spatial Databases. Inf. Sci. 90, 121–136 (1996)
13. Kieler, B.: Derivation of Semantic Relationships between Different Ontologies with the Help of Geometry. In: Workshop at AGILE 2008, Girona, Spain (2008)
14. Jones, C.B., Abdelmoty, A.I., Fu, G.: Maintaining Ontologies for Geographical Information Retrieval on the Web. In: Meersman, R., Tari, Z., Schmidt, D.C. (eds.) CoopIS 2003, DOA 2003, and ODBASE 2003. LNCS, vol. 2888, pp. 934–951. Springer, Heidelberg (2003)
15. Klien, E., Lutz, M.: The Role of Spatial Relations in Automating the Semantic Annotation of Geodata. In: Cohn, A.G., Mark, D.M. (eds.) COSIT 2005. LNCS, vol. 3693, pp. 133–148. Springer, Heidelberg (2005)
16. Smart, P.D., Abdelmoty, A.I., El-Geresy, B.A., Jones, C.B.: A framework for combining rules and geo-ontologies. In: First International Conference on Web RR Systems (2007)
17. Kammersell, W., Dean, M.: Conceptual Search: Incorporating Geospatial Data into Semantic Queries. In: Terra Cognita - Directions to the Geospatial Semantic Web (2006)
18. OO jDREW, http://www.jdrew.org/oojdrew/
19. JTS Topology Suite, http://www.vividsolutions.com/jts/jtshome.htm

Rules and Norms:
Requirements for Rule Interchange Languages
in the Legal Domain

Thomas F. Gordon[1], Guido Governatori[2], and Antonino Rotolo[3]

[1] Fraunhofer FOKUS, Berlin, Germany
thomas.gordon@fokus.fraunhofer.de
[2] NICTA, Queensland Research Laboratory, Brisbane, Australia
guido.governatori@nicta.com.au
[3] CIRSFID, University of Bologna, Bologna, Italy
antonino.rotolo@unibo.it

Abstract. In this survey paper we summarize the requirements for rule interchange languages for applications in the legal domain and use these requirements to evaluate RuleML, SBVR, SWRL and RIF. We also present the Legal
Knowledge Interchange Format (LKIF), a new rule interchange format developed
specifically for applications in the legal domain.

1 Introduction

An extensive research has been devoted in the last years for developing rule languages
in the legal domain. Interesting efforts has been carried out especially in the field of e-
contracting, business processes and automated negotiation systems. This led to devise
new languages, or adjust existing ones, specifically designed for documenting and modeling the semantics of business vocabularies, facts, and rules. Significant examples are
SBVR [39], the case handling paradigm [43], OWL-S [29], ContractLog [30], Sadiq
et al.'s constraint specification framework [37], the Web Service Modeling Ontology
(WSMO) [33], the ConDec language [31], PENELOPE [14], and RuleML for business
rules [21,16].

But legal rules are not only pervasive in modeling e-transactions—where formalizing and handling legal rules and contract clauses is required for providing tools to
support legally valid interactions and/or to legally ground contractual transactions—but
their sound and faithful representation is obviously crucial for representing legislative
documents, regulations, and other sources of law (for instance, in the domains of e-
governance and e-government).

Since the seminal work of Sergot et al. [41], which formalized the British Nationality
Act in a logic programming setting, the AI & Law community has devoted an extensive
effort for modeling many aspects of legal rules and regulations[1]. However, there are
still a few works which address the problem of devising rule interchange languages,

---

[1] The interested reader may consult a large number of relevant works published in the Artificial
Intelligence and Law journal and in the proceedings of conferences such as ICAIL and JURIX.

G. Governatori, J. Hall, and A. Paschke (Eds.): RuleML 2009, LNCS 5858, pp. 282–296, 2009.

properly speaking, for the legal domain. LKIF is probably the first systematic attempt in this regard (LKIF will be discussed here in Section 3.5).

Significant experiences for representing legislative documents throughout XML languages, for instance, are CEN MetaLex [7], SDU BWB (see [27]), LexDania (see [27]), NormeinRete [26], AKOMA NTOSO [2]. Other XML standards in legal domain are CHLexML [10], EnAct [3], Legal RDF [28], eLaw (see [27]), Legal XML (see [27]), LAMS [24], JSMS (see [27]), and UKMF (see [27]). Much of these XML-based attempts are ambitious, valuable and effective. Yet, they mostly focused on representing *legal documents* rather than modeling directly *legal rules*. In addition, some are focused on specific application areas, others model a few aspects of the many concerns that exist in reality, or, if developed in order to be sufficiently general, exhibit some limitations since they are not based on robust or comprehensive conceptual models for representing legal rules to be applied in the legal domain (for a detailed evaluation of modeling, e.g., legislation, see [27]).

In general, many of the drawbacks affecting many existing languages are perhaps due to the fact that there has not yet been an overall and systematic effort to establish a general list of requirements for rule interchange languages in the legal domain or because there is not yet an agreement in particular among the practitioners working in this field. This survey paper is meant to offer a list of minimal requirements to a large audience of computer scientists, legal engineers and practitioners who are willing to model legal rules. These requirements are then discussed with regard to some existing rule interchange languages.

Note that a remarkable, and additional difficulty is that it is sometimes not immediate to adjust and extend existing standards for rule interchange languages when we need to use them in the legal domain. Indeed, although the legal domain has several features which are shared by other domains, some aspects are very specific for the law. Consider the notion of information retrieval. In the law, question-answering "seems more relevant than information retrieval", since "question requires some deduction or inference before an appropriate answer can be given" and "regulations may contain many different articles about the same topic and one can only assess whether something is permitted or not by understanding the full documentation". "A rather detailed understanding is required, in particular, because regulations generally contain complex structures of exceptions" [6, p. 9]. The peculiarity of the legal domain thus poses specific problems for developing suitable and faithful representation languages.

The layout of the paper is as follows. In Section 2 we provide a rather comprehensive list of requirements for devising rule interchange languages. The subsequent sections discuss these requirements to evaluate RuleML, SBVR, SWRL, RIF, and LKIF. Some brief conclusions end the paper.

2 Requirements

The law is a complex phenomenon, which can be analyzed into different branches according to the authority who produces legal norms and according to the circumstances and procedures under which norms are created. But, independently of these aspects, it is possible to identify some general features that norms should enjoy.

First of all, it is widely acknowledged in legal theory and AI & Law that norms have basically a conditional structure like [23,38]

$$\text{if } A_1, \dots, A_n \text{ then } B \tag{1}$$

where $A_1, \dots, A_n$ are the applicability conditions of the norm and B denotes the legal effect which ought to follow when those applicability conditions hold[2].

This very general view highlights an immediate link between the concepts of norm and rule. However, there are many types of rules. The common sense, dictionary meaning of rule is "One of a set of explicit or understood regulations or principles governing conduct within a particular sphere of activity." [1]. In classical logic, rules can be inference rules or material implications. In computer science, rules can be production rules, grammar rules, or rewrite rules.

When we use the term 'rule' in the legal field, we usually mean rule in the regulatory sense. But rules express not only regulations about how to act. For example, von Wright [45] classified norms into the following main types (among others):

1. determinative rules, which define concepts or constitute activities that cannot exist without such rules. These rules are also called in the literature 'constitutive rules'.
2. technical rules, which state that something has to be done in order for something else to be attained;
3. prescriptions, which regulate actions by making them obligatory, permitted, or prohibited. These norms, to be complete, should indicate
   - who (the norm-subjects)
   - does what (the action-theme)
   - in what circumstances (the condition of application) and
   - the nature of their guidance (the mode).

Notice that the notion of norm proposed by von Wright is very general and extends well over the notion of norm in legal reasoning; but in some cases the component of a rule have to modified. For example, legal systems can have provisions to handle changes in the systems itself. Thus, it is possible to have norms about how to change other norms. These rules have again a prescriptive character, but we have to adjust the element, in particular these rules should specify, *what* (the content to be modified), *how* (the new content), in what circumstances, and the nature of the modifications (e.g.,, substitution, derogation, abrogation, annulment, ...)

Many of these aspects have been acknowledged in the field of artificial intelligence and law, where there is now much agreement about the structure and properties of rules [15,32,22,44,38]. Important requirements for legal rule languages from the field of AI & Law include the following:

Isomorphism [5]. To ease validation and maintenance, there should be a one-to-one correspondence between the rules in the formal model and the units of natural language text which express the rules in the original legal sources, such as sections

---

[2] Indeed, norms can be also unconditioned, that is their effects may not depend upon any antecedent condition. Consider, for example, the norm "everyone has the right to express his or her opinion". Usually, however, norms are conditioned. In addition, unconditioned norms can formally be reconstructed in terms of (1) with no antecedent conditions.

of legislation. This entails, for example, that a general rule and separately stated exceptions, in different sections of a statute, should not be converged into a single rule in the formal model.

Reification [15]. Rules are objects with properties, such as

Jurisdiction. The limits within which the rule is authoritative and its effects are binding (of particular importance are spatial and geographical references to model jurisdiction).

Authority [32]. Who produced the rule, a feature which indicates the ranking status of the rule within the sources of law (whether the rule is a constitutional provision, a statute, is part of a contract clause or is the ruling of a precedent, and so on).

Temporal properties [19]. Rules usually are qualified by temporal properties, such as:

1. the time when the norm is in force and/or has been enacted;
2. the time when the norm can produce legal effects;
3. the time when the normative effects hold.

Rule semantics. Any language for modeling legal rules should be based on a precise and rigorous semantics, which allows for correctly computing the legal effects that should follow from a set of legal rules.

Defeasibility [15,32,38]. When the antecedent of a rule is satisfied by the facts of a case, the conclusion of the rule presumably holds, but is not necessarily true. The defeasibility of legal rules breaks down into the following issues:

Conflicts [32]. Rules can conflict, namely, they may lead to incompatible legal effects. Conceptually, conflicts can be of different types, according to whether two conflicting rules

– are such that one is an exception of the other (i.e., one is more specific than the other);
– have a different ranking status;
– have been enacted at different times;

Accordingly, rule conflicts can be resolved using principles about rule priorities, such as:

– *lex specialis*, which gives priority to the more specific rules (the exceptions);
– *lex superior*, which gives priority to the rule from the higher authority (see 'Authority' above);
– *lex posterior*, which gives priority to the rule enacted later (see 'Temporal parameters' above).

Exclusionary rules [32,38,15]. Some rules provide one way to explicitly undercut other rules, namely, to make them inapplicable.

Contraposition [32]. Rules do not counterpose. If some conclusion of a rule is not true, the rule does not sanction any inferences about the truth of its premises.

Contributory reasons or factors [38]. It is not always possible to formulate precise rules, even defeasible ones, for aggregating the factors relevant for resolving a legal issue. For example: "The educational value of a work needs to be taken into consideration when evaluating whether the work is covered by the copyright doctrine of fair use."

Rule validity [19]. Rules can be invalid or become invalid. Deleting invalid rules is not an option when it is necessary to reason retroactively with rules which were valid at various times over a course of events. For instance:

1. The *annulment* of a norm is usually seen as a kind of repeal which invalidates the norm and removes it from the legal system as if it had never been enacted. The effect of an annulment applies *ex tunc*: annulreled norms are prevented from producing any legal effects, also for past events.

2. An *abrogation* on the other hand operates *ex nunc*: The rule continues to apply for events which occured before the rule was abrogated.

Legal procedures. Rules not only regulate the procedures for resolving legal conflicts (see above), but also for arguing or reasoning about whether or not some action or state complies with other, substantive rules [16]. In particular, rules are required for procedures which

1. regulate methods for detecting violations of the law;

2. determine the normative effects triggered by norm violations, such as reparative obligations, namely, which are meant to repair or compensate violations[3].

Normative effects. There are many normative effects that follow from applying rules, such as obligations, permissions, prohibitions and also more articulated effects such as those introduced, e.g., by Hohfeld (see [38]). Below is a rather comprehensive list of normative effects [35]:

Evaluative, which indicate that something is good or bad, is a value to be optimised or an evil to be minimised. For example, "Human dignity is valuable", "Participation ought to be promoted";

Qualificatory, which ascribe a legal quality to a person or an object. For example, "*x* is a citizen";

Definitional, which specify the meaning of a term. For example, "Tolling agreement means any agreement to put a specified amount of raw material per period through a particular processing facility";

Deontic, which, typically, impose the obligation or confer the permission to do a certain action. For example, "*x* has the obligation to do *A*";

Potestative, which attribute powers. For example, "A worker has the power to terminate his work contract";

Evidentiary, which establish the conclusion to be drawn from certain evidence. For example, "It is presumed that dismissal was discriminatory";

Existential, which indicate the beginning or the termination of the existence of a legal entity. For example, "The company ceases to exist";

Norm-concerning effects, which state the modifications of norms such as abrogation, repeal, substitution, and so on.

Persistence of normative effects [20]. Some normative effects persist over time unless some other and subsequent event terminate them. For example: "If one causes damage, one has to provide compensation.". Other effects hold on the condition and only while the antecedent conditions of the rules hold. For example: "If one is in a public office, one is forbidden to smoke".

---

[3] Note that these constructions can give rise to very complex rule dependencies, because we can have that the violation of a single rule can activate other (reparative) rules, which in turn, in case of their violation, refer to other rules, and so forth.

Values [4]. Usually, some values are promoted by the legal rules. Modelling rules sometimes needs to support the representation of *values* and *value preferences*, which can play also the role of meta-criteria for solving rule conflicts. (Given two conflicting rules r_1 and r_2, value v_1, promoted by r_1, is preferred to value v_2, promoted by r_2, and so r_1 overrides r_2.)

An interesting question is whether rule interchange languages for the legal domain should be expressive enough to fully model all the features listed above, or whether some of these requirements can be meet at the reasoning level, at the level responsible for structuring, evaluating and comparing legal arguments constructed from rules and other sources. The following sections will consider this issue when discussing these requirements in the context of some existing rule interchange formats: RuleML, SBVR, SWRL, RIF, and LKIF.

3   Overview of Some Rule Interchange Languages

3.1   The Rule Markup Language (RuleML)

RuleML[4] is an XML based language for the representation of rules. It offers facilities to specify different types of rules from derivation rules to transformation rules to reaction rules. It is capable of specifying queries and inferences in Web ontologies, mappings between Web ontologies, and dynamic Web behaviours of workflows, services, and agents [8]. RuleML was intended as the canonical web language for rules, based on XML markup, formal semantics and efficient implementations. Its purpose is to allow exchange of rules between major commercial and non-commercial rules systems on the Web and various client-server systems located within large corporations to facilitate business-to-customer (B2C) and business-to-business (B2B) interactions over the Web.

RuleML provides a way of expressing business rules in modular stand-alone units. It allows the deployment, execution, and exchange of rules between different systems and tools. It is expected that RuleML will be the declarative method to describe rules on the Web and distributed systems [47]. RuleML arranges rule types in an hierarchical structure comprising reaction rules (event-condition-action-effect rules), transformation rules (functional-equational rules), derivation rules (implicational-inference rules), facts ('premiseless' derivation rules, i.e., derivation rules with empty bodies), queries ('conclusionless' derivation rules, i.e., derivation rules with empty heads) and integrity constraints (consistency-maintenance rules). Each part of a rule is an expression that has specific functions in the rule. The RuleML Hierarchy first directly branches out into two categories: Reaction Rules and Transformation Rules. Transformation Rules then break down into Derivation Rules, that, in turn, subdivide into Facts and Queries. Finally, Queries break down into Integrity Constraints [36].

The way RuleML achieves flexibility and extensibility is based on the use and composition of modules. Each module is meant to implement a particular feature relevant for a specific language or application (e.g., modules for various types of negation, for example, classical negation, and negation as failures). Each module is intended to refer to a semantic interpretation of the feature implemented in the module. However,

---

[4] http://www.ruleml.org

RuleML does not have a mechanism to specify semantic structures on which to evaluate elements of the language.

The key strength of Rules is its extensibility. Thus despite that currently there is no dialect specifically intended for the representation of legal rules a few works proposed extension and interpretation for this area, in particular for the representation of (business) contracts [21,16,18].

The contribution of [21] by Grosof was the proposal of adopting courteous logic programming (a variant of defeasible logic) as execution model for RuleML rule-base corresponding to the clauses of a contract. Accordingly, Grosof's proposal meets the defeasiblity key requirement for modelling legal rules. Technically [21] uses derivation rules, but then a courteous logic program implemented as Sweet Jess rules constitutes executable specifications, where the conclusion of a rule can be executed by a computer program producing effects. Thus the approach bridges the gap among the various types of rules in the RuleML family.

The limitation of [21] is that it does not consider normative effects (i.e., it is not possible to differentiate between obligations and permissions). This limitation has been addresses by Governatori [16], where defeasible logic is extended with the standard deontic operators for obligations, permissions and prohibitions as well as a new special deontic operator to model violations and penalties for the violations. Furthermore [16] distinguishes between constitutive and prescriptive rules. It provides a RuleML compliant DTD for representing the various deontic elements, and discusses various options for the modelling of such notions in defeasible logic. [18] implements [16], in a Semantic Web framework with support for RDF databases, to provide an environment to model, monitor and perform business contracts.

The modelling approach proposed in [16] has proven successful for various legal concepts (for example the legal notion of trust [34]) and it has been extended to cover temporal aspects [20] norm dynamics [19], and it has been applied to the study of business process compliance [17].

3.2   Semantics of Business Vocabulary and Business Rules (SBVR)

SBVR [39] is a standard proposed by the Object Managament Group (OMG) for the representation and formalisation of business ontologies, including business vocabularies, business facts and business rules. The main purpose of SBVR is to give the basis for formal and detailed natural language declarative specifications of business entities and policies. It provides a way to represent statements in controlled natural language as logic structures called semantic formulations. The formal representation is based on several logics including first order logic, alethic modal logic and deontic logic, furthermore it adopts model theoretic interpretations for semantic formulations. It is worth noticing that the focus of SBVR is on modelling not providing a framework for executing the rules.

The two most relevant and salient features of SBVR for the modelling of norms are the introduction of deontic operators to represent obligations and permissions and the use of controlled natural languages for modelling norms. These two features, combined with the underlying formalisation, make SBVR a conceptual language able to capture some of the requirements discussed in Section 2. In particular the requirements about the structural isomorphism and the ability to capture some normative effects.

Unfortunately, the semantics for the deontic modalities is left underspecified and the proposed interpretation suffers from some drawbacks to model norms. First of all it assumes formulas like Barcan formula and its converse that allow for the permutation of universal quantifiers and alethic modalities (i.e., $\Box \forall x \phi(x) \equiv \forall x \Box \phi(x)$). The main consequence for this is that it then forces the used of possible world models with constant domains. While this assumption seems to be harmless, it has some important consequences for the modelling of norms. Recent literature on deontic logic (see, among others, [40]) agrees that normal deontic logics –that is logics that admit necessitation (i.e., from $\vdash \phi$ derive $\vdash O\phi$, where O is the deontic modality for obligation)– are not suitable to model norms. However, any deontic logic based on possible world semantics with constant domains, and having at least one genuine obligation is a normal deontic logic [46]. Thus an adequate model theoretic semantics for the deontic modalities seems problematic. Another problem caused by standard deontic logic is that of contrary-to-duty obligations, i.e., obligations arising from violations of other obligations. It is well known that these cannot be handled properly by standard deontic logics [9]. However, as [16] points out these are frequent in legal documents, and contracts in particular. [39] recognises this limitation and the issue of handling this is left for future versions of the specifications.

SBVR suggests the equivalence $(\phi \rightarrow O\psi) \equiv O(\phi \rightarrow \psi)$ to transform semantic expressions having the deontic operator not as main operator into an expression where the deontic operator is the main operator. The proposed transformation imposes additional non-standard constraints on possible world semantics; moreover the proposed transformation poses some concerns on its conceptual soundness since typically the deontic modality applies just to the conclusion of the rule or to the conditional corresponding to the rule (see the discussion about prescriptive rules in Section 2).

The final drawback of the proposed semantics for SBVR is that, being based on classical first order logic it is not suitable to handle conflicts. But as we have highlighted in the discussion of the requirements, handling conflicts is one of the key features if one wants to use rule to reason with legal rules.

3.3   The Semantic Web Rule Language (SWRL)

The Semantics Web Rule Language (SWRL) is a W3C proposal for a rule interchange format which combines ontologies represented in the Description Logic (DL) subset of OWL with an XML format for rules in the Unary/Binary Datalog subset of the Rule Markup Language (RuleML).[5] While both OWL-DL and Datalog, separately, are decidable subsets of first-order logic, the union OWL-DL and Datalog, as in SWRL, is undecidable.

Three approaches to implementing inference engines for SWRL have been tried. In Hoolet, SWRL files are translated into a language for full first-order logic and a general purpose first-order theorem prover is used to derive inferences, with all the undesirable computational properties this entails. In Bossam, SWRL files are translated into rules for a forward-chaining production rule system[6]. This procedure translates OWL-DL axioms into rules, but with a loss of information, since some information expressable in

---

[5] http://www.w3.org/Submission/SWRL/

[6] http://owl.man.ac.uk/hoolet/

OWL-DL axioms cannot be represented in Bossam's production rule language. Thus the resulting inference engine with this approach is incomplete. Finally, a third approach, taken by Pellet, is to start with tableaux theorem-prover for OWL-DL and extend this to support the "DL-safe" subset of SWRL [42].

Let us now try to evaluate SWRL with respect to the requirements we have identified for modeling and reasoning with legal rules. Since SWRL rules are Horn clauses, it is not possible to model legal rules in an isomorphic way. Most legal rules would need to be modeled using several SWRL rules. Morever, the lack of negation in Horn clause logic is a problem, since both the condtions and conclusions of legal rules are often negated. Perhaps this can be overcome in SWRL to some extent by defining complementary predicates using OWL classes. Since rules are represented in XML in SWRL, they can be reified by giving them identifiers using XML attributes. Similarly, the various properties of legal rules, such as their validity, could presumably also be represented using XML attributes. But since these attributes would be at a meta-level, outside the formal syntax and semantics of the SWRL logic, and since SWRL inherits the monotonic semantics of classical first-order logic, it is not clear how these measures could be used to resolve conflicts among legal rules, using principals like *lex superior* or to reason with exclusionary rules. A further problem is that SWRL provides no standard way to annotate the conditions of rules with information about the distribution of the burden proof, but it should be possible to extend SWRL, again using XML attributes, to provide this information. Semantically, unlike legal rules SWRL rules do contrapose, since they are interpreted as material conditionals of classical logic, but in practice SWRL reasoners are too weak to derive the undesired conclusions. Morever, even if the reasoners were stronger, without some way to represent negative facts, *modus tollens* would never be applicable. If we separate the syntax of SWRL from its semantics, it might be possible to develop a nonmontonic logic which solves some of these problems, while retaining SWRL's syntax, but with some additional XML attributes for annotating rules. But it is difficult to imagine how this approach could satisfy the isomorphism requirement.

3.4   The Rule Interchange Format (RIF)

The Rule Interchange Format (RIF) Working Group of the World-Wide-Web Consortium was established in 2005, about a year afer the SWRL proposal was submitted, with the goal of developing an extensible rule interchange format for the Web, building on prior experience in related initiatives and W3C submissions, included RuleML, SWRL, Common Logic and SBVR, among others.[7]

Like RuleML, RIF is intended to be an extensible framework for a whole family of rule languages, possibly with different semantics. Currently, RIF consists of draft reports for several components, including the following:

RIF Core. Defines an XML syntax for definite Horn rules without function symbols, i.e Datalog, with a standard first-order semantics.
RIF Basic Logic Dialect (RIF-BLD). Defines a language, building on RIF Core, for definite Horn rules with equality and a standard first-order semantics.

---

[7] http://www.w3.org/2005/rules/wg/charter.html

RIF Production Rule Dialect (RIF-PRD). Extends RIF Core to define a language for production rules, i.e condition-action rules in which the actions supported are limited to modifications of the facts in working memory.

RIF RDF and OWL Compatibility. Defines semantics for the integrated use of RIF, RDF and OWL in applications.

RIF Framework for Logic Dialects (RIF-FLD). Defines a framework which may be used to configure RIF dialects. For example, one can choose whether negation is interpreted classically or negation-as-failure, as in logic programming, or whether rules are interpreted as material implications or inference rules.

For the purpose of representing legal rule, RIF Core and RIF-BLD both appear to suffer from the same problems as SWRL, and for the same reasons. The production rule dialect of RIF does not seem relevant, since production rules, with their ability to delete information from working memory, are a procedural programming paradigm which may or may not be useful for implementing a legal reasoning support system, but which are not suitable as a language for modeling legal norms. An interesting question is whether the RIF Framework for Logic Dialects (RIF-FLD) could be used to configure a RIF dialect which is more suitable for modeling legal norms. Although this question requires further research, our initial impression is that the space of configurations possible is limited subsets of first-order logic and well-known logic programming paradigms. Languages suitable for modeling legal norms presumably fall outside of this space, since neither first-order logic nor common logic programming languages provide sufficient support isomorphic modeling of legal, allocating the burden of proof among the parties in a legal dispute, or modeling principals for resolving conflicts, such as *lex superior* for resolving rule conflicts.

3.5   The Legal Knowledge Interchange Format (LKIF)

The Legal Knowledge Interchange Format (LKIF) was developed in a three-year European research project, ESTRELLA[8], which completed its work at the end of 2008 [12,11]. The goal of the ESTRELLA project, with respect to LKIF, was to develop an interchange format for formal models of legal norms which is sufficient for modeling legal knowledge in a broad range of application scenarios, builds on existing standards, especially in the context of the Semantic Web, and informed by the state-of-the-art of the field of Artificial Intelligence and Law.

LKIF is an XML Schema for representing theories and arguments constructed from theories. A theory in LKIF consists of a set of axioms and defeasible inference rules. The language of individuals, predicate and function symbols used by the theory can be imported from an ontology represented in the Web Ontology Language (OWL). Importing an ontology also imports the axioms of the ontology. All symbols are represented using Universal Resource Identifiers (URIs). Other LKIF files may also be imported, enabling complex theories to modularized.

Axioms are named formulas of full first-order logic. The heads and bodies of inference rules are sequences of first-order formulas. All the usual logical operators are

---

[8] IST-4-027655.

supported and may be arbitrarily embedded: disjunction ($\wedge$), conjunction ($\vee$), negation ($\neg$), material implication ($\rightarrow$) and the biconditional ($\leftrightarrow$). Both existential ($\exists$) and universal ($\forall$) quantifiers are supported. Free variables in inference rules represent schema variables.

Terms in formulas may be atomic values or compound expressions. Values are represented using XML Scheme Definition (XSD) datatypes. Atomic formulas are reified and can be used as terms, allowing some meta-level propositions to be expressed.

The schema for atomic formulas has been designed to allow theories to be displayed and printed in plain, natural language, using Cascaded Style Sheets (CSS). An atomic proposition may first be represented in propositional logic, using natural language, and later enriched to become a first-order model, by marking up the variables and constants of the proposition and specifying its predicate using an XML attribute. This feature of LKIF is essential for enabling domain experts, not just computer specialists, to write and validate theories.

Support for allocating the burden of proof when constructing arguments from theories in dialogues is provided. An *assumable* attribute is provided for atomic formulas, to indicate they may be assumed true until they have been challenged or questioned. An *exception* attribute is provided for negated formulas, to indicate that P may be presumed not true *unless* P has been proven. This is similar to negation-as-failure in logic programming, in that the failure to find a proof for P is sufficient to prove $\neg P$, if $\neg P$ is an exception.

Arguments in LKIF link a sequence of premises to a conclusion, where both the premises and the conclusion are atomic formulas. Attributes are provided for stating the direction of the argument (pro or con), the argumentation scheme applied and the role of each premise in an argument. Arguments can be linked together to form argument graphs. The legal proof standard each proposition at issue must satisfy, such as "preponderance of the evidence" or "beyond reasonable doubt" may be specified. Attributes are provided for recording the relative weight assigned to each argument by the finder of fact, such as the jury, or some other audience, as well as the status of each issue in the proceeding.

All of the main elements of an LKIF file may be assigned Universal Resource Identifiers, allowing them to be referenced in other documents, anywhere on the World Wide Web. Cross references between elements of legal source documents and the elements of the LKIF document which model these sources may be included within the LKIF file, using a sequence of *source* elements. The scheme allows m to n relationships between legal sources and elements of the LKIF model to be represented.

LKIF builds on and uses many existing World Wide Web standards, including the XML, Universal Resource Identifiers, XML Namespaces, the Resource Description Framework (RDF) and the Web Ontology Language (OWL). However, for a variety of reasons it does not use other XML schemas for modeling legal rules, such as Common Logic, RuleML, the Semantic Web Rule Language (SWRL), or the Rule Interchange Format (RIF). Common Logic is an ISO standard for representing formulas of first-order classical logic. While LKIF includes a sublanguage for first-order logic, LKIF has been designed to allow formulas of first-order logic to be represented in human readable form in natural language, to ease development, maintenance and validation by domain experts. Moreover, the ISO Common Logic standard does not look like it will be widely adopted within the World Wide Web community, which has its own standards body, the World Wide Web Consortium. RuleML, SWRL and RIF, among other efforts, are competing to become the Web standard for rules. At the beginning of the ESTRELLA project, SWRL was the leading candidate. In the meantime, during the development of LKIF in Estrella, RIF has become the leading contender. But neither SWRL nor RIF are currently expressive enough for the legal domain. Legal rules can be understood as domain-dependent defeasible inference rules. They cannot be adequately modeled as material implications in first-order logic. However, an LKIF theory can in principal import a first-order theory represented in any XML format, to be used as part of the axioms of the theory. This feature of LKIF enables a part of the legal theory to be represented in first-order logic, using whatever format eventually becomes the World Wide Web standard.

A reference inference engine for LKIF, called Carneades, was developed in ES-TRELLA [13]. Carneades is written in a functional style, using the Scheme programming language, and is available as Open Source software.[9]. Carneades places some restrictions on LKIF rules: The heads of rules are limited to literals (postive or negated atomic formulas) and the biconditional ($\leftrightarrow$) operator and first-order quantifiers are not supported. (Free variables, represented schema variables, are supported.) Since LKIF is a very expressive language, the computational complexity of various reasoning tasks can be high, depending on which features of the language have been used in a model. Carneades allows programmers to choose a search strategy (depth-first, breadth-first, iterative deepening), and to develop and plug-in custom, heuristic search strategies. Resource bounds can be set to assure that every search for arguments about an issue terminates in a predictable period of time.

Because of the open-ended nature of legal reasoning, no formal model of a legal domain, in any logic, can guarantee that inferences are legally correct in some absolute sense. The formal model may be incorrect or incomplete. Or the search space may be so large as to make the legal problem undecidable or intractable. Thus legal reasoning and argumentation necessarily has a procedural component. Legal procedures are designed to assure that justifiable decisions can be made in finite time, expending limited resources, as in Loui's conception of resource-bounded, non-demonstrative reasoning [25]. LKIF and Carneades are designed for use in such procedures. The ability of Carneades to generate arguments, making the reasoning of the system transparent and auditable, are essential for documenting and justifying legal decisions.

4 Conclusions

In this paper we outlined a comprehensive list of requirements for rule interchange languages for applications in the legal domain. We used these requirements to assess the suitability of some rule interchange languages, such as RuleML, SBVR, SWRL and RIF, for modeling legal rules. We finally presented the Legal Knowledge Interchange Format (LKIF), a new rule interchange format developed specifically for legal applications.

---

[9] http://carneades.berlios.de

Currently, there is no language, among those that we have examined here, which satisfy all the requirements we have listed in Section 2: all languages have thus their pros and cons. It should be noted, however, that not all those requirements play the same role in the legal domain. While the concept of defeasibility, for example, is almost ubiquitous in the law, others, such as the representation of some temporal properties (in particular, the time when a rule is in force) are definitely more important when we are dealing, e.g., with legislation.

Accordingly, it seems to us that some languages are not currently expressive enough for the legal domain. In particular, RIF and SWRL fail to meet the defeasibility requirement, which is quite fundamental: legal rules are often defeasible and cannot be correctly represented through material implications in first-order logic. Hence, LKIF and RuleML look suitable and more flexible in this regard. Another requirement, among others, which seems crucial for modeling legal rules is the correct representation of the many different types of normative effects, and the need to capture, for example, the deontic concepts. Here, SBVR and RuleML [16], though with some limitations, show how to do that in a rather satisfactory way.

Finally, it should remarked that, for specific types of application, some (but not all) of these requirements can be somehow relaxed. For example, strict isomorphism is not always compulsory if we have to devise a system for monitoring norm compliance but we do not develop additional in-house rules for normalising legal norm, namely, for identifying formal loopholes, deadlocks and inconsistencies, and making hidden conditions (such as chains of reparative obligations) explicit. Without such a mechanism, it may hard to guarantee that a given process is compliant, because we do not know if all relevant norms have been considered. Anyway, we do believe that most of the requirements seem fundamental for representing legal rules.

Acknowledgements

We would like to thank Harold Boley and Monica Palmirani for their valuable comments on earlier versions of this paper.

NICTA is funded by the Australian Government as represented by the Department of Broadband, Communications and the Digital Economy and the Australian Research Council through the ICT Centre of Excellence program.

References

1. Abate, F., Jewell, E.J. (eds.): New Oxford American Dictionary. Oxford University Press, Oxford (2001)
2. Architecture for Knowledge-Oriented Management of African Normative Texts using Open Standards and Ontologies (2009), http://www.akomantoso.org/
3. Arnold-Moore, T.: Automatic generation of amendment legislation. In: Proc. ICAIL 1997. ACM, New York (1997)
4. Bench-Capon, T.: The missing link revisted: The role of teleology in representing legal argument. Artificial Intelligence and Law 10(1-3), 79–94 (2002)
5. Bench-Capon, T., Coenen, F.: Isomorphism and legal knowledge based systems. Artificial Intelligence and Law 1(1), 65–86 (1992)

6. Benjamins, V.R., Casanovas, P., Breuker, J., Gangemi, A. (eds.): Law and the Semantic Web: Legal Ontologies, Methodologies, Legal Information Retrieval and Applications. Springer, Heidelberg (2005)

7. Boer, A., Hoekstra, R., Winkels, R.: Metalex: Legislation in XML. In: Proc. JURIX 2002. IOS Press, Amsterdam (2002)

8. Boley, H., Tabet, S., Wagner, G.: Design rationale for RuleML: A markup language for Semantic Web rules. In: Cruz, I.F., Decker, S., Euzenat, J., McGuinness, D.L. (eds.) Proc. SWWS 2001, The first Semantic Web Working Symposium, pp. 381–401 (2001)

9. Carmo, J., Jones, A.J.: Deontic logic and contrary to duties. In: Gabbay, D., Guenther, F. (eds.) Handbook of Philosophical Logic, 2nd edn., vol. 8, pp. 265–343. Kluwer, Dordrecht (2002)

10. ChLexML (2009), http://www.svri.ch/

11. ESTRELLA Project. Estrella user report. Deliverable 4.5, European Commission (2008)

12. ESTRELLA Project. The legal knowledge interchange format (LKIF). Deliverable 4.3, European Commission (2008)

13. ESTRELLA Project. The reference LKIF inference engine. Deliverable 4.3, European Commission (2008)

14. Goedertier, S., Vanthienen, J.: A declarative approach for flexible business. In: Eder, J., Dustdar, S. (eds.) BPM Workshops 2006. LNCS, vol. 4103, pp. 5–14. Springer, Heidelberg (2006)

15. Gordon, T.F.: The Pleadings Game, An Artificial Intelligence Model of Procedural Justice. Springer, New York (1995), Book version of 1993 Ph.D. Thesis; University of Darmstadt (1995)

16. Governatori, G.: Representing business contracts in RuleML. International Journal of Cooperative Information Systems 14(2-3), 181–216 (2005)

17. Governatori, G., Milosevic, Z., Sadiq, S.: Compliance checking between business processes and business contracts. In: Proc. EDOC 2006, pp. 221–232. IEEE, Los Alamitos (2006)

18. Governatori, G., Pham, D.H.: Dr-contract: An architecture for e-contracts in defeasible logic. International Journal of Business Process Integration and Management 5(4) (2009)

19. Governatori, G., Rotolo, A.: Changing legal systems: Legal abrogations and annulments in defeasible logic. The Logic Journal of IGPL (forthcoming)

20. Governatori, G., Rotolo, A., Sartor, G.: Temporalised normative positions in defeasible logic. In: Proc. ICAIL 2005, pp. 25–34. ACM Press, New York (2005)

21. Grosof, B.: Representing e-commerce rules via situated courteous logic programs in RuleML. Electronic Commerce Research and Applications 3(1), 2–20 (2004)

22. Hage, J.C.: Reasoning with Rules – An Essay on Legal Reasoning and its Underlying Logic. Kluwer Academic Publishers, Dordrecht (1997)

23. Kelsen, H.: General theory of norms. Clarendon, Oxford (1991)

24. Legal and Advice Sectors Metadata Scheme (LAMS5), http://www.lcd.gov.uk/consult/meta/metafr.htm

25. Loui, R.P.: Process and policy: resource-bounded non-demonstrative reasoning. Computational Intelligence 14, 1–38 (1998)

26. Lupo, C., Batini, C.: A federative approach to laws access by citizens: The Normeinrete system. In: Traunmüller, R. (ed.) EGOV 2003. LNCS, vol. 2739, pp. 413–416. Springer, Heidelberg (2003)

27. Lupo, C., Vitali, F., Francesconi, E., Palmirani, M., Winkels, R., de Maat, E., Boer, A., Mascellani, P.: General XML format(s) for legal sources. Technical report, IST-2004-027655 ESTRELLA European project for Standardised Transparent Representations in order to Extend Legal Accessibility: Deliverable 3.1 (2007)

28. McClure, J.: Legal-rdf vocabularies, requirements and design rationale. In: Proc. V Legislative XML Workshop, Florence. European Press (2006)

29. The OWL services coalition: OWL-S 1.2 pre-release (2006),
    http://www.ai.sri.com/daml/services/owl-s/1.2/
30. Paschke, A., Bichler, M., Dietrich, J.: Contractlog: An approach to rule based monitoring and
    execution of service level agreements. In: Adi, A., Stoutenburg, S., Tabet, S. (eds.) RuleML
    2005. LNCS, vol. 3791, pp. 209–217. Springer, Heidelberg (2005)
31. Pesic, M., van der Aalst, W.: A declarative approach for flexible business. In: Eder, J., Dust-
    dar, S. (eds.) BPM Workshops 2006. LNCS, vol. 4103, pp. 169–180. Springer, Heidelberg
    (2006)
32. Prakken, H., Sartor, G.: A dialectical model of assessing conflicting argument in legal rea-
    soning. Artificial Intelligence and Law 4(3-4), 331–368 (1996)
33. Roman, D., Keller, U., Lausen, H., de Bruijn, J., Lara, R., Stollberg, M., Polleres, A., Feier,
    C., Bussler, C., Fensel, D.: Web service modeling ontology. Applied Ontology 1(1), 77–106
    (2005)
34. Rotolo, A., Sartor, G., Smith, C.: Good faith in contract negotiation and performance. Inter-
    national Journal of Business Process Integration and Management 5(4) (2009)
35. Rubino, R., Rotolo, A., Sartor, G.: An OWL ontology of fundamental legal concepts. In:
    Proc. JURIX 2006, pp. 101–110 (2006)
36. RuleML. The Rule Markup Initiative August 20 (2009), http://www.ruleml.org
37. Sadiq, S., Orlowska, M., Sadiq, W.: Specification and validation of process constraints for
    flexible workflows. Information Systems 30(5), 349–378 (2005)
38. Sartor, G.: Legal reasoning: A cognitive approach to the law. In: Pattaro, E., Rottleuthner,
    H., Shiner, R., Peczenik, A., Sartor, G. (eds.) A Treatise of Legal Philosophy and General
    Jurisprudence, vol. 5. Springer, Heidelberg (2005)
39. OMG: Semantics of business vocabulary and business rules (SBVR) (2008),
    http://www.businessrulesgroup.org/sbvr.shtml
40. Sergot, M.: A computational theory of normative positions. ACM Transactions on Computa-
    tional Logic 2(4), 581–622 (2001)
41. Sergot, M., Sadri, F., Kowalski, R., Kriwaczek, F., Hammond, P., Cory, H.: The British Na-
    tionality Act as a logic program. Communications of the ACM 29(5), 370–386 (1986)
42. Sirin, E., Parsia, B., Grau, B., Kalyanpur, A., Katz, Y.: Pellet: A practical OWL-DL reasoner.
    Web Semantics 5(2), 51–53 (2007)
43. van der Aalst, W., Weske, M., Grünbauer, D.: Case handling: a new paradigm for business
    process support. Data Knowledge Engineering 53(2), 129–162 (2005)
44. Verheij, B.: Rules, Reasons, Arguments. Formal Studies of Argumentation and Defeat. Ph.d.,
    Universiteit Maastricht (1996)
45. von Wright, G.H.: Norm and Action. Routledge, London (1963)
46. Waagbø, G.: Quantified modal logic with neighborhood semantics. Zeitschrift für Mathema-
    tische Logik und Grundlagen der Mathematik 38, 491–499 (1992)
47. Wagner, G., Antoniou, G., Tabet, S., Boley, H.: The abstract syntax of RuleML – towards a
    general web rule language framework. In: Proc. Web Intelligence 2004, pp. 628–631. IEEE,
    Los Alamitos (2004)

A Java Implementation of Temporal Defeasible Logic

Rossella Rubino and Antonino Rotolo

CIRSFID/Law School, University of Bologna, Italy
{rossella.rubino,antonino.rotolo}@unibo.it

Abstract. In this paper we report on a Java implementation of a variant of Temporal Defeasible Logic, an extension of Defeasible Logic developed to capture the concept of temporal persistence. The system consists of three elements: a graphical user interface for selecting defeasible theories, and for visualizing conclusions; a parser, which translates sets of rules in TXT or RuleML formats; and the inference engine to compute conclusions.

1 Introduction

Defeasible Logic (DL) is based on a logic programming-like language and, over the years, proved to be a flexible formalism able to capture different facets of non-monotonic reasoning (see [2]). Standard DL has a linear complexity [15] and has also several efficient implementations (e.g., [6]).

DL has been recently extended to capture the temporal aspects of several phenomena, such as legal positions [12] and modifications (e.g., [11]), and deadlines [8]. The resulting logic, called Temporal Defeasible logic (TDL), has been developed to model the concept of temporal persistence within a non-monotonic setting and, remarkably, it preserves the nice computational properties of standard DL [10]. In addition, this logic distinguishes between permanent and transient (non-permanent) conclusions, which makes the language suitable for applications, for example, in the legal domain, where normative effects may persist over time unless some other and subsequent events terminate them (example: "If one causes damage, one has to provide compensation"), while other effects hold on the condition and only while the antecedent conditions of the rules hold (example: "If one is in a public office, one is forbidden to smoke").

We believe that TDL and the present implementation should be interesting for the RuleML community, since Courteous Logic Programming, which is one of the many variants of DL [4], has been advanced as the inferential engine, e.g., for business contracts represented in RuleML (see, e.g., [14]). In short, when we need to deal with time (especially in normative contexts), we think that TDL could be adopted as a model for developing suitable inferential mechanisms for RuleML, whose syntax can be extended to represent the TDL features related with the notion of temporal persistence [16].

So far, no implementation of TDL has been developed. To the best of our knowledge, this work reports on the first attempt. The layout of the paper is as follows. In Section 2 we briefly outline a variant of TDL. In Section 3 we describe the system implementing the logic, which consists of a graphical user interface, a parser for translating rules expressed in TXT and RuleML formats, and the inference engine. For space reasons, we will only provide some details on the third and last component (Section 3.1). Some conclusions end the paper.

G. Governatori, J. Hall, and A. Paschke (Eds.): RuleML 2009, LNCS 5858, pp. 297–304, 2009.

2 Temporal Defeasible Logic

The language of propositional TDL is based on the concept of *temporal literal*, which is an expression such as l^t (or its negation, $\neg l^t$), where l is a literal and t is an element of a discrete totally ordered set $\mathscr{T}$ of instants of time $\{t_1, t_2, \dots\}$: l^t intuitively means that l holds at time t. Given a temporal literal l the complement $\sim l$ is $\neg p^t$ if $l = p^t$, and p^t if $l = \neg p^t$.

A *rule* is an expression $lbl : A \hookrightarrow^x m$, where lbl is a unique label of the rule, A is a (possibly empty) set of temporal literals, $\hookrightarrow \in \{\rightarrow, \Rightarrow, \rightsquigarrow\}$, m is a temporal literal and x is either π or τ signaling whether we have a *persistent* or *transient* rule. *Strict rules*, marked by the arrow $\rightarrow$, support indisputable conclusions whenever their antecedents, too, are indisputable. *Defeasible rules*, marked by $\Rightarrow$, can be defeated by contrary evidence. *Defeaters*, marked by $\rightsquigarrow$, cannot lead to any conclusion but are used to defeat some defeasible rules by producing evidence to the contrary. A *persistent* rule is a rule whose conclusion holds at all instants of time after the conclusion has been derived, unless interrupting events occur; *transient* rules establish the conclusion only for a specific instant of time. Thus $ex_1 : p^5 \Rightarrow^\pi q^6$ means that if p holds at 5, then q defeasibly holds at time 6 and continues to hold after 6 until some event overrides it. The rule $ex_2 : p^5 \Rightarrow^\tau q^6$ means that, if p holds at 5, then q defeasibly holds at time 6 but we do not know whether it will persist after 6. Note that we assume that defeaters are only transient: if a persistent defeasible conclusion is blocked at t by a transient defeater, such a conclusion no longer holds after t unless another applicable rule reinstates it.

We use some abbreviations. Given a rule r and a set R of rules, $A(r)$ denotes the antecedent of r while $C(r)$ denotes its consequent; R^π denotes the set of persistent rules in R, and $R[\psi]$ the set of rules with consequent ψ. R_s, R_{sd} and R_{dft} are respectively the sets of strict rules, defeasible rules, and defeaters in R.

There are in TDL three kinds of features: facts, rules, and a superiority relation among rules. Facts are indisputable statements, represented by temporal literals. The superiority relation describes the relative strength of rules, i.e., about which rules can overrule which other rules. A *TDL theory* is a structure $(F, R, \prec)$, where F is a finite set of facts, R is a finite set of rules and $\prec$ is an acyclic binary superiority relation over R.

TDL is based on a constructive inference mechanism based on tagged conclusions. Proof tags indicate the strength and the type of conclusions. The strength depends on whether conclusions are indisputable (the tag is Δ), namely obtained by using facts and strict rules, or they are defeasible (the tag is ∂). The type depends on whether conclusions are obtained by applying a persistent or a transient rule: hence, conclusions are also tagged with π (persistent) or τ (transient).

Provability is based on the concept of a derivation (or proof) in a TDL theory D. Given a TDL theory D, a *proof* P from D is a finite sequence of tagged temporal literals such that: (1) each tag is either $+\Delta^\pi$, $-\Delta^\pi$, $+\partial^\pi$, $-\partial^\pi$, $+\Delta^\tau$, $-\Delta^\tau$, $+\partial^\tau$, or $-\partial^\tau$; (2) the proof conditions *Definite Provability* and *Defeasible Provability* given below are satisfied by the sequence P^1.

---

[1] Given a proof P we use $P(n)$ to denote the n-th element of the sequence, and $P[1..n]$ denotes the first n elements of P.

The meaning of the proof tags is a follows:

- $+\Delta^\pi p^{t_p}$ (resp. $+\Delta^\tau p^{t_p}$): we have a definite derivation of p holding from time t_p onwards (resp. p holds at t_p);
- $-\Delta^\pi p^{t_p}$ (resp. $-\Delta^\tau p^{t_p}$): we can show that it is not possible to have a definite derivation of p holding from time t_p onwards (resp. p holds at t_p);
- $+\partial^\pi p^{t_p}$ (resp. $+\partial^\tau p^{t_p}$): we have a defeasible derivation of p holding from time t_p onwards (resp. p holds at t_p);
- $-\partial^\pi p^{t_p}$ (resp. $-\partial^\tau p^{t_p}$): we can show that it is not possible to have a defeasible derivation of p holding from time t_p onwards (resp. p holds at t_p).

The inference conditions for $-\Delta$ and $-\partial$ are derived from those for $+\Delta$ and $+\partial$ by applying the Principle of Strong Negation [3]. For space reasons, in what follows we show only the conditions for $+\Delta$ and $+\partial$.

Definite Provability
If $P(n+1) = +\Delta^x p^{t_p}$, then
1) $p^{t_p} \in F$ if $x = \tau$; or
2) $\exists r \in R_s^x[p^{t'_p}]$ such that
$\quad \forall a^{t_a} \in A(r) : +\Delta^y a^{t_a} \in P[1..n]$
where:

(a) $y \in \{\pi, \tau\}$;
(b) if $x = \pi$, then $t'_p \le t_p$;
(c) if $x = \tau$, then $t'_p = t_p$.

Defeasible Provability
If $P(n+1) = +\partial^x p^{t_p}$, then
1) $+\Delta^x p^{t_p} \in P[1..n]$ or
2) $-\Delta^x \sim p^{t_p} \in P[1..n]$ and
2.1) $\exists r \in R_{sd}^x[p^{t'_p}]$ such that
$\quad \forall a^{t_a} \in A(r) : +\partial^y a^{t_a} \in P[1..n]$, and
2.2) $\forall s \in R^y[\sim p^{t \sim p}]$ either
$\quad$ 2.2.1) $\exists b^{t_b} \in A(s), -\partial^y b^{t_b} \in P[1..n]$ or
$\quad$ 2.2.2) $\exists w \in R^y[p^{t \sim p}]$ such that
$\quad\quad \forall c^{t_c} \in A(w) : +\partial^y c^{t_c} \in P[1..n]$ and
$\quad\quad s \prec w$

where:

1. $y \in \{\pi, \tau\}$;
2. if $x = \pi$, then $t'_p \le t_{\sim p} \le t_p$;
3. if $x = \tau$, then $t'_p = t_{\sim p} = t_p$.

Consider the conditions for definite provability. If the conclusion is transient (if $x = \tau$), the above conditions are the standard ones for definite proofs in DL, which are just monotonic derivations using forward chaining. If the conclusion is persistent ($x = \pi$), p can be obtained at t_p or, by persistence, at any time t'_p before t_p. Finally, notice that facts lead to strict conclusions, but are taken not to be persistent.

Defeasible derivations run in three phases. In the first phase we put forward a supported reason (rule) for the conclusion we want to prove. Then in the second phase we consider all (actual and potential) reasons against the desired conclusion. Finally in the last phase, we have to rebut all the counterarguments. This can be done in two ways: we can show that some of the premises of a counterargument do not obtain, or we can show that the counterargument is weaker than an argument in favour of the conclusion. If $x = \tau$, the above conditions are essentially those for defeasible derivations in DL. If $x = \pi$, a proof for p can be obtained by using a persistent rule which leads to p holding at t_p or at any time t'_p before t_p. In addition, for every instant of time between the t'_p and t_p, p should not be terminated. This requires that all possible attacks were not triggered (clause 2.2.1) or are weaker than some reasons in favour of the persistence of p (clause 2.2.2).

3 The Implementation

The system implementing TDL consists of three elements: (a) a parser, which translates sets of rules in TXT or RuleML formats to generate a corresponding TDL theory; (b) a Graphical User Interface for selecting defeasible theories, and for visualizing conclusions and the execution time of the algorithm; (c) an inference engine which implements the algorithm of [10] to compute conclusions of the generated TDL theory.

The parser translates sets of rules in TXT or RuleML formats to generate a corresponding theory to be processed by the inference engine. Any of such theories consists of a set of rules and a temporal interval within which to compute the theory conclusions. (Facts and the superiority relation can be safely removed to obtain an equivalent TDL theory [10].) In particular, TDL rules can be represented in RuleML as derivation rules [7]. However, standard RuleML does not support temporal concepts in its syntax. So, in addition to introducing some attributes for representing the rule strength (whether a rule is strict, defeasible, or a defeater) and the rule labels [7], the syntax of the RuleML elements Atom and Imp is extended in order to express the time parametrizing the literals and the duration (whether the rule is persistent or transient) [16].

The Graphical User Interface allows the user to select a set of rules in RuleML or TXT format and to decide the time interval within which to compute their conclusions. Rules are then elaborated by assigning a unique label to each rule and the signs $+/-$ to the literals according to whether they are positive or negative: the rules are visualized accordingly. The data are then elaborated in such a way as to create and return the hash tables on which is based the algorithm [10]:

- a hash table keeps track of the rules in which an atom occur; in particular, for each atom α, the table will point to the rules where $+\alpha$ occurs in their heads and their bodies, and where $-\alpha$ occurs in their heads and their bodies;
- a hash table keeps track of the times of the literals occurring in the head of the rules;
- a hash table keeps track of the times of the literals occurring in the heads of transient rules;
- a hash table keeps track of the least times associated with the literals occurring in the head of persistent rules[2].

After the data are inputed into the system, it is possible to compute the definite and defeasible conclusions, which are stored in a hash table containing pairs string-ArrayList of integers. The strings represent the atoms of the theory while the integers are the time instants when the atoms hold. For space reasons, in the remainder of this paper we will only illustrate some details of the inference engine.

3.1 Inference Engine

The inference engine of the system implements in Java the algorithm for TDL developed by Governatori and Rotolo in [10][3]. The algorithm computes the extension of any TDL

---

[2] These data are used, in particular, to remove the rules which have in their body literals parametrized by times preceding those stored in this hash table. See Section 3.1 for more details.

[3] See http://www.defeasible.org/implementations/TDLJava/index.html for the full code and the Javadoc documentation. See [16] for more details.

theory D, where the concept of extension is defined as follows: if HB_D is the Herbrand Base for D, the extension of D is the 4-tuple $(\Delta^+, \Delta^-, \partial^+, \partial^-)$, where $\#^\pm = \{p^t | p \in HB_D, D \vdash \pm\#^x p^t, t \in \mathcal{T}\}$, $\# \in \{\Delta, \partial\}$, and $x \in \{\pi, \tau\}$.

Δ^+ and Δ^- are the positive and negative definite extensions of D, while ∂^+ and ∂^- are the positive and negative defeasible extensions.

The computation of the extension of a TDL theory runs in three steps [10][4]:

(i) in the first step the superiority relation is removed by creating an equivalent theory where $\prec = \emptyset$; any fact a^t, too, is removed by replacing it with a rule $\rightarrow^\tau a^t$;

(ii) in the second step the theory obtained from the first phase is used to compute the definite extension;

(iii) in the third step the theory from the first step and the definite extension are used to generate the theory to be used to compute the defeasible extension.

The Java class implementing the algorithm is TDLEngine. This has, as its main attributes, the theory (a set of rules and atoms), the theory conclusions, the time interval within which to compute these conclusions, the execution time of the algorithm, and a log manager.

The methods of the class TDLEngine are of two types: (i) those that are proper of the algorithm; (ii) those that are functional to the algorithm execution. Here, we will only describe the former ones.

It is worth noting that the computation makes use of time intervals to give a compact representation for sets of contiguous instants. The algorithm works both with proper intervals such as $[t, t']$, i.e., intervals with start time t and end time t', and punctual intervals such as $[t]$, i.e., intervals corresponding to singletons.

Following the idea of [15], the computation of the definite and defeasible extensions is based on a series of (theory) transformations that allow us (1) to assert whether a literal is provable or not (and the strength of its derivation) (2) to progressively reduce and simplify a theory. The key ideas depend on a procedure according to which, once we have established that a literal is positively provable we can remove it from the body of rules where it occurs without affecting the set of conclusions we can derive from the theory. Similarly, we can safely remove rules from a theory when one of the elements in the body of the rules is negatively provable. The methods of TDLEngine for this purpose are computeDefiniteConclusions and computeDefeasibleConclusions.

The method computeDefiniteConclusions works as follows. At each cycle, it scans the set of literals of the theory in search of temporal literals for which there are no rules supporting them (namely, supporting their derivation). This happens in two cases: (i) there are no rules for a temporal literal l^t or (ii) all the persistent rules having the literal in their head are parametrized by a greater time than t. For each of such temporal literals we add them to the negative definite extension of the theory, and remove all rules where at least one of these literals occurs. Then, the set of rules is scanned in search of rules with an empty body. In case of a positive match we add the conclusion of the rule to the positive definite extension (with an open ended interval for a persistent rule and with a punctual interval otherwise). Finally we remove such temporal literals matching

---

[4] Governatori and Rotolo [10] proved that, given a TDL theory D, the extension of D can be computed in linear time, i.e., $O(|R| * |H_D| * |\mathcal{T}_D|)$, where R are the rules of D and $\mathcal{T}_D$ is the set of distinct instants occurring in D. It is also shown that the proposed algorithm is correct.

the newly added conclusions from the body of rules. The cycle is repeated until (1) there are no more literals to be examined, or (2) the set of strict rules is empty, or (3) no addition to the extension happened in the cycle.

The method `computeDefeasibleConclusions` is more complex. As regards the scanning of the set of literals of the theory–in search of temporal literals for which there are no rules supporting them–the procedure is basically the same of `computeDefiniteConclusions` (with the difference that when we eliminate a rule we update the state of the extension instead of waiting to the end as in the case of the definite extensions). Then we search for rules with empty body. Suppose we have one of such rules, say a rule for l^t, and we know that the complement of l, i.e., $\sim l$, cannot be proved at t. So we add $(\sim l, [t])$ to ∂^-. At this stage we still have to determine whether we can insert l in ∂^+ and the instant/interval associated to it. We have a few cases. The rule for l is a defeater: defeaters cannot be used to prove conclusions, so in this case, we are done. If the rule is transient, then it can prove the conclusion only at t, and we have to see if there are transient rules for $\sim l^t$ or persistent rules for $\sim l^{t'}$ such that $t' \leq t$. If there are we have to wait to see if we can discard such rules. Otherwise, we can add $(l, [t])$ to ∂^+. Finally, in the last case the rule is persistent. What we have to do in this case is to search for the minimum time t' greater or equal to t in the rules for $\sim l$, and we can include $(l, [t, t'])$ in ∂^+.

The method `computeDefeasibleConclusions` basically calls three subroutines: `proved`, `discard`, and `persistence`.

The subroutine corresponding to `persistence` updates the state of literals in the extension of a theory after we have removed the rules in which we know at least one literal in the antecedent is provable with $-\partial^x$. Consider, for example, a theory where the rules for p and $\neg p$ are: $r: \Rightarrow^\pi p^1$, $s: q^5 \Rightarrow^\tau \neg p^{10}$, $v: \Rightarrow^\pi \neg p^{15}$. In this theory we can prove $+\partial^\pi p^t$ for $1 \leq t < 10$, no matter whether q is provable or not at 5. Suppose that we discover that $-\partial^x q^5$. Then we have to remove rule s. In the resulting theory from this transformation can prove $+\partial^\pi p^t$ for $1 \leq t < 15$. Thus we can update the entry for l from $(l, [1, 10])$ to $(l, [1, 15])$.

Secondly, `discard` adds a literal to the negative defeasible extension and then removes the rules for which we have already proved that some literal in the antecedent of these rules is not provable. The literal is parametrised by an interval. Then it further calls `persistence` that updates the state of the extension of a theory.

Third, `proved` allows to establish if a literal is proved with respect to a given time interval I. As a first step, it inserts a provable literal in the positive defeasible extension of the theory. Then it calls `discard` with the complementary literal. The next step is to remove all the instances of the literal temporalised with an instant in the interval I from the body of any rule. Finally, the rule is removed from the set of rules.

4 Discussion and Conclusions

There are two mainstream approaches to reasoning with and about time: a point based approach, as TDL, and an interval based approach [1]. Notice, however, that TDL is able to deal with constituents holding in an interval of time: an expression $\Rightarrow a^{[t_1, t_2]}$, meaning that a holds between t_1 and t_2, can just be seen as a shorthand of the pair of rules $\Rightarrow^\pi a^{t_1}$ and $\leadsto^\tau \neg a^{t_2}$. Non-monotonicity and temporal persistence are covered by a number of different formalisms, some of which are quite popular and mostly based on variants of Event Calculus or Situation Calculus combined with non-monotonic logics (see, e.g., [17,18]). TDL has some advantages over many of them. In particular, while TDL is sufficiently expressive for many purposes, it is possible in TDL to compute the set of consequences of any given theory in linear time to the size of the theory. To the best of our knowledge, no logic with a comparable expressive power of TDL is so efficient. Temporal and duration based defeasible reasoning has been also developed by [5,13]. [13] focuses on duration and periodicity and relationships with various forms of causality. In particular, [5] proposed a sophisticated interaction of defeasible reasoning and standard temporal reasoning (i.e., mutual relationships of intervals and constraints on the combination of intervals). In these cases no complexity results are available, but these systems cannot enjoy the same nice computational properties of TDL, since both are based on more complex temporal structures.

On account of the feasibility of TDL, in this paper we reported on the first implementation in the literature of this logic, which has been developed in Java because of the good performance and flexibility results of another implementation proposed for a modal (non-temporal) extension of DL [9]. We are still at a preliminary stage for testing the system. In particular, we have not yet done a systematic performance evaluation using tools generating scalable test defeasible logic theories: this study is a matter of further research. We have evaluated only some theory types (exemplified below):

- **Backward Persistence Transient**
  - Rules: $a^x \Rightarrow^\tau a^{x-1}, \Rightarrow^\tau a^{100}$
  - Output: $(a, [0, 100])$
- **Backward Persistence Persistence**
  - Rules: $a^x \Rightarrow^\pi a^{x-1}, \Rightarrow^\tau a^{100}$
  - Output: $(a, [0, 100])$
- **Backward Opposite Persistence**
  - Rules: $a^{100} \Rightarrow^\pi \neg a^0, \Rightarrow^\tau a^{100}$
  - Output: $(a, [100]), (\neg a, [0, 99])$

- **Backward Persistence Lazy**
  - Rules: $a^{100} \Rightarrow^\pi a^0, \Rightarrow^\tau a^{100}$
  - Output: $(a, [0, 100])$
- **Persistence via Transient**
  - Rules: $a^x \Rightarrow^\tau a^{x+1}, \Rightarrow^\tau a^0$
  - Output: $(a, [0, 100])$
- **Persistence**
  - Rules: $\Rightarrow^\pi a^0$
  - Output: $(a, [0, 100])$

Our preliminary experiments, reported in Table 1, were performed on an Intel Core Duo (1,80 GHz) with 3 GB main memory. We focused, among others, on some types of backward persistence, where conclusions persist from times which precede the ones when rules leading to such conclusions apply. These reasoning patterns occur, e.g., in the legal domain, where the retroactivity of normative effects is a common phenomenon [11]. As expected, all cases of backward persistence where conclusions are re-used to derive persistent literals ("Backward Persistence Transient" and "Backward Persistence Persistence") are more computationally demanding, while the other cases, where literals persist by default, are comparable to standard persistence (the last row in Table 1). We also tested the system with some real-life scenarios, which cannot be described here [16][5]. Two of them are particularly significant, one formalizing the regulation on road traffic restrictions of the Italian town of Piacenza, another corresponding to a real E-commerce scenario. We verified that the system behaved correctly in all cases.

---

[5] See also http://www.defeasible.org/implementations/TDLJava/index.html

Table 1. Performances on Persistence

Theory	Rules	Atoms	Time	Execution time
Backward Persistence Transient	11	1	[0,100]	1110 ms
Backward Persistence Persistence	11	1	[0,100]	984 ms
Backward Opposite Persistence	2	1	[0,100]	15 ms
Backward Persistence Lazy	2	1	[0,100]	32 ms
Persistence via Transient	11	1	[0,100]	1250 ms
Persistence	1	1	[0,100]	16 ms

References

1. Allen, J.: Towards a general theory of action and time. Artificial Intelligence 23 (1984)
2. Antoniou, G., Billington, D., Governatori, G., Maher, M.J.: Representation results for defeasible logic. ACM Transactions on Computational Logic 2, 255–287 (2001)
3. Antoniou, G., Billington, D., Governatori, G., Maher, M.J.: Embedding defeasible logic into logic programming. Theory and Practice of Logic Programming 6, 703–735 (2006)
4. Antoniou, G., Maher, M.J., Billington, D.: Defeasible logic versus logic programming without negation as failure. Journal of Logic Programming 42, 47–57 (2000)
5. Augusto, J., Simari, G.: Temporal defeasible reasoning. Knowledge and Information Systems 3, 287–318 (2001)
6. Bassiliades, N., Antoniou, G., Vlahavas, I.: A defeasible logic reasoner for the Semantic Web. International Journal on Semantic Web and Information Systems 2, 1–41 (2006)
7. Governatori, G.: Representing business contracts in RuleML. International Journal of Cooperative Information Systems 14(2-3), 181–216 (2005)
8. Governatori, G., Hulstijn, J., Riveret, R., Rotolo, A.: Characterising deadlines in temporal modal defeasible logic. In: Orgun, M.A., Thornton, J. (eds.) AI 2007. LNCS (LNAI), vol. 4830, pp. 486–496. Springer, Heidelberg (2007)
9. Governatori, G., Pham, H.D.: A semantic web based architecture for e-contracts in defeasible logic. In: Adi, A., Stoutenburg, S., Tabet, S. (eds.) RuleML 2005. LNCS, vol. 3791, pp. 145–159. Springer, Heidelberg (2005)
10. Governatori, G., Rotolo, A.: Temporal defeasible logic has linear complexity. Technical report, NICTA, Queensland Research Laboratory, Australia (2009),
http://www.governatori.net/publications.html
11. Governatori, G., Rotolo, A.: Changing legal systems: Legal abrogations and annulments in defeasible logic. The Logic Journal of IGPL (forthcoming)
12. Governatori, G., Rotolo, A., Sartor, G.: Temporalised normative positions in defeasible logic. In: ICAIL 2005, pp. 25–34. ACM Press, New York (2005)
13. Governatori, G., Terenziani, P.: Temporal extensions to defeasible logic. In: Orgun, M.A., Thornton, J. (eds.) AI 2007. LNCS (LNAI), vol. 4830, pp. 476–485. Springer, Heidelberg (2007)
14. Grosof, B.N.: Representing e-commerce rules via situated courteous logic programs in RuleML. Electronic Commerce Research and Applications 3(1), 2–20 (2004)
15. Maher, M.: Propositional defeasible logic has linear complexity. Theory and Practice of Logic Programming 1, 691–711 (2001)
16. Rubino, R.: Una implementazione della logica defeasible temporale per il ragionamento giuridico. PhD thesis, CIRSFID, University of Bologna (2009)
17. Shanahan, M.: Solving the Frame Problem: A Mathematical Investigation of the Common Sense Law of Inertia. MIT Press, Cambridge (1997)
18. Turner, H.: Representing actions in logic programs and default theories: A situation calculus approach. Journal of Logic Programming 31(1-3), 245–298 (1997)

Fill the Gap in the Legal Knowledge Modelling

Monica Palmirani[1], Giuseppe Contissa[2], and Rossella Rubino[1]

[1] CIRSFID - University of Bologna
{monica.palmirani,rossella.rubino}@unibo.it
[2] CODEX - Stanford University
contissa@stanford.edu

Abstract. There is a gap between the legal text description in XML trends and the legal knowledge representation of the norms that from the text starts. This gap affects the effectiveness of the legal resources exploitation and the integrity of the legal knowledge on the Web. This paper presents a legal document model for managing the legal resources in integrated way and linking all the different levels of representation[1].

1 Introduction

The AI & Law community dedicated the last twenty years to model the legal norms using different logics and formalisms [14]. The methodology used starts from a re-interpretation of the legal text by a Legal Knowledge Engineer that extracts the norms, applies models and theory using a logic and finally represents them with a particularly formalism. In the last ten years several Legal XML standards were arisen for describing the legal text [9] and rules (RuleML, RIF, SWRL, etc.). In the mean-time the Semantic Web, in particular the Legal Ontology research, combined with the NLP extraction of the semantic [13], has borne a great impulse to the modelling of the legal concepts [3]. In this scenario there is the urgent need to close the gap be-tween the text description, definitely using XML techniques, and the norms modelling in order to realise an integrated and self-contained representation of the legal re-sources available on the Web.

The reasons of this urgent need are primarily four.

- the legal knowledge is now presented in a disjointed way by the original text that inspired the logical modelling. This disconnection between legal document man-agement and logic representation of the norms affects strongly the real usage of the legal resources in many applications of the law in favour of the citizens, public administrations, enterprises;
- the change management over the time of the legal document, especially the legisla-tive one that for its nature is variable and subject to frequent modifications, significantly affects the coordination between the text and the rules that should be remodelled;

[1] This work was partially performed inside of the Estrella IST6-project n. 2004-027655. We would like to thanks Thomas Gordon for encouraging us in this vision and also for providing to the authors useful comments during several discussions on the LKIF.

G. Governatori, J. Hall, and A. Paschke (Eds.): RuleML 2009, LNCS 5858, pp. 305–314, 2009.
© Springer-Verlag Berlin Heidelberg 2009

- the legal validity of the text as published by the official authority should be preserved by any manipulation. On the other hand it is important to connect legal document resources, that includes itself many legality values, with the multiple interpretations coming from the legal knowledge modelling;
- a theory of legal document modelling able to separate clearly the many layers of representation of the resource: content (text), structure of the text, metadata on the document, ontology on the legal concepts expressed in the text, legal content modelling (normative part of the text) is fundamental for preserving over the time the digital legal text enriched by many semantic annotations, including also logic representation of the norms.

This paper aims to present a new approach for joining two fields of the legal research, AI&Law and Legal Document Management, for realising the structure of the Semantic Web in favour to a concrete application of the legal knowledge information in the retrieval and in the legal reasoning field. A pilot case and the related methodology is presented for better explaining the model presented. Finally the conclusions comments the results and critical issues.

2 Layers of Legal Document Modelling

The state of the art of the last ten years produced plenty of Legal XML standards for describing the document as legal resource. We can divide these standards in four categories:

- the *first generation* of Legal document XML standard[2], was oriented mostly to describe the legal text and its structure with an approach near to the database entities or the typography-word processing paradigms;
- the *second generation* posed more attention to the document modelling and to the description of text, structure and metadata[3]. Nevertheless the descriptiveness of the elements was not preceded by an abstract analysis of the classes of data and the result is a very long list of tags, a complex inclusions of DTDs or XML-schema, with a frequent overlapping between metadata and text definition and a weak instruments for linking the text to any other layers;
- the *third generation* is based on pattern. The pattern defines the properties of the class and its grammar, content model, behaviour and hierarchy respect the other classes, so any additional tag belongs to an existing abstract class and in this way it is preserved the consistency over the time. A strong attention to divide the text, structure, metadata and ontology is a primary principle in order to track in robust way any new layer put on the top of the pure text. Because the pattern defines general rules that no longer impose real constraints in the mark-up action, so the clarity of design scarifies the prescriptiveness [4];
- the *four generation* uses the pattern jointly with co-constraint grammar like, among the others, RELEX NG [10], Schematron [6], DSD [8], etc. for resolving above mentioned problem of lack of prescriptiveness.

---

[2] Like EnAct or Formex.
[3] As NiR or Lexdania.
[4] Akoma Ntoso [15] and CEN/Metalex are examples of this approach.

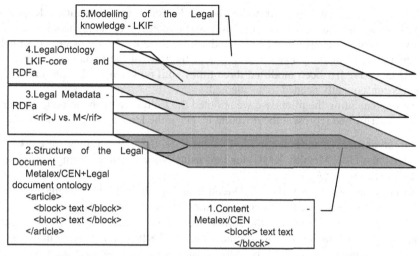

Fig. 1. Layers of representation in the Legal Document Modelling

For now we are using CEN/Metalex, as a transformation of NiR, jointly with LKIF [5] that supports RELEX NG.

The aim is to define a general legal document architecture able to describe all the following five layers (see Fig. 1) with a unique syntax or with reasonable hooks for integrating, in a cleanest way, all the different layers without confusion:

- **text:** part of the document officially approved by the authority with the legal power;
- **structure:** of the text: part of the document that states an organisation of the text;
- **metadata:** any information that was not approved by the authority in the deliberative act;
- **ontology:** any information about the reality in which the document act a role (e.g. for a judgement the juridical system concepts) or any concept called from the text that needs a modelling;
- **legal knowledge representation:** the part of the interpretation and modelling of the meaning of the text under legal perspective. Several XML standards are present in the state of the art for managing rules (RIF, RuleML, SWRL), nevertheless LKIF seems to provide a flexible language able to describe different possible theories or logic models (propositional, predicative, argumentative, non-monotonic, deontic, defeasible, etc.) more fitted for the legal domain.

3 Pilot Case Scenario

A pilot case scenario is used for explaining in concrete the Legal Document Modelling. It is based on the Italian Saving Tax Law (Legislative decree No 48 of 18 April 2005) that implements the EU Directive (Council Directive 2003/48/EC) in the same domain. One of the most important problems in the national legislative system of the EU member states is to implement, by transposition, the EU Directive taking in consideration the mandatory norms of the supra-national regulation and in the same time to be coherent with the national legislation.

The Italian pilot case aims to model the EU Directive and the national legislation concerning the Saving Tax Law with LKIF and to detect the inconsistencies between the two legislative norms using LKIF standard. On the other hand this pilot case helps the legal drafting activity and versioning over the time of the Italian corpus because the EU Directive has been modified three times until now and the Italian Saving Tax Law has consequently needed to be adapted.

For implementing the pilot case we adopted the following methodology in order to model, describe and represent the different levels of the legal knowledge information:

- **text, structure, metadata:** the EU Directive and the Italian Law were marked-up in NiR and CEN/Metalex XML standard using Norma-Editor[5] during the mark-up and validation actions;
- **ontology:** an ontology of the EU Directive and the Italian Law was made in order to model and define the macro-concepts specific for the Tax domain;
- **rules:** the EU Directive and the Italian Law were modeled in LKIF-rule syntax manually by a legal knowledge engineering. The LKIF-rule files were imported inside of an inference engine[6] properly customized by CIRSFID with a specific dialog interface for checking the consistency and for managing the comparison between the EU directive and the Italian Law.

Finally all the legal resources and the results were stored and delivered on the Web with a native XML database[7]. In this way all the legal resources (text, ontology, logic representation) were interconnected and presented using Internet interface and API for fostering the legal knowledge through information retrieval engine, reasoning engine, application layer.

4 Legal Text Description and Representation

The document texts were marked up firstly with the national Italian standard NormeinRete for grasping as much as possible the full descriptiveness of the legal document using a specific national reached mark-up. After that we translated the NIR-XML in CEN/Metalex using Norma-Editor [11] for working with a standard based on pattern and more flexible to manage the other layers linking.

Two mechanisms are present in CEN/Metalex for realizing the connection between multi-layers division above presented:

- RDFa assertions in the XML, for linking the structural part (second level) with the full ontology of classes (fourth level), with clear distinction and no overlap of relations and literal values;
- URIs naming convention, based on FRBR [1], is defined to identify the legal resources. This naming permits to link the LKIF-rules or arguments assertion directly to the part of the text involved in (connection between second level with the fifth level).

---

[5] Norma-Editor is a specialized legal drafting editor developed by CIRSFID and based on Microsoft-Word. It is able to convert the final mark-up action in different XML formats.

[6] RuleBurst engine by RuleBurst Europe Ltd., now a branch of Oracle Corporation UK Ltd.

[7] eXistrella native XML database presented in Jurix2008 LegalXML Workshop, Palmirani, Cervone.

A fragment of XML text in the Table 2 demonstrates the twofold mechanism: (i) `rdfa:property` says that the current metadata is-a member of the `lkif-document:URI` ontology class; (ii) the `rdfa:content` value specifies the URI of the resources that will be used in the LKIF XML file.

Table 1. Fragment of CEN/Metalex Saving Tax Law: identification of the document

```
<meta xmlns:rdfa="http://www.w3.org/TR/xhtml-rdfa-primer" .
   name="manifestationURI"
   id="metalex_d1e5"
   value="it/act/2005-04-18/84/2005_04_18_dlgs_84.xml"
   rdfa:property="[lkif-document:URI]"
   rdfa:content="/it/act/2005-04-18/84/ita@/2005_04_18_dlgs_84.xml"/>
```

5   LKIF Ontology of the Italian Savings Tax Law

The aim of the ontological level is to model any concepts (definition and properties) that is useful for completing the understanding of the rules or arguments represented in LKIF. Secondary the OWL syntax permits to make some consistency check on the concepts using some existing engine like Pellet. The rationale is to define in the ontology any static concepts (T-Box) of the context useful for the understanding of the rule and arguments. Some classes of this ontology[8] are presented hereinafter:

- **income.** According to the LKIF ontology the *income* can be classified as a `top:Mental_Concept`. We consider, here, only the subcategory of **savings income** in the form of interest payments. The form of an income has been modelled as the property *form* of the class *savings income*.

Fig. 2. Savings_income class

- Every **payment**, which is an `action:Action` in the LKIF ontology, has a subject, a payer, a beneficiary, the amount of the payment, the medium adopted to make the payment and finally the date and the place where the payment has been performed. An **interest payment** is a payment whose object is an interest.

Fig. 3. Interest_payment class

---

[8] They are an extension to the LKIF-core ontology [4].

- The **beneficial owner** is the individual, that is a j.1:Natural_Person in the LKIF ontology, who receives an interest payment or for whom an interest payment is secured. We defined some properties which are useful: the tax identification number, the place and the date of birth, the name of the beneficial owner, his address and his country.

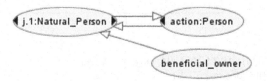

Fig. 4. Beneficiary_owner class

- The **paying agent** is any economic operator who makes an interest payment to the beneficial owner or secures a savings income payment for the beneficial owner. The **economic operator** is the individual or body which actively takes part - from the side of demand and of offer - in a specific market.

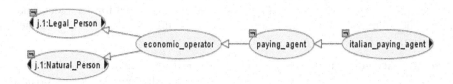

Fig. 5. Paying_agent class

These OWL classes are linked both in the CEN/Metalex XML file and in the LKIF-rules modelling using two different syntaxes base on the same principles in order to maintain consistent and self-contained the legal document representation in the Web:

- inside of the CEN/Metalex XML we focused our attention on linking the fragment of the text with the ontology class using RDFa assertion by the rdfa:rel and rdfa:href attributes.

Table 2. Fragment of CEN/Metalex Saving Tax Law: assertion in RDFa using the class interest_payment

```
<htitle name="rubrica" id="metalex_title_art2 " rdfa:about="" rdfa:rel="rdf:type"
rdfa:href="[savings:interest_payment]">Interest payments.</htitle>
```

- inside of the LKIF-rule we specify the `pred` attribute's value with the name of the predicate that is defined inside of the ontology class `saving:paying_agent`. This mechanism permits to use classes, properties and definitions uniquely modelled inside of the ontology with a great benefit for the consistency and interoperability.

Table 3. Fragment of LKIF Saving Tax Law modelling: paying_agent predicate

`<s pred="savings:paying_agent"><v>X</v>` shall communicate the information concerning the payment to Revenue Agency under 1-1b`</s>`

6 LKIF Modelling

The main goal was to represent in LKIF the rules for fulfilling the requirements of the Italian Pilot, which involves a comparison of EU and Italian norms in a legal drafting perspective. A high degree of granularity was requested and the knowledge representation should have been isomorphic to the maximum possible extent, as described by Bench-Capon [2] and Karpf [7]. Each legal source is represented separately, preserving its structure and the traditional mutual relations, references and connections with other legal sources, thanks to the fact that structural elements in the source texts correspond to specific elements in the representation. The representation of the legal sources and their mutual relations should also be separated from all other parts of the model, in particular the representation of queries and facts management.

Therefore, each norm from the legal source text was represented in a correspondent rule (or set of rules). No deviations were made, as long it was possible, from the original structure of the text, even when it was redundant or confusing.

The representation of the Italian Savings Tax Law was carried out as follows: the law was analyzed to find the core part to be modelled, keeping a correspondence with the modelled part of the Directive. The following excerpt is a fragment representation of Art. 1 comma 1, first paragraph[9].

The rules are represented with `head` and `body`, the `body` is a set of sentences (`<s>`) concatenated with Boolean operators. In case of `exception` we should specify a new rule with `pred` value "execption": this produce a duplication of the rule and a verbose representation and a not efficient method for expressing the exception of exception.

The `<sources>` block specifies the URIs where to find the corresponding part of the rule into the text with a granularity since the word.

---

[9] *"Art. 1. Subjects held to the communication - 1. Banks, investment firms, Poste italiane S.p.a., security investment fund management companies, financial companies and trust companies, resident in Italy, shall communicate to the Revenue Agency the information concerning the paid interest or the interest for which the immediate payment is secured for the immediate benefit of individuals, that are beneficial owners, resident in another Member State; for this purpose the individuals shall be considered beneficial owners of interest if they received the payment as final recipient."*

Table 4. Fragment of LKIF-XML modelling rules in Italian Saving Tax Law

```xml
<?xml version="1.0" encoding="utf-8"?>
<?oxygen RNGSchema="LKIF2.rnc" type="compact"?>
<?xml-stylesheet type="text/css" href="LKIF2.css"?>
<lkif xmlns:savings="savingsTax_ita.owl">
  <sources>
    <source element="#ita_savings"      uri="/it/act/2005-04-18/84/eng@/main.xml"/>
    <source element="#s1_ita"           uri="/it/act/2005-04-18/84/eng@/main.xml#art1"/>
    <source element="#s1-1a-ita"        uri="/it/act/2005-04-18/84/eng@/main.xml#art1-com1"/>
    <source element="#s1-1a_01-ita"     uri="/it/act/2005-04-18/84/eng@/main.xml#art1-com1"/>
    <source element="#s1-1a_02-ita"     uri="/it/act/2005-04-18/84/eng@/main.xml#art1-com1"/>
    <source element="#s1-1a_03-ita"     uri="/it/act/2005-04-18/84/eng@/main.xml#art1-com1"/>
  </sources>
  <theory id="ita_savings">
    <rules>
      <rule id="s1-1a-ita">
        <head>
          <s pred="savings:paying_agent"><v>X</v> shall communicate the information concerning
the payment to Revenue Agency under 1-1a</s>
        </head>
        <body>
          <and>
            <s><v>X</v> is a subject listed in 1-1</s>
            <s><v>X</v> is resident in Italy</s>
            <s><v>X</v> pays or secures interest to <v>Y</v></s>
            <s><v>Y</v> is an individual</s>
            <s><v>Y</v> is a beneficial owner</s>
            <s><v>Y</v> is resident in another member state</s>
            <s><v>X</v> operates as debtor or charged to pay</s>
          </and>
        </body>
      </rule>
      <rule id="s1-1a_01-ita">
        <head>
          <s><v>X</v> is a subject listed in 1-1</s>
        </head>
        <body>
          <or>
            <s><v>X</v> is a bank</s>
            <s><v>X</v> is an investment firm</s>
            <s><v>X</v> is poste italiane spa</s>
            <s><v>X</v> is a security investment fund management company</s>
            <s><v>X</v> is a financial company</s>
            <s><v>X</v> is a trust company</s>
          </or>
        </body>
      </rule>
      <rule id="s1-1a_02-ita">
        <head>
          <s><v>X</v> pays or secures interest to <v>Y</v></s>
        </head>
        <body>
          <or>
            <s><v>X</v> pays interest to <v>Y</v></s>
            <s><v>X</v> secures interest payment for immediate benefit of<v>Y</v></s>
          </or>
        </body>
      </rule>
```

Table 5. Fragment of LKIF-XML modelling rules in EU Saving Tax Directive

```
<rule id="par-2-1-exception-b2-EUSD">
  <head>
   <s pred="rule:excluded"> <c>par-2-pa1-EUSD </c>excluded<s>beneficial-owner<v>x</v>
    </s></s>
   <s pred="rule:excluded"> <c>par-2-pa2-EUSD</c> excluded<s>beneficial-owner<v>x</v>
    </s></s>
  </head>
  <body>
   <and>
    <s>individual<v>x</v>
    </s>
    <s>provides-evidence<v>x</v>
     <s>entity-taxed-on-its-profits<v>y</v>
     </s></s>
    <s>provides-evidence<v>x</v>
     <s>acts-on-behalf-of<v>x</v><v>y</v>
     </s></s>
   </and>
  </body>
 </rule>
```

7 Conclusions

The pilot case underlines several considerations concerning the gap between the five levels of information that we need to model for describing a legal document:

- the granularity of the XML document marked-up is not isomorphic to the rules and statements modeling in term of part of text (word, paragraphs, etc.);
- the relationship between rules and text is a N:M cardinality, so both the LKIF-rules and the CEN/Metalex need to improve their syntax for implementing a smart mechanism able to capture, without duplication, the multiple referencing;
- the interaction between rules in LKIF and concepts expressed in the ontology, since the definitions can be considered too complex to be inserted into the ontology, also because some of them are rules by themselves;
- the relationship between the XML document and the ontology it is useful even if it is not in contrast with the usage of the same classes in the LKIF-rules;
- the legal document change over the time as well as the ontology definition so in the LKIF-rule it is necessary a mechanism for managing the dynamicity over the time as well as into ontology;
- the non-monotonic dimension of the law, the exception of exception convinced us to include some extension in LKIF for implementing the defeasible logic paradigm strictly linked with the text.

From these conclusions we learnt the lesson that we need a strong architecture of the legal document divided in five levels. For permitting a coordination between these layers we need standards designed with mechanisms for interconnecting them in asynchronous way. Our future work will go in the direction to fill the gap between the different markup granularity and temporality among the different layers. Finally we aim to implement the defeasible logic extending the LKIF schema.

References

1. Bekiari, C., Doerr, M., Le Boeuf, P.: International Working Group on FRBR and CIDOC CRM Harmonization. FRBR object-oriented definition and mapping to FRBRER (v. 0.9 draft) (2008), http://cidoc.ics.forth.gr/docs/frbr_oo/frbr_docs/FRBR_oo_V0.9.pdf (accessed August 20 2009)
2. Bench-Capon, T., Coenen, F.: Isomorphism and legal knowledge based systems. Artificial Intelligence and Law 1(1), 65–86 (1992)
3. Boer, A., Radboud, W., Vitali, F.: MetaLex XML and the Legal Knowledge Interchange Format. In: Casanovas, P., Sartor, G., Casellas, N., Rubino, R. (eds.) Computable Models of the Law. LNCS (LNAI), vol. 4884, pp. 21–41. Springer, Heidelberg (2008)
4. Breuker, J., Boer, A., Hoekstra, R., Van Den Berg, C.: Developing Content for LKIF: Ontologies and Framework for Legal Reasoning. In: Legal Knowledge and Information Systems, JURIX 2006, pp. 41–50. ISO Press, Amsterdam (2006)
5. Gordon Thomas, F.: Constructing Legal Arguments with Rules in the Legal Knowledge Interchange Format (LKIF). In: Computable Models of the Law, Languages, Dialogues, Games, Ontologies 2008, pp. 162–184 (2008)
6. Jelliffe, R.: The Schematron Assertion Language 1.5, http://www.ascc.net/xml/resource/schematron/Schematron2000 (accessed June 20, 2009)
7. Karpf, J.: Quality assurance of Legal Expert Systems, Jurimatics No 8, Copenhagen Business School (1989)
8. Klarlund, N., Møller, A., Schwartzbach, M.I.: DSD: A Schema Language for XML. In: Proceedings of the third workshop on Formal methods in software practice, Portland (2000)
9. Lupo, C., Vitali, F., Francesconi, E., Palmirani, M., Winkels, R., de Maat, E., Boer, A., Mascellani, P.: General xml format(s) for legal sources - Estrella European Project IST-2004-027655. Deliverable 3.1, Faculty of Law. University of Amsterdam, Amsterdam (2007)
10. Murata, M.: RELAX (REgular LAnguage description for XML) (2000), http://www.xml.gr.jp/relax (accessed June 20, 2009)
11. Palmirani, M., Brighi, R.: An XML Editor for Legal Information Management. In: Proceeding of the DEXA 2003, Workshop su E-Government, Praga, September 1-5, pp. 421–429. Springer, Heidelberg (2003)
12. Peruginelli, G., Ragona, M. (eds.): Law via the Internet Free Access, Quality of Information, Effectiveness of Rights, EPAP, p. 494 (2009)
13. Proceeding of the 12th International Conference on Artificial Intelligence and Law, Barcelona, June 8-12, p. 243. ACM, New York (2009)
14. Sartor, G.: Legal Reasoning: A Cognitive Approach to the Law. In: Treatise on Legal Philosophy and General Jurisprudence, vol. 5. Springer, Berlin (2005)
15. Vitali, F.: Akoma Ntoso Release Notes, http://www.akomantoso.org (accessed August 20, 2009)

The Making of SPINdle

Ho-Pun Lam[1,2] and Guido Governatori[2]

[1] School of Information Technology and Electrical Engineering
The University of Queensland, Brisbane, Australia
[2] NICTA*, Queensland Research Laboratory, Brisbane, Australia

Abstract. We present the design and implementation of SPINdle – an open source Java based defeasible logic reasoner capable to perform efficient and scalable reasoning on defeasible logic theories (including theories with over 1 million rules). The implementation covers both the standard and modal extensions to defeasible logics. It can be used as a standalone theory prover and can be embedded into any applications as a defeasible logic rule engine. It allows users or agents to issues queries, on a given knowledge base or a theory generated on the fly by other applications, and automatically produces the conclusions of its consequences. The theory can also be represented using XML.

Keywords: Defeasible Logic, Modal Defeasible Logic, Reasoning.

1 Introduction

Defeasible logic (DL) is a non-monotonic formalism originally proposed by Nute [1]. It is a simple rule-based reasoning approach that can reason with incomplete and contradictory information while preserving low computational complexity [2]. Over the years, the logic has been developed notably by [3,4,5]. Its use has been advocated in various application domains, such as business rules and regulations [6], agent modeling and agent negotiations [7], applications to the Semantic Web [8] and business process compliance [9]. It is suitable to model situations where conflicting rules may appear simultaneously.

In this paper we report on the implementation of SPINdle which implements reasoners to compute the consequences of theories in defeasible logic. The implementation covers both standard defeasible logic and modal defeasible logic.

The most important features of SPINdle are the following:

- It supports all rule types of defeasible logic, such as fact, strict rules, defeasible rules, defeaters and superiority.
- It supports Modal Defeasible Logics [10] with modal operator conversions.
- It supports negation and conflicting (mutually exclusive) literals.
- A theory can be represented using XML and plain text (with pre-defined syntax), and a theory and its extension can also be exported using XML.
- A visual theory editor is developed for editing standard defeasible logic theory.

* NICTA is funded by the Australian Government as represented by the Department of Broadband, Communications and the Digital Economy and the Australian Research Council through the ICT Centre of Excellence program.

G. Governatori, J. Hall, and A. Paschke (Eds.): RuleML 2009, LNCS 5858, pp. 315–322, 2009.

As a result, SPINdle is a powerful tools supporting:

- rule, facts and ontologies;
- monotonic and nonmonotonic (modal) rules reasoning with inconsistencies and incomplete information

In the rest of this paper, section 2 gives a brief introduction to the syntax and semantics of both standard defeasible logic and modal defeasible logic. Section 3 describes the implementation details of SPINdle, the algorithm that it used and the data structure that it proposed to enhance the performance of the inference process. Section 4 presents the performance statistics from a study undertaken using various types and sizes of defeasible logic theories. Section 5 gives our conclusions and poses future research/development work on SPINdle.

2   Defeasible Logic

2.1   Basics of Defeasible Logic

A *defeasible theory D* is a triple $(F, R, >)$ where F and R are finite set of facts and rules respectively, and $>$ is a superiority relation on R. Here SPINdle only considers rules that are essentially propositional. Rules containing free variables are interpreted as the set of their ground instances.

Facts are indisputable statements, represented either in form of states of affairs (literal and modal literal) or actions that have been performed. Facts are represented by predicates. For example, "John is a human" is represented by $human(John)$.

A *rule*, on the other hand, describes the relations between a set of literals (premises) and a literal (conclusion). We can specify the strength of the rule relation using the three kinds of rules supported by DL, namely: *strict rules*, *defeasible rules* and *defeaters*; and can specify the mode the rules used to connect the antecedent and the conclusion. However, in such situations, the conclusions derived will be *modal literals*.

Strict rules are rules in the classical sense: whenever the premises are indisputable (e.g. facts) then so is the conclusion. An example of a strict rule is: "human are mammal", written formally:

$$human(X) \rightarrow mammal(X)$$

Defeasible rules are rules that can be defeated by contrary evidence. An example of such a rule is "mammal cannot flies"; written formally:

$$mammal(X) \Rightarrow \neg flies(X)$$

The idea is that if we know that X is a mammal then we may conclude that it cannot fly *unless there is other, not inferior, evidence suggesting that it may fly* (for example that mammal is a bat).

Defeaters are a special kind of rules that cannot be used to draw any conclusions. Their only use is to prevent some conclusions. That is, they are used to defeat some defeasible rules by producing evidence to the contrary. For example the rule

$$heavy(X) \rightsquigarrow \neg flies(X)$$

states that an animal is heavy is not sufficient enough to conclude that it does not fly. It is only evidence against the conclusion that a heavy animal flies. In other words, we don't wish to conclude that ¬*flies* if *heavy*, we simply want to prevent a conclusion *flies*.

DL is a "skeptical" nonmonotonic logic, meaning that it does not support contradictory conclusions. Instead DL seeks to resolve conflicts. In cases where there is some support for concluding A but also support for concluding ¬A, DL does not conclude neither of them. However, if the support for A has priority over the support for ¬A then A is concluded.

As we have alluded to above, no conclusion can be drawn from conflicting rules in DL unless these rules are prioritised. The *superiority relation* is used to define priorities among rules, that is, where one rule may override the conclusion of another rule. For example, given the following facts:

$$\rightarrow bird \qquad\qquad \rightarrow brokenWing$$

and the defeasible rules:

$$r: \qquad bird \Rightarrow fly$$
$$r': \quad brokenWing \Rightarrow \neg fly$$

which contradict one another, no conclusion can be made about whether a bird with a broken wing can fly. But if we introduce a superiority relation $>$ with $r' > r$, then we can indeed conclude that the bird cannot fly.

We now give a short informal presentation of how conclusions are drawn in DL. Let D be a theory in DL (as described above). A *conclusion* of D is a tagged literal and can have one of the following four forms:

$+\Delta q$ meaning that q is definitely provable in D (i.e. using only facts and strict rules).
$-\Delta q$ meaning that we have proved that q is not definitely provable in D.
$+\partial q$ meaning that q is defeasible provable in D.
$-\partial q$ meaning that we have proved that q is not defeasible provable in D.

Strict derivations are obtained by forward chaining of strict rules while a defeasible conclusion p can be derived if there is a rule whose conclusion is p, whose prerequisites (antecedent) have either already been proved or given in the case at hand (i.e. facts), and any stronger rule whose conclusion is ¬p has prerequisites that fail to be derived. In other words, a conclusion p is (defeasibly) derivable when:

- p is a fact; or
- there is an applicable strict or defeasible rule for p, and either
  - all the rules for ¬p are discarded (i.e. not applicable) or
  - every rule for ¬p is weaker than an applicable rule for p.

A full definition of the proof theory can be found in [3]. Roughly, the rules with conclusion p form a team that competes with the team consisting of the rules with conclusion ¬p. If the former team wins p is defeasibly provable, whereas if the opposing team wins, p is non-provable.

2.2 Modal Defeasible Logic

Modal logics have been put forward to capture many different notions somehow related to the intensional nature of agency as well as many other notions. Usually modal logics are extensions of classical propositional logic with some intensional operators. Thus any modal logic should account for two components: (1) the underlying logical structure of the propositional base and (2) the logical behavior of the modal operators. Alas, as is well-known, classical propositional logic is not well suited to deal with real life scenarios. The main reason is that the descriptions of real-life cases are, very often, partial and somewhat unreliable. In such circumstances, classical propositional logic is doomed to suffer from the same problems.

On the other hand, the logic should specify how modalities can be introduced and manipulated. Some common rules for modalities are, e.g., **Necessitation** (from $\vdash \phi$ infer $\vdash \Box \phi$) and **RM** (from $\vdash \phi \rightarrow \psi$ infer $\vdash \Box \phi \rightarrow \Box \psi$). Both dictate conditions to introduce modalities purely based on the derivability and structure of the antecedent. These rules are related to well-known problem of omniscience and put unrealistic assumptions on the capability of an agent. However, if we take a constructive interpretation, we have that if an agent can build a derivation of ψ then she can build a derivation of $\Box \psi$.

To this end, SPINdle follows the semantics proposed by [10] on reasoning with modal defeasible logic. However, due to the limited space, readers interested in understand the semantics, modal operator conversions, conflict detections, conflict resolutions, and algorithm implemented in SPINdle please refer to the paper for details.

3 Implementation

3.1 SPINdle System Architecture

SPINdle, written in Java, consists of three major components (Fig. 1): the Rule Parser, the Theory Normalizer and the Inference Engine.

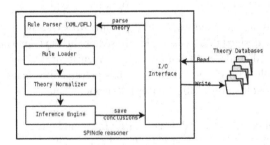

Fig. 1. Main components of the SPINdle reasoner

SPINdle accepts defeasible logic theories represented using XML or plain text (with pre-defined syntax). The *Rule Parser* is used to transform the theory from a saved theory document into a data structure that can be processed in the next module. The nature of the rule parser is rather similar to the *Logic Program Loader* module of the DR-Device family of applications developed by [8].

After loading a theory into SPINdle, users can then modify/manipulate the theory according to their applications need. The *I/O Interface* module in Fig. 1 provides useful methods for helping users to load and save (modified) theory (also the derived conclusions) to and from the database. Theories can also be exported using XML for agent communication, which is a common scenario in the Semantic Web.

3.2 The Inference Process

The whole inference process has two phases: A *pre-processing* phase where we transform the theory using the techniques described in [3] into an equivalent theory *without superiority relation and defeaters*, which later helps to simplify the reasoning process in the reasoning engine. The *Theory Normalizer* module in Fig. 1 carry out this process by further breaking it down into three linear transformations: one to transform the theory to regular form, one to empty the superiority relation and one to empty the defeaters. In addition to this, the theory normalizer also transform rules with *multiple heads* into an equivalent sets of rules with single head. It is expected that the transformed theory will produce the same sets of conclusions in the language of the theory they transform.

Following [2] (a.k.a. the *Delores* algorithm), the *conclusions generation* phase (the *Inference Engine* module) is based on a series of (theory) transformations that allow us (1) to assert whether a literal is provable or not (and the strength of its derivation); (2) to progressively reduce and simplify a theory. The reasoner will, in turn:

- Assert each fact (as an atom) as a conclusion and remove the atom from the rules where the atom occurs positively in the body, and "deactivates" (remove) the rule(s) where the atom occurs negatively in the body. The atom is then removed from the list of atoms.
- Scan the list of rules for rules with empty head. It takes the head element as an atom and search for rule(s) with conflicting head. If there are no such rules then the atom is appended to the list of facts and the rule will be removed from the theory. Otherwise the atom will append to the pending conclusion list until all rule(s) with conflicting head can be proved negatively.
- Repeats the first step.
- The algorithm terminates when one of the two steps fails or when the list of facts is empty. On termination the reasoner output the set of conclusions.

Finally, the conclusions are either exported to the users, or saved in the theory database for future use. Since each atom/literal in a theory is processed exactly once and every time we have to scan the set of rules, the complexity of the above algorithm is $O(|\mathcal{L}| * |\mathcal{R}|)$, where $\mathcal{L}$ is the size of the distinct modal literals and $\mathcal{R}$ is the number of rules in the theory [10]. This complexity result can be improved through the use of proper data structure (Fig. 2). For each literal p a linked-list (the dashed arrow) of the occurrences of p in rule(s) can be created during the theory paring phase. Each literal occurrence has a link to the record for the rule(s) it occurs in. Using this data structure, whenever a literal update occurs, the list of rules relevant to that literal can be retrieved easily. Thus instead of scanning through all the rules for empty head (step 2), only rules relevant to that particular literal will be scanned. The complexity of the inference process will therefore reduced to $O(|\mathcal{L}| * |\mathcal{M}|)$, where $\mathcal{M}$ is the maximum number of rules a literal associated with in the theory,

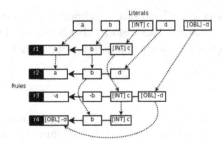

Fig. 2. Data structure for literals-rules association

It is important to note that the aforementioned algorithm is a generalized version that is common in inferencing both standard defeasible logic and modal defeasible logic. In the case of modal defeasible logic, due to the modal operator conversions, an additional process adding extra rules to the theory is needed. In addition, for a literal p with modal operator $\Box_i$, besides its complement (i.e., $\Box_i \neg p$), the same literal with modal operator(s) in conflict with $\Box_i$ should also be included in the conflict literal list (step 2) and only literal with strongest modality will be concluded.

4   Performance Evaluation

SPINdle has been extensively tested for correctness, scalability and performance using different forms of theories generated by a tool that comes with Deimos [11]. In this section we are going to describe the tests we have performed and the result we obtained.

In the test, SPINdle is compiled using the SUN Java SDK 1.6 without any optimization flags. The test begins with a heap size of 512MB and gradually increased to 2GB according to the theory size. Performance has been measured on a Pentium 4 PC (3GHz) with Linux (Ubuntu) and 2GB main memory. Figure 3 shows the performance and memory usage of SPINdle in running theories of various types and sizes.

The various theory types were designed to explore various aspects of defeasible inference. For example, in theory simple-chain of size n, a fact a_0 is at the end of a chain of n rules $a_i \Rightarrow a_{i+1}$. A defeasible proof of a_n will use all the rules and the fact. In theory *tree(n,k)*, a_0 is the root of a k-branching tree of depth n, in which every literals occur once. More details about the various theory types can be found in [11]. Notices that in our performance evaluation the theory sizes refer to the number of rules (both strict rules and defeasible rules) that were stored in the theory. The statistics show only the time used for the inference process (excluding the time used for loading theories into the reasoner) and the total memory used. Figure 3 (a,b) show the performance and memory usage with large theory $(10000 < n < 180000)$[1]; while (c,d) show the same information in theory *tree* with $k = 3$ and n={2188, 6562, 19684, 59050 and 177148}.

Results (Fig. 3) show that SPINdle can handle inferences with thousands of rules in less than three seconds. The reasoning time growth (in most cases) almost linearly

---

[1] SPINdle does not impose any limitation on the theory size. SPINdle can reason on any theory as long as it can be put onto the memory. The largest theory size tested so far is with $n = 1,000,000$ (1 million).

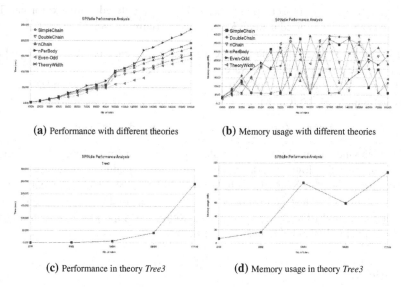

(a) Performance with different theories   **(b)** Memory usage with different theories

(c) Performance in theory *Tree3*   **(d)** Memory usage in theory *Tree3*

Fig. 3. Performance and Memory usage with different theories

proportional to the size of the theories tested, which is coherent with the complexity described in section 3.2.

On the contrary, the memory usage fluctuates as the theories sizes increase, regardless the semantics of the input theories. This may be due to the dynamic nature of hash table that we used in handling the literal and theory objects as well as the temporary variables that we created during the inference process. Nevertheless, for the theories tested, our results show that SPINdle can handle 10,000 rules with less than 50MB memory and 180,000 rules with less than 400MB.

Last but not least, apart from the default configuration, we can also optimize the performance (also the memory usage) of SPINdle through different setting. For example, a *defeasible rule only* engine can be used (instead of the ordinary one) if the input theory contains defeasible rules only. Our results (not shown here) show that the performance can be improved by up to 3% and with about 20% drop in memory usage (on average) if we configure SPINdle properly.

5   Conclusion

We have presented SPINdle, a reasoner that support reasoning on both the standard defeasible logic and modal defeasible logic. It features include:

– Full implementation of defeasible logic, including fact, strict and defeasible rules, defeaters, and superiority relation among rules.
– It supports reasoning with Modal Defeasible Logic theory.
– It can reason with incomplete and inconsistent theories.
– It is efficient (due to the low computational complexity), and with low memory consumption.

- It supports theory represented in both XML and pre-defined plain text format. Theory can also be exported using XML for agent communication.

We have also showed that the complexity of the inference process can be reduced through the use of proper data structure.

SPINdle is freely available for downloading and experimentation under the LGPL license agreement, at the following address:

`http://spin.nicta.org.au/spindleOnline`

To further enhance the usability of SPINdle, a visual editor for writing rules and theory in standard defeasible logic was developed and is also available for download.

Future directions for SPINdle include both algorithm and technological improvement, which include interface to support direct import from the web and processing of OWL/RDF data and RDF schema ontologies. Reasoning support for *Temporal Defeasible Logic* [12] will also be included in the future.

References

1. Nute, D.: Defeasible logic. In: Gabbay, D., Hogger, C. (eds.) Handbook of Logic for Artificial Intelligence and Logic Programming, vol. III, pp. 353–395. Oxford University Press, Oxford (1994)
2. Maher, M.J.: Propositional defeasible logic has linear complexity. Theory and Practice of Logic Programming 1(6), 691–711 (2001)
3. Antoniou, G., Billington, D., Governatori, G., Maher, M.J.: Representation results for defeasible logic. ACM Transactions on Computational Logic 2(2), 255–286 (2001)
4. Antoniou, G., Billington, D., Governatori, G., Maher, M.J.: Embedding defeasible logic into logic programming. Theory and Practice of Logic Programming 6(6), 703–735 (2006)
5. Billington, D.: Defeasible logic is stable. J. Logic Computation 3(4), 379–400 (1993)
6. Antoniou, G., Billington, D., Governatori, G., Maher, M.J.: On the modeling and analysis of regulations. In: Proceedings of the Australian Conference Information Systems, pp. 20–29 (1999)
7. Governatori, G., Rotolo, A.: Defeasible logic: Agency, intention and obligation. In: Lomuscio, A., Nute, D. (eds.) DEON 2004. LNCS (LNAI), vol. 3065, pp. 114–128. Springer, Heidelberg (2004)
8. Bassiliades, N., Antoniou, G., Vlahavas, I.: A defeasible logic reasoner for the semantic web. International Journal of Semantic Web and Information Systems 2(1), 1–41 (2006)
9. Governatori, G., Milosevic, Z., Sadiq, S.: Compliance checking between business processes and business contracts. In: 10th International Enterprise Distributed Object Computing Conference (EDOC 2006), pp. 221–232. IEEE Computing Society, Los Alamitos (2006)
10. Governatori, G., Rotolo, A.: BIO logical agents: Norms, beliefs, intentions in defeasible logic. Journal of Autonomous Agents and Multi Agent Systems 17, 36–69 (2008)
11. Maher, M.J., Rock, A., Antoniou, G., Billington, D., Miller, T.: Efficient defeasible reasoning systems. International Journal of Artificial Intelligence Tools 10, 483–501 (2001)
12. Governatori, G., Rotolo, A., Sartor, G.: Temporalised normative positions in defeasible logic. In: 10th International Conference on Artificial Intelligence and Law (ICAIL 2005), pp. 25–34. ACM Press, New York (2005)

Approaches to Uncertain or Imprecise Rules - A Survey

Matthias Nickles and Davide Sottara

[1] Department of Computer Science,
University of Bath
Bath, BA2 7AY, United Kingdom
nickles@gmx.net
[2] DEIS, Facolta di Ingegneria,
Universita di Bologna, Bologna, Italy
dsotty@gmail.com

Abstract. With this paper we present a brief overview of selected prominent approaches to rule frameworks and formal rule languages for the representation of and reasoning with uncertain or imprecise knowledge. This work covers selected probabilistic and possibilistic logics, as well as implementations of uncertainty and possibilistic reasoning in rule engine software.

Keywords: Rules, Uncertainty Reasoning, Imperfect Knowledge, Semantic Web, Knowledge Engineering.

1 Introduction

This survey paper presents a short overview of selected more or less prominent approaches to rule frameworks and formal rule languages for the representation of and reasoning with various kinds of uncertain, imprecise or ambiguous information. These properties of information are quite different and knowledge may be affected by one or more of them at the same time. After Smets ([42]), the term "imperfection" will be used as a general concept encompassing all kinds of them, since "uncertainty", which is also sometimes used in a general way, is actually a specific type of imperfection.

The knowledge an agent has about the world can often be conveniently encoded using formulas expressed in some logic-based language. In classical logic, a formula is either true or false. However, in many practical cases the truth of a formula might be unknown, unclear or uncertain, in which case we cannot assign it a definite truth value.

There are many possible reasons for such *imperfect knowledge*. Several attempts to outline the differences and classify them in a standard framework exist: among them, the already cited survey by Smets [42] but also, for example [26,31,33]. More recently, the W3C *Incubator Group on Uncertainty Reasoning for the Web* has defined an ontology (see [10]) for the representation of imperfection of information on the Web which is to some degree resembled in this

G. Governatori, J. Hall, and A. Paschke (Eds.): RuleML 2009, LNCS 5858, pp. 323–336, 2009.

paper. Even if there is much philosophical debate, it is generally accepted that such imperfection can take on the following major forms (among others):

Uncertainty derives from a lack of knowledge about a fact or an event, be it past, present or future, even if the actual state of the world is known to belong to some set of alternatives. Uncertainty may be *aleatory*, i.e. intrinsically present in some random phenomenon so that the knowledge gap cannot be filled, or *epistemic*, i.e. due to a partial ignorance of the agent, for example because of missing, questionable or inconsistent data. Often, but not always, the degree of uncertainty can be measured in some statistical, *objective* way. In the epistemic case, or whenever a subjective judgement is adopted, the degree of uncertainty is typically denoted as the *degree of belief* of some agent in a statement. But the subjective belief of an agent can of course also comprise statistical information.

Imprecision arises when knowledge is as complete as it can be, but the terms used to denote it do not allow to identify the entity that are being referred in a precise and univocal way. This imprecision may lead to *ambiguity*, when there is more than one possible interpretation, *approximation*, when a class of entities is collapsed into one representative, and *vagueness*, when the boundaries of the definition of a concept are relaxed - usually *fuzzified* - in some way.

Inconsistency is a property of a knowledge base with *conflicting* information, such that there is no possible world it can describe. Conflicts have to be resolved, usually removing, ignoring or modifying part or all the conflicting information. Inconsistency may be a symptom of *incorrect* or noisy information. But of course there is possibly also erroneous information which does not lead to any (apparent) inconsistency. In this survey, we do not consider formal frameworks for inconsistency handling.

Notice that an imperfect representation of knowledge may be more concise, robust and less expensive to obtain than its perfect version. Consider for example the age-related version of the *Sorite Paradox*: if a person is young on one day, they will also be young the day after, until the day when they will be old. This is a paradox in classical logic unless a different adjective is defined for every day in the life of a person, but is perfectly acceptable in fuzzy logic, where one single property, "young", has a truth degree that varies continuously with age. Given the variety of sources, most knowledge based systems are likely to have to deal with some type of imperfection in the data they process: moreover, given the robustness/conciseness trade-off, a system handling imperfection without ignoring is more applicable and powerful than an idealized system which simply prunes away uncertainty and imprecision. Of all the possible applications, we will discuss an example class, chosen because it is usually a domain of traditional (from a logical point of view), i.e., "perfect" rule-based systems:

Complex Event Processing (CEP) is an emerging approach [36] based on the concept of event, that is here, a message notifying that some state has changed in a system at a certain time. Many real-world systems (e.g. a stock markets in finance, a plant in chemistry, a human body in medicine, . . .) generate dozens of such events of different types at high frequencies, but only a few are usually relevant at a given time. The challenge, then, is to filter, sort and analyze the events, possibly aggregating them in higher order events at different abstraction levels, so to extract only the relevant information. Uncertain formal rules can be used to reason about such events, and *reaction rules* are particularly suitable in the context of CEP since they can be used to trigger actions (including the generation of new events) when the current events match their preconditions. Event processing, however, may be affected by imperfection in different ways:

- If some events are partially unobservable, there may be uncertainty due to the missing data. Moreover, the conditions used to detect a complex event from simpler ones may not be certain (i.e. when detecting the insurgence of a disease from its symptoms). Finally, all event-based predictions are intrinsically uncertain.
- Events may be reported with imprecision, e.g. because the measurements are unreliable. It may also be convenient, if not necessary, to express some constraints - especially the temporal ones - with some degree of vagueness (e.g., event A happens more or less contemporarily with event B).
- There may be unexpected, conflicting combinations of events, especially in case of failures.

Being a relatively novel discipline, not many real-world applications of *imperfect* CEP exist yet, possibly because commercial rule-based systems are efficient at handling imprecision, especially in the fuzzy case, but still have serious limitation when processing uncertainty.

A further set of examples and case studies related to uncertain and imprecise information on the Web can be found in [10], along with a discussion on possible solution approaches.

The remainder of this paper is structured as follows: Section 2 deals with different models of imperfection, such as fuzzy sets and probability theory, focusing on how they have been applied to extend traditional logics in the formalization of rules. Section 3 discusses whether and how mainstream rule-based systems, both academic and commercial, support imperfect rules.

2 Probabilistic and Possibilistic Reasoning with Rules

The different types of imperfection are the domains of different theories: in particular, uncertainty is typically handled using probabilistic approaches, while possibilistic ones are used for imprecision. The theoretical backgrounds are far beyond the scope of this work, so the main techniques will just be recalled briefly, to focus more properly on their application in logic.

2.1 Probabilistic Approaches

Probability Theory

Probability theory is the mathematical theory of *random events*. The probability of an event A is represented by a real number ranging from 0 (impossible event) to 1 (certain event), and is usually denoted as $Pr(A)$. If uncertainty is aleatory, this probability is normally estimated using a "frequentist" approach, by repeating observations of events in long-run experiments and choosing the ratio of favorable outcomes over the total number of experiments as the probability. In contrast to this view, with the *Bayesian* or epistemic interpretation of probability, which is the underlying theory for most approaches in Artificial Intelligence (especially in Machine Learning), probability is a measure of the belief in some hypothesis which is not necessarily grounded in any physical properties or empirical observations (but could be). Under this view, for a rational agent, probability is typically grounded in terms of betting behavior: $Pr(A)$ is here the amount of money that a rational agent would be indifferent to betting on the occurrence of event A. In order to calculate probabilities, an agent typically starts with some personal prior belief, which is then progressively updated using the application of *Bayes' rule* as new information is acquired.

Various logics with support for probabilistic reasoning (purely statistical approaches as well as Bayesian reasoning) have been developed. In the following, we will present a subset of those first-order languages which are able to represent logical rules in the sense of Horn clauses (see below). Our survey does not aim at a complete or representative list of such formal frameworks, but just at a list of hopefully interesting example approaches and as a starting point for further reading.

One of the most simple languages for rules, and the core of *RuleML* [8], is *Datalog* [22]. A *Datalog program* allows to contain rules, that is, clauses (disjunctions of literals) with at most one positive literal (*Horn clauses*), with certain restrictions (see below). Rules can thus be written in the form $H \longleftarrow B_1 \wedge ... \wedge B_n$, where H and the B_i are atoms (the Datalog notation follows that of Prolog and is thus slightly different).

So-called *extensional predicates* are fully defined in an extensional manner by lists of *facts* (positive ground unit clauses). *Intensional predicates* are in contrast defined entirely by rules. Extensional predicates correspond to relations ("tables") in extensional databases, intensional predicates correspond to intensional database relations. The difference between these two type of predicates in Datalog becomes very important for probabilistic extensions of Datalog, as explained further below.

Datalog can be seen both as a *deductive database system* (a database system which can derive new data (facts) using logical inference) and as a Prolog-like logic language. But in contrast to Prolog, functions within terms are not allowed, and every variable in the head of a clause must also appear in the body of this clause (so-called *safe rules* or *range-restricted rules*). Variants of Datalog such as *stratified Datalog* imply further restrictions. Datalog allows for very effective query evaluation, and queries are ensured to terminate. Its expressivity covers relational algebra (roughly: the core of SQL), but goes beyond it, by means of so-called *recursive queries.*

Probabilistic Datalog (Datalog$_P$) [28] is an extension of Datalog which additionally allows for the probabilistic weighting of facts (but not - extensionally - of rules). Informally, the idea here is that each ground fact corresponds to an event in the sense of probability theory, and rules allow for boolean combinations of events and their probabilities. However, naively applied, this approach would lead to inconsistencies and other problems, since probability theory is not *truth-functional*, that is, values of complex expressions are not necessarily functions of the values of the constituents of these expressions. This is a problem which all formal approaches which aim for a combination of probability theory and logical calculi need to be aware of.

The semantics of *Datalog$_P$* programs is a probabilistic *possible-world semantics.* "Possible worlds" correspond to subsets of the least Herbrand model of the respective Datalog program, and a probability structure provides a probability distribution over all possible worlds. This uncertainty enabled possible-world semantics is typically - but not always - used for probabilistic logics, and reflects the aforementioned view of probabilities as degrees of belief. Essentially, a possible-world semantics assigns probabilities to propositions, and the probability of a certain proposition is the probability of the set of possible worlds where this particular proposition is true.

In order to deal with the absence of truth functionality, *Datalog$_P$* follows two alternative directions: basic *Datalog$_P$* yields probability intervals instead of "point" probabilities in case of derived event expressions (which are not given explicitly as probabilistically annotated ground facts) in order to reflect incomplete knowledge about event independence, which is the actual cause of lack of truth-functionality of probabilistic calculi. Alternatively, *Datalog$_{PID}$* ("*Datalog$_P$* with independence assumptions") makes the quite strong assumption of universal event independence, that is $Pr(e_1 \wedge e_2) = Pr(e_1) \cdot Pr(e_2)$ for any events e_1 and e_2. Under this assumption boolean combinations of the constituents of probabilistic event expressions become possible. Please find technical details in [28].

Approaches which combine Description Logics for the Semantic Web with Datalog/Prolog-like logic programming (and thus full rules as defined above) are described in the next section.

Whereas the logics described so far are subsets of first-order logic, in [30], Halpern presents three different probabilistic (full) first-order languages $\mathcal{L}_1$, $\mathcal{L}_2$, and $\mathcal{L}_3$. Basic *Datalog$_P$* is a subset of $\mathcal{L}_2$. All three languages have a very expressive syntax, allowing for, e.g., probabilistically-weighted arbitrary first-order formulas (including rules in the sense of Horn clauses) and conditional probabilities. However, they are undecidable and their relevance is largely theoretical and important mainly because of their influence on historically newer and practically more relevant languages.

Formulas of $\mathcal{L}_1$ take the basic form of $w_x(\varphi) \geq pr$, where φ is a first-order formula and pr is a probability. This syntax is more or less identical to the syntax of Bacchus' logic L_p [20]. The semantics is quite similar, but not identical. The informal meaning of the statement form above is "The probability that a randomly chosen object x in the domain satisfies φ is at least pr". This means that we do here (and also for L_p) *not* encounter a possible-world semantics here, but instead a "statistical semantics" (or more appropriate: *domain-frequency semantics*) which puts a probability distribution over the *domain of discourse*. This semantics is an implementation of the empirical interpretation of probabilities mentioned at the beginning of this section. Probabilities reflect here "objective" statistical proportions or frequencies.

In contrast, $\mathcal{L}_2$ uses a possible-world semantics. The syntactical form of $\mathcal{L}_2$ formulas is $lb \leq \varphi \leq up$, where lp is the lower bound for the probability of φ and up is the upper bound. The semantics works much like that of $Datalog_P$, only that φ is not restricted to facts.

Both approaches can be converted into each other - however, to understand better what the practical difference between these two languages respectively their semantics is, and why a degree-of-belief semantics is not appropriate in certain cases, and domain-frequency semantics is not adequate in other case, we look at the following example (taken from [30]):

Suppose one would like to formalize the two statements "The probability that a randomly chosen bird is greater than 0.9" and "The probability that Tweety (which is a bird) can fly is greater than 0.9", using some first-order probabilistic languages. For the second statement, a possible-world semantics which assigns a probability of 0.9 or higher to the set of those worlds in which Tweety can fly seems appropriate. However, the first statement cannot simply be formalized using a possible-world semantics, at least not in a straightforward way [30,20]. A seeming way (among others which do not work very well) would be to attach probability 0.9 to the worlds in which $\forall x Bird(x) \rightarrow Flies(x)$ holds. However, if in all worlds there is at least a single bird which does not fly, and so the probability of this statement is zero, it could still be the case that the probability that a randomly chosen bird flies is greater than 0.9. In contrast to the possible-world semantics, a domain-frequency semantics can represent the first statement without problems, but would on the other hand have in certain contexts problems with the representation of statements like the second statement above [30,20].

In order to allow for a simultaneous reasoning with both views of probability, Halpern introduces the language $\mathcal{L}_3$. It allows statements such as $w(w_x(Flies(x)|Bird(x)) > .99) < 2 \wedge w(w_x(Flies(x)|Bird(x)) > .9) > .95$, hence combining the syntactical features of $\mathcal{L}_1$ and $\mathcal{L}_2$. What essentially happens here is that agents can hold degrees of beliefs about statistically uncertain statements.

While *Bayesian networks* (also called *belief networks*) are able to represent a sort of event-conditional rules, and certain variations of Bayesian networks encode causal rules, they work only on a propositional level, and thus do not fall into our scope of interest. However, several formal approaches exist which extend Bayesian networks with first-order capabilities (relations). *Bayesian logic programs* (BLP) [32] for example can be seen as a generalization of Bayesian networks and logic programming, implementing a possible-world semantics. The logic component of BLP consists of so-called *Bayesian clauses*. A Bayesian clause is a rule of the form $A|A_1, ..., A_n$, where each A_i is a universally quantified *Bayesian atom*. The main difference between Bayesian clauses and ordinary clauses (apart from the use of | instead of : − as in Prolog or Datalog rules) is that the Bayesian atoms have values from a finite domain instead of boolean values. In addition to Bayesian clauses, a BLP consists of a set of conditional probability distributions over Bayesian clauses c (encoding $Pr(head(c)|body(c))$) and so-called *combining rules* in order to retrieve a *combined conditional probability distribution* from the combination of the multiple different conditional probability distributions. From a BLP a Bayesian network can be easily computed and then queried using standard Bayesian inference. Another prominent example for relational extensions of Bayesian Networks are *Probabilistic Relational Models* (PRMs). However, they cannot express arbitrary quantified first-order rules.

Multi-entity Bayesian networks (MEBNs) [34] are another example for a formal framework which integrates first-order logic with Bayesian probability theory. In contrast to most other "relational Bayesian" approaches, they have full first-order representation power.

Stochastic Logic Programs (SLPs) [38] are sets of rules in form of range-restricted clauses labeled with probabilities. The resulting annotated rules are called *stochastic clauses*. The semantics of SLPs assigns a probability distribution to the atoms of each predicate in the Herbrand base of a program. SLPs are a generalization of *Hidden Markov Models* as well as stochastic grammars to first-order logic programming, and are expressive enough to encode (undirected) Bayesian networks. SLPs cannot only be manually constructed, but also learned using a combination of Inductive Logic Programming (ILP) and stochastic parameter estimation [38].

Stochastic Logic Programs encode a sort of domain-frequency semantics. It can be shown, however, that SLPs (respectively, BLPs) can be translated into BLPs (respectively, extended SLPs) [40].

A quite recent approach to the combination of first-order logic and probabilistic theory are *Markov Logic Networks* (MLNs) [41]. A MLN is a set of (unrestricted) first-order formulas with a weight (not a probability) attached to each formula. MLNs are used as "templates" from which *Markov networks* are constructed. Markov networks are graphical models for the joint distribution of a set of random variables, allowing to express conditional dependencies Bayesian networks cannot represent, and vice versa. The (ground) Markov network generated from the MLN then determines a probability distribution over possible worlds, and is used to compute probabilities of restricted formulas using probabilistic inference. Machine learning algorithms allow to learn the formula weights in MLNs from relational databases.

2.2 Dealing with Imprecision: Possibilistic Theories

Fuzzy Sets

Fuzzy sets [44] are sets S whose elements x have gradual degrees of membership, evaluated by a membership function $\mu_S : X \mapsto [0, 1]$. Fuzzy sets can be used to define vague concepts, such as "old" and "tall", and reason with and over them. The concept of fuzzification, i.e. extension with vagueness, has been applied to many contexts: just to cite some, fuzzy mathematics, fuzzy system theory, fuzzy classification and even meta-fuzzy systems (if a membership degree can't be estimated precisely, higher order fuzzy sets can be defined, which meta-membership degrees are, in turn, fuzzy sets over the domain of membership degrees themselves, $[0, 1]$), even if we are obviously interested in fuzzy logic, first defined by Zadeh himself in [45]. Even if a complete discussion of fuzzy sets goes beyond the scope of this paper, an interesting property must be recalled (see also [25] for a complete discussion). The membership of an object x in a fuzzy class S is usually a qualitative, vague but certain feature - e.g. a man is tall: the quantitative aspect of the matching is contained in the truth degree returned by the membership function. This degree may be defined a priori, but more often it is evaluated on one or more quantitative features of x. For example, the function μ_{Tall} is likely to rely on the height of the men it tests for membership, so it should be more explicitly written $\mu_{Tall}(x|h)$. In a different scenario, the only available information is that x is a member of S, i.e. a man is tall, in a degree greater or equal than some value α. Here, the x's height is unknown: there is uncertainty about its actual value, but S (or, more properly, its α-cut) defines a possibility distribution stating, for each candidate value h^*, how coherent it would be should it have to be used as the real value. In this case, one gets a distribution $\pi_{Tall}(h|x)$: more generally, according to Zadeh's coherence principle, it has to be that $\mu_S(x|h) = \pi_S(h|x)$. In a way, this duality is reflected by two different approaches to fuzzy logic.

Fuzzy Logic in a Narrow Sense

The fuzzification of first-order logic can be obtained directly by extending the underlying algebraic structure. In boolean FOL `true` and `false` are the only allowed truth values for a formula: one can obtain a fuzzy version of this logic assuming that the truth value is a member of a lattice (a partially ordered set in which any two elements always have a unique supremum and infimum), which in practice is usually the unit interval $[0, 1]$. Like in classical logic, the formulas are still built from atoms using connectives and quantifiers, but their evaluation is delegated to complex operators which generalize the boolean ones. To preserve an axiomatic structure, only a minimal set of operators is defined primitively (e.g. the implication $\rightarrow$ and the negation $\neg$), while the others are derived according to the canonical definitions (e.g. $\neg x \lor y \Leftrightarrow x \rightarrow y$). The choice of the implication operator is strictly connected to the choice of the conjunction operator $\star$, implemented using a triangular norm, a commutative, associative, monotonic binary map on $[0, 1]$ having 1 as neutral element: in fact, it holds that $x \star z \leq y \Leftrightarrow z \leq (x \rightarrow y)$. Interestingly, there is not a unique choice for $\star$, so, given the mutual dependencies, there exist families of operators and thus different logics according to which operators and how they are defined. However, it has been shown that all alternatives can be reduced to three basic t-norms, namely minimum, product and Lukasiewicz. The operators, in any case, are truth functional, which makes evaluation computable efficiently. The truth value of the atoms may be given as a fact, or evaluated using the equivalent of a membership function operating on its arguments: afterwards, operators are applied to determine the truth value of formulas. For these reasons, these "mathematical"fuzzy logics are special instances of many-valued logic. A complete discussion can be found in [29]. This kind of fuzzy logic is arguably the easiest to implement in a rule-based system, since it is an immediate generalization of boolean logic. Moreover, fuzzy rules are more robust than their crisp counterparts (e.g. all thresholds can be substituted using gradual variations) and, in many cases, less rules are required to express equivalent concepts. The drawback is that the rules become more complex to write and manage, so few systems actually exist that exploit the full expressiveness of this type of logic.

Fuzzy Logic in a Broader Sense

When Zadeh defined fuzzy logic ([45]), he was more interested in adopting a formalism close to the way people think and express concepts than in defining a formal extension of mathematical logic. This vision led to the definition of linguistic variables, which values are fuzzy sets over a specific domain. For example, the variable age may have *young, mature* and *old* as values, all fuzzy sets over, say, the integer interval 0..100. This allows to write conjunctive, horn-like rules having the form "if X_1 is A_1 and ... and X_n is A_n then Y is B". Tipically, several rules entailing information on Y are written and then combined disjunctively. In order to apply the rules, the quantitative inputs x (e.g. the age of a person) are fuzzified using, for example, a membership function. The fuzzified variables are matched against the qualitative constraints in the rules to entail the fuzzy conclusions Y, again fuzzy sets, which are defuzzified to obtain a quantitative output value. A set of such fuzzy rules is equivalent to a local, declarative approximation of a relation $\Pi_{j=1}^n X_j \mapsto Y$. For example, stating that if X1 is A and X2 is B then Y is C and providing a definition for A,B and C which has full membership in x_1^0, x_2^0 and y^0 respectively is but a way to define $f(x_1^0, x_2^0) = y^0$ with implicit continuity: an input similar to (x_1^0, x_2^0) should be mapped to a point close to y^0 (even if domain and range may use different metrics, the distances should be correlated). A more complete discussion, including the corresponding interpretation in narrow fuzzy logic, can be found in [29].

For this reason, this type of fuzzy logic has been widely used in many real-world applications for the monitoring and control of dynamical systems, especially non linear.

Possibility theory and Possibilistic logic.

In possibility theory [24], a possibility distribution associates a value $\pi(x) \in [0, 1]$ to each element x in a set of alternatives X. This value measures the compatibility of x with the actual state of the world. Unlike probability theory, where relative odds are considered, in possibility theory each individual is considered separately: in fact, the possibility $\Pi(A)$ associated to a set A is given by $\max(\pi(x)) : x \in A$. The dual concept of possibility is necessity, defined by negating the possibility of the complementary:

$N(A) = 1 - \Pi(A)$. Possibility can be physical or epistemic, depending whether objective or subjective factors are taken into account when defining π. As already discussed, the membership function of a fuzzy set can be considered a possibility distribution on its domain, given the only knowledge that the actual state of the world belongs to the set. Possibility and/or Necessity degrees can be used in a possibility-annotated logic (if they are restricted to $0, 1$, the resulting logic is actually a form of three-valued logic). The main drawback of this approach with respect to mathematical fuzzy logic is that the operators, albeit simpler, are not truth functional: in some cases, it is only possible to entail a bound, lower or upper, for the actual degree of a formula instead of an exact value. The complete theory can be found in [35].

3   Mainstream Software Tools and Standards

The (Semantic) Web is a heterogeneous software environment in which many systems and frameworks coexist.

We have shown that there exist different families of logic, with different expressiveness, which can be used to define and reason with rules. On the other hand, there exist many *rule engines*, with different characteristics, that a user can adopt in their projects. In modern applications, such rule engines are currently mainly used for two following tasks:

Processing business rules and, as an emerging field, reasoning over Semantic Web ontologies enhanced with rules [19,8].

The main goal of our survey is to see whether existing rule engines support imperfect rules natively, i.e. if the rule engine handles all the inference procedures typical of the respective logic.

Inevitably, however, other aspects have also to been taken into account: whether a tool is commercial or freely available, whether it is for an academic or industrial use, whether it is just a rule engine or a full fledged *Business Rule Management System* (BRMS), and its execution environment (e.g. Java vs. .NET).

Even limiting our research to mainstream projects[1], i.e. general purpose tools supported by a some community (to whatever extent), thus discarding several student projects and many ad-hoc engines built for specific applications, we have noticed that there are more than two dozens of different alternatives, but only a few with support for imperfection, and most of these work with fuzzy logic.

Note that we disregard in this section logical reasoners and other implementations of probabilistic and possibilistic frameworks which are not rule engines in the usual sense.

Imperfection in Rule-based Systems

Fuzzy logic is perhaps the easiest type of imperfect logic to implement in a rule-based system. Inference in fuzzy logic is a generalization of the boolean case.

---

[1] Inevitably, we can't consider all existing tools, so we preventively apologize for not citing or discussing some software.

Most importantly, the operators are truth functional, i.e. they just aggregate the degrees associated to their operands, so the complexity remains limited. However, this still means that a standard rule engine can't evaluate fuzzy rules natively without an extension. Many mainstream BRMSs, both commercial and free, include an engine, typically based on the RETE algorithm ([27]), and several additional tools giving support for rule management (remote and local), event handling, editing and reporting.

Among them, we can cite *InRule* [12], *ObjectConnections Common Knowledge* [14], *Microsoft BizTalk* [5], *Fair Isaac's Blaze Advisor* [4], *ILOG JRules* [11], *OpenRules* [16], *PegaSystems PegaRules* [6], *Open Lexicon* [15], *XpertRule KnowledgeBuilder* [18] and *JBoss Drools* [13].

Adding fuzzy logic (or any other type of logic) to any of these systems would require a refactoring of the internal rule engine and, possibly, the rule language, neither of which is a simple operation, although such a process is being attempted in Drools.

The mainstream fuzzy-capable systems, instead, are open source rule shells, typically originated in an academic context, without many of the additional features of BRMSs. The most famous are possibly *FuzzyShell* [3], *FuzzyClips* [1] and *FuzzyJess* [2]; we also know of a commercial data mining tool, Scientio XMLMiner / MetaRule, which has fuzzy capabilities [17]. FuzzyJess is one of the most used given its Java-oriented nature: it is actually a rewriting of FuzzyCLips, itself an extension of the CLIPS engine. FuzzyClips, moreover, has the merit of supporting two types of imperfection: fuzzy logic and confidence, in the form of certainty factors.

The first rule-based system to introduce uncertainty in automatic reasoning, *MYCIN* ([21]), adopted imperfect rules annotated with certainty factors to model a sort of quality score. The way of handling the factors was not theoretically very sound, so later systems used more structured approaches, even if the idea of using confidence was further developed (see for example [43]). In FuzzyClips, however, they have been introduced in a more coherent way, again truth-functional, and their evaluation proceeds in parallel with the evaluation of the fuzzy truth degrees of the formulas. Notice that all these fuzzy shells support fuzzy logic in the broader sense of the term.

In contrast, no commercial mainstream rule engine that we are aware of supports probabilistic logic. Whereas Bayesian networks have become a very popular tool for handling uncertainty and many mature software packages exist which implement Bayesian networks, hardly any product already supports any of the various probabilistic logics, even if recently at least two projects have been started, namely *Balios* and *BLOG*.

Interoperability

One of the limitations of the different engines is their using proprietary languages to write logic formulas and rules in particular. However, current Web standards for knowledge representation are not able to represent imperfect probabilistic or possibilistic rules.

To achieve a good degree of interoperation, standards on rule representation and interchange are being proposed in the last few years. The *Rule Interchange Format* (RIF) [7] is a proposed W3C standard format for rule representation and interchange, based on XML. *RuleML* [8] is an initiative which develops a XML- and RDF- based markup language for rules, with Datalog-rules as the core. RuleML uses a modular approach to support different rule-based logics with different types of complexity and expressiveness, in order to promote rule interoperability between industry standards. RuleML supports various kinds of reasoning engines (e.g., forward vs backward chaining, RETE vs Prolog, ...) and leaves knowledge engineers the choice of implementation for entities and facts (e.g., objects, plain symbols, XML trees, ...). RuleML is supported by various rules engines, such as jDREW and Mandarax. A combination of the current standard ontology language OWL and RuleML is proposed to the W3C in form of the *Semantic Web Rule Language* (SWRL) [9].

The issues related to the introduction of imperfection in rule languages have recently been discussed in [23], where a module for uncertainty and fuzzy reasoning with rules is defined. This work is remarkable since it shows that most types of imperfect logic can be encoded simply by allowing truth degrees and operators to be customized using appropriate tags (degree) and attributes (kind). Such knowledge, however, should be processed by an engine capable of changing its configuration at run-time, a task that requires more than the creation of a language translator. The extensions proposed in [23] can be integrated into RuleML, but also in a preliminary version of the W3C Rule Interchange Format (RIF) [7]. Another approach to the integration of rules based on proposed standards and fuzzy logic is *f-SWRL* [39].

As for probability theory, candidates for future standard languages will possible integrate a description logic with rules (in the sense of logic programming) and uncertainty reasoning, such as the formal framework proposed in [37].

References

1. Fuzzy clips,
   http://ai.iit.nrc.ca/irpublic/fuzzy/fuzzyclips/fuzzyclipsindex.html
2. Fuzzy jess, http://www.nrc-cnrc.gc.ca/eng/ibp/iit.html
3. Fuzzyshell, http://cobweb.ecn.purdue.edu/rvl/projects/fuzzy/
4. http://www.fico.com/en/products/dmtools/pages/
   fico-blaze-advisor-system.aspx
5. http://www.microsoft.com/biztalk/en/us/default.aspx
6. http://www.pega.com/products/rulestechnology.asp
7. Rule interchange format (rif) working group,
   http://www.w3.org/2005/rules/wiki/rif_working_group
8. Ruleml, http://www.ruleml.org
9. Swrl: A semantic web rule language combining owl and ruleml,
   http://www.w3.org/submission/swrl/
10. W3c uncertainty reasoning for the web incubator group,
    http://www.w3.org/2005/incubator/urw3/xgr-urw3
11. http://www.ilog.com/products/jrules

12. http://www.inrule.com
13. http://www.jboss.org/drools
14. http://www.objectconnections.com
15. http://www.openlexicon.org
16. http://www.openrules.com
17. http://www.scientio.com
18. http://www.xpertrule.com
19. Antoniou, G., Damásio, C.V., Grosof, B., Horrocks, I., Kifer, M., Maluszynski, J., Patel-Schneider, P.F.: Combining Rules and Ontologies. A survey (2005)
20. Bacchus, F.: l_p, a logic for representing and reasoning with statistical knowledge. Computational Intelligence 6, 209–231 (1990)
21. Buchanan, B.G., Shortliffe, E.H.: Rule-based Expert Systems: the MYCIN experiments of the Stanford Heuristic Programming Project. Addison-Wesley, Reading (1984)
22. Ceri, S., Gottlob, G., Tanca, L.: What you always wanted to know about datalog (and never dared to ask). IEEE Transactions on Knowledge and Data Engineering 1(1), 146–166 (1989)
23. Damsio, C.V., Pan, J.Z., Stoilos, G., Straccia, U.: Representing uncertainty in RuleML. Fundam. Inf. 82(3), 265–288 (2008)
24. Dubois, D.: Possibility theory and statistical reasoning. Computational Statistics & Data Analysis 51(1), 47–69 (2006)
25. Dubois, D., Esteva, F., Godo, L., Prade, H.: Fuzzy-set based logics an history-oriented presentation of their main developments. In: Gabbay, D.M., Woods, J. (eds.) Handbook of the History of Logic. The Many Valued and Non-monotonic Turn in Logic, vol. 8, pp. 325–449. Elsevier, Amsterdam (2007)
26. Dubois, D., Prade, H.: Possibility theory, probability theory and Multiple-Valued logics: A clarification. Annals of Mathematics and Artificial Intelligence 32(1-4), 35–66 (2001)
27. Forgy, C.: Rete: A fast algorithm for the many patterns/many objects match problem. Artif. Intell. 19(1), 17–37 (1982)
28. Fuhr, N.: Probabilistic datalog - a logic for powerful retrieval methods. In: Proceedings of the 18th Annual International ACM SIGIR Conference on Research and Development in Information Retrieval (1995)
29. Hájek, P.: Metamathematics of Fuzzy Logic. Trends in Logic: Studia Logica Library, vol. 4. Kluwer Academic Publishers, Dordrecht (1998)
30. Halpern, J.Y.: An analysis of first-order logics of probability. Artificial Intelligence 46, 311–350 (1990)
31. Halpern, J.Y.: Reasoning about Uncertainty. MIT Press, Cambridge (2003)
32. Kersting, K., Raedt, L.D.: Bayesian logic programs. In: Proceedings of the 10th International Conference on Inductive Logic Programming (2000)
33. Klir, G.J.: Generalized information theory. Fuzzy Sets Syst. 40(1), 127–142 (1991)
34. Laskey, K.B., Costa, P.C.: Of klingons and starships: Bayesian logic for the 23rd century. In: Proceedings of the Twenty-first Conference on Uncertainty in Artificial Intelligence (2005)
35. Logic, P., Dubois, D., Prade, H.: Possibilistic logic
36. Luckham, D.C.: The Power of Events: An Introduction to Complex Event Processing in Distributed Enterprise Systems. Addison-Wesley Longman Publishing Co., Inc., Boston (2001)
37. Lukasiewicz, T.: Probabilistic description logic programs. International Journal of Approximate Reasoning 45(2), 288–307 (2007)

38. Muggleton, S.: Learning stochastic logic programs. Electronic Transactions in Artificial Intelligence (2000)
39. Pan, J.Z., Stamou, G.B., Tzouvaras, V., Horrocks, I.: f-swrl: A fuzzy extension of swrl. In: Duch, W., Kacprzyk, J., Oja, E., Zadrożny, S. (eds.) ICANN 2005. LNCS, vol. 3697, pp. 829–834. Springer, Heidelberg (2005)
40. Puech, A., Muggleton, S.: A comparison of stochastic logic programs and bayesian logic programs. In: IJCAI 2003 workshop on learning statistical models from relational data, IJCAI (2003)
41. Richardson, M., Domingos, P.: Markov logic networks. Machine Learning 62(1-2), 107–136 (2006)
42. Smets, P.: Imperfect Information: Imprecision and Uncertainty, pp. 225–254 (1996)
43. Wang, P.: Confidence as higher order uncertainty. null (1994)
44. Zadeh, L.A.: Fuzzy sets. Information and Control 8(3), 338–353 (1965)
45. Zadeh, L.A.: Fuzzy logic and approximate reasoning, pp. 238–259 (1996)

Fuzzy Reasoning with a Rete-OO Rule Engine

Nikolaus Wulff[1] and Davide Sottara[2]

[1] University of Applied Sciences Münster, 48149 Münster, Germany
[2] DEIS, Facolta di Ingegneria, Universita di Bologna, 40131 Bologna, Italy

Abstract. Rules and rule engines play an important role in automated decision making processes like business workflows or system monitoring. Classical inference machines evaluate rules until a final "yes" or "no" decision: this crisp classification schema can turn into a deficiency when they have to deal with uncertain or inprecise knowledge. To circumvent some of these limitations we have built the "Java Expert Fuzzy Inference System" (*Jefis*) and implemented factory methods to deploy the *Jefis* library as an extension for the classical rule engine JBoss Drools. We outline the new features and give examples of uncertain formulated rules executing within the *Jefis* Drools extender.

Keywords: inference engine, fuzzy logic, uncertain reasoning.

1 Introduction

Recently D. Sottara et al. proposed [1] to enrich the JBoss[1] rule engine with annotated logic, allowing different imperfect reasoning schemas, implemented as the sub-module *Drools:Chance*. These novel enhancements offer new possibilities and challenges for the application programmer developing expert systems and indicate first steps towards convergence with the fuzzy RuleML inititative as proposed in [2]. The modifications necessary within the *Drools* framework are described elsewhere, whereas this paper concentrates on the steps to develop a fuzzy based plugin architecture. The first part of the paper explains the basic concepts of the "Java Expert Fuzzy Inference System" *Jefis*, whereas the second part concentrates on the integration and usage within *Drools*.

There exist many rule engines supporting fuzzy logic, both commercial and open source, such as FuzzyCLIPS [3] and FuzzyShell [4], not to mention the countless ad-hoc solutions implemented in many projects, academic or not. Our attention, turned on Drools, which we had already completed some successful third-party projects with. Since the project has started only recently, the results we have are preliminary, so this paper will be focused mainly on the creation and customization of fuzzy rules, leaving the discussion on applications, performance and optimization for future works. To begin with let us explain how we converted from a classical rule engine user into a fuzzy one.

As an example take a rule based cooling control system using the fan current as control variable. Classical rule based systems encode knowledge in the form of statements like:

---

[1] M. Proctor et al.: *Drools* "http://www.jboss.org/drools"

G. Governatori, J. Hall, and A. Paschke (Eds.): RuleML 2009, LNCS 5858, pp. 337–344, 2009.

```
IF temperature == 60 degree THEN fan current = 6 mA
```

This is a rule easy to formulate within the language of any expert system. The phrases 'IF' and 'THEN' are capitalized to indicate the semantic parts of a rule separating premise and conclusion. One drawback of this approach is, that it does not state what happens, if the temperature is 59° C or 61° C, since the rule has been formulated for an exact temperatur value. Obviously specifying many fine granular rules for every possible temperature – or temperature interval – is not feasible. What helps is to formulate rules which express the original intention of the initiator: *if the temperature is hot cool more, if not cool less.* This type of rule is much closer to the way of human thinking and to our natural language. Expressing logical assertions using human-like language is a concept first introduced by Lotfi Zadeh [5] in the late sixties within the context of *fuzzy logic.* Using this approach will result in a small number of rules such as:

```
R1: IF temperature IS cold   THEN current IS low
R2: IF temperature IS normal THEN current IS medium
R3: IF temperature IS hot    THEN current IS high
```

The possible temperature and current ranges have been partitioned using linguistic variables, with the labels "cold", "normal" and "hot" for temperature and "low", "medium" and "high" for current respectively. These linguistic variables denote concepts which have to be modeled appropriately and will be represented by *fuzzy sets* forming two fuzzy partitions as depicted in Fig. 1. A fuzzy set $\mathcal{A}$ is defined as an ordered pair

$$\mathcal{A} = \left\{ \left(x, \mu_{\mathcal{A}}(x)\right) \mid x \in \Omega \text{ with } \mu_{\mathcal{A}}(x) : \Omega \to [0,1] \right\} \tag{1}$$

over some set Ω called the universe of discourse, where $\mu_{\mathcal{A}}(x)$ is the membership function describing in how far the value x belongs to the set $\mathcal{A}$.

The new keyword 'IS' serves as indicator for the fuzzy nature of these rules. The IS operator will activate every rule where the fuzzy membership of the left hand site $\mathcal{P}$ is greater than zero and does not require an exact match. Depending on the degree of the match the conclusion $\mathcal{C}$ will partly be activated by

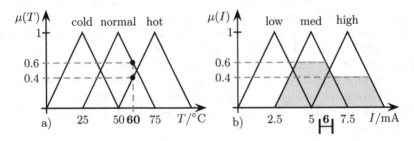

Fig. 1. a) Fuzzification of the temperature 60° C and b) rule based activated fuzzy sets of the output partition, defuzzified as 6 mA fan current

the implication operator. Figure 1 a) demonstrates the fuzzification of the crisp temperature 60° C via the fuzzy sets of the input partition giving the membership vector $\boldsymbol{\mu}(60°\,\text{C}) = \{0, 0.6, 0.4\}^T$ for the sets labeled "cold", "normal" and "hot".

Using the above example the rules R2 and R3 will partly "fire", activating two conclusions for the current, which has to be "medium"/0.6 and "high"/0.4 at the same time, depicted as the shaded area of Fig. 1 b), i.e. there is no clear decision for the "medium" or "high" set, as both are valid to a certain degree.

From the activated output sets a crisp value can be calculated, for example with the weighted mean $I = 0.6 \cdot 5\,\text{mA} + 0.4 \cdot 7.5\,\text{mA} = 6\,\text{mA}$ for the fan current. Note that the cases 59° C or 61° C will not require any additional rules, but can be deduced using the same method, causing in the input partition slightly modifed membership values and thus yielding a current in the vicinity of 6 mA, marked by the range indicator.

2   Basic *Jefis* Fuzzy Set Operations

The building block of *Jefis* are fuzzy sets $\mathcal{A}$ as defined in formula 1. Instead of the membership function $\mu_{\mathcal{A}}(x)$ we will frequently write $\mathcal{A}(x)$ to shorten formulae. It is important to notice, that even if theoretically the domain Ω can be "anything", for ease of computation within the *Jefis* framework it is restricted to be a subset of the real line. The actual membership $\mathcal{A}(x)$ for some value x is returned by the call `set.containment(x)` for any instantiated *Jefis* fuzzy set, no matter if it is a "composite set" – resulting from some join or union operations – or a "primitive leaf set". The library makes extensive use of mature design patterns – as described by E. Gamma et al. [6], which are partly visualized in the UML diagram Fig. 2 showing some of the most important static associations and relations. The Java interface `FuzzySet` offers the three

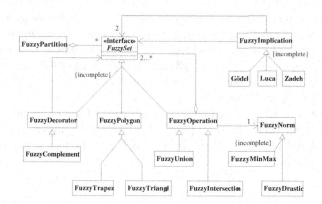

Fig. 2. Partial UML diagram of the basic *Jefis* fuzzy set abstractions and relations

Table 1. Short overview of some t- and s-norms implemented within *Jefis*

name	t-norm ($\cap$)	s-norm ($\cup$)
MinMax	min(a,b)	max(a,b)
Bounded	$\max(0, a+b-1)$	$\min(1, a+b)$
Drastic	$\begin{cases} \min(a,b) & \text{if } \max(a,b) = 1 \\ 0 & \text{otherwise} \end{cases}$	$\begin{cases} \max(a,b) & \text{if } \min(a,b) = 0 \\ 1 & \text{otherwise} \end{cases}$
Algebraic	ab	$a + b - ab$
Hamacher	$\begin{cases} \frac{ab}{a+b-ab} & \text{if } a,b \neq 0 \\ 0 & \text{otherwise} \end{cases}$	$\begin{cases} \frac{a+b-2ab}{1-ab} & \text{if } a,b \neq 1 \\ 1 & \text{otherwise} \end{cases}$
Einstein	$\frac{ab}{1+(1-a)(1-b)}$	$\frac{a+b}{1+ab}$

important classical set operations $A \cup B$ (s-norm), $A \cap B$ (t-norm) and $\neg A$ (c-operation) extended for fuzzy sets A and B, via the `join`, `intersection` and `complement` methods, implemented with aid of the composite and decorator pattern. The *Jefis* implementations for the $\cup$ and $\cap$ operations instantiate a concrete subclass of the `FuzzyNorm` abstraction providing the t- and s-norm implementations, like `FuzzyMinMax` or `FuzzyProduct`, by use of the strategy pattern. This allows a user defined transparent exchange of the t- and s-norm operators at application startup or during runtime. The implemented operators are partially listed in table 1, together with their definitions.

Notice that, giving the pairing between a t-norm and its corresponding s-norm, they are implemented together within the same strategy instance of a concrete FuzzyNorm class, which also fixes within the *Jefis* library the complement operation to be $c(A(x)) \equiv 1 - A(x)$, provided by the FuzzyComplement decorator class, as out of the three operations only two are independent.

The ability to formulate rules in a human readable natural language is one of the advantages of fuzzy logic. Consider the introductive example and Fig. 1 a). The temperature is a *linguistic variable* where the classifiers "cold", "normal" and "hot" are *linguistic terms*. The former is modeled within *Jefis* as a `FuzzyPartition`, where the later are modeled as named `FuzzySets`. Technical the sets are contained within a fuzzy partition which serves as a container. FuzzyPartitions are mostly defined in terms of triangular membership functions, which are easy to implement and calculate[2], but *Jefis* offers also other parabola shaped fuzzy sets like `FuzzyS`, `FuzzyZ` and `FuzzyP`. Linguistic variables are the natural way in *Jefis* to build n-array relations $R : \Omega_1 \times \Omega_2 \times \cdots \times \Omega_n \to [0,1]$ over n universes Ω_k of discourse.

To make things concrete consider the temperatur current example. To monitor the cooling system there are nine possible rules in the temperature $\times$ current space, which build a rule set $\Theta = (T_j \times I_k, S_l)$ described by a relation matrix $R : T \times I \to S$ as in table 2.

---

[2] For polygon shaped Fuzzy Sets like triangle and trapezoid *Jefis* provides direct join and intersection operations and performs a closed analytical calculation of the center of gravity without numerical integration for Mamdani or Larsen inference.

Table 2. Possible monitoring relation for the cooling system

	low	medium	high
cold	green	yellow	yellow
normal	yellow	green	yellow
hot	red	yellow	green

This mapping of temperatur and current for all (t, i)-tuples to the status output can be formulated with corresponding rules as:

```
R11: IF temperature IS cold AND current IS low
        THEN status IS green
R12: IF temperature IS cold AND current IS medium
        THEN status IS yellow

  ...

R33: IF temperature IS hot AND current IS high
        THEN status IS greend
```

The cooling status S is fuzzyfied as a traffic light with labels "green", "yellow" and "red" using a `FuzzyPartition` with three `FuzzySets` in a similar way as for the temperatur and current.

Starting point of this paper was the rule of type "if $\mathcal{P}$ then $\mathcal{C}$": in order to assert a fuzzy conclusion, an implication operator $I(p \to q) : [0,1] \times [0,1] \to [0,1]$ extended for fuzzy logic is requiered. A rule $R_{kj} : A_k \to B_j$ with some system input $\tilde{A}(x) \subset A_k$ will usally activate an variational output set

$$\tilde{B}_j(y) = \sup_{x \in \Omega} t\big(\tilde{A}(x)\,,\, I(A_k(x), B_j(y))\big) \qquad (2)$$

where the degree of variation between $\tilde{B}_j$ and B_j will depend on the implication operator and the t-norm in use. The literature counts meanwhile over seventy different implication operator definitions – for a nice overview see [7] and [8]. Therefore we allow to choose between different implementations via the strategy pattern and offer a possibility to implement and inject custom ones via a parametrizable factory. Table 3 lists some of the more wellknown implication operators. In addition also the most common used engineering implications[3] of Mamdani [9] and Larsen [10] are included.

Or-ing the different activated output sets $\tilde{B}_j$ of the rules R_{kj} via an appropriate s-norm join operation –, the final output fuzzy set $\tilde{B}(y)$ is obtained

$$\tilde{B}(y) = \bigcup_{j=1}^{n} \tilde{B}_j(y) \qquad (3)$$

which has to be converted into a crisp output if desired via the center of gravity or weighted mean method during defuzzyfication.

---

[3] Called approximations as the requirements $I(0,0) = 1$ and $I(0,1) = 1$ demanded for an implication operator are not held.

Table 3. Short overview of some implication operators implemented within *Jefis*

name	$I(p, q)$
Zadeh	$\max(1 - p, \min(p, q))$
Łucasiwiecz	$\min(1, 1 - p + q)$
Kleene-Dienes	$\max(1 - p, q)$
Yager	q^p
Gödel	$\begin{cases} 1 & \text{if } p \leq q \\ q & \text{otherwise} \end{cases}$
Goguen	$\begin{cases} 1 & \text{if } p \leq q \\ q/p & \text{otherwise} \end{cases} \simeq \min(1, q/p)$
Mamdani	$\min(p, q)$
Larsen	$p * q$

3   The *Jefis* Interface to *Drools*

So far we have presented the plain Java implementation of *Jefis* without interfacing the *Drools* rule engine. To do the later we exploit the novel *Drools* enhancements capable to handle uncertain reasoning first announced in [1]. To get the main idea we shortly summarize the rule execution. The "left hand side" (LHS) of a rule, modelling a premise $\mathcal{P}$ is transformed by *Drools* implementation of the RETE algorithmn into a network which nodes evaluate the different constraints present in $\mathcal{P}$. Given one or more objects to be tested, the individual constraints are evaluated by modules implementing the Evaluator interface, and return information on whether each object satisfies the constraints or not. Usually the result of an Evaluator will be a boolean "true" or "false", but with Drools novel uncertainty mode turned on, the result of the evaluation can be a simple fuzzy truth degree, modeled by a real value in [0,1]. Moreover, evaluators are pluggable and can be defined by the user in external libraries, thus increasing significantly the flexibility of the system. Different constraints can be used to build complex logic formulas using logic connectives, such as AND (&&) and OR (||), which encapsulate operators combining the degrees of their

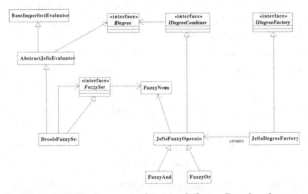

Fig. 3. Partial UML diagram of some *Jefis* to *Drools* adapter classes

operands. Finally, a generalized Modus Ponens operator allows to entail a degree for the conclusion $\mathcal{C}$. With help of a Factory fuzzy *Drools* evaluators decorate *Jefis* fuzzy sets, see the UML diagram Fig. 3. Whenever a fuzzy constraint in the LHS of a rule has to be evaluated the call will be forwarded to the corresponding fuzzy set A and its containment degree μ_A will concur to define the global degree of the premise $\mathcal{P}$, activating the RHS THEN-part of the rule to a certain degree.

```
rule "R11"
    when $c: Cooler(temperature is "cold" && fanCurrent is "low" )
    then $c.setStatus("green", drools.getConsequenceDegree());
```

Listing 1.1. A fuzzy rule base formulated within the *Drools* rule language

Listing 1.1 shows a short extract of a fuzzy rule base. The Cooler instance is a simple Java bean, with getter methods for the temperature and the fan current. 'IS' is a custom evaluator, which takes the *Jefis* fuzzy set evaluators identified by the labels "cold" and "low", which will in turn calculate the membership degrees $\mu_{\text{cold}}(t)$, $\mu_{\text{low}}(i)$ for the present temperature t and current i as returned by the cooler instance. Moreover the '&&'-operator and the implication are not longer boolean but fuzzy. The degree at which a conclusion is entailed is accessible in a rule's RHS and can be used, for example, to express a fuzzy consequence such as "status is green to some degree". Unless configured differently during the session setup, in the default case it will be implemented using the *Jefis* Min t-Norm. The underlying implementation of each operation can further be exchanged within a rule by annotating an '&&'-operator like in the next example 1.2.

```
rule "R31"
    when $c: Cooler(temperature is "hot" &&@(kind="Drastic") fanCurrent is "low")
    then // do what ever is needed ...
```

Listing 1.2. An annotated fuzzy AND operator

The proposed syntax has been inspired by the fuzzy RuleML initiative [2] adopted for the *Drools* DRL, although it is not formulated in XML. Just as well it is possible to annotate the '||'-operator using a different s-norm for the OR operation. Likewise, the implication operator of formula (2), combining the premise's and the implication's degree to yield the degree of the conclusion, can be selected on a rule by rule base using the rule attribute "kind" as in the next example.

```
rule "R31"
    kind "Goedel" // select the used implication operator
    when $c: Cooler(temperature is "hot" && fanCurrent is "low" )
    then // do what ever is needed ...
```

Listing 1.3. Explicit implication operator selection

Like with '&&'- and '||'-operations, all annotated rules belonging to the same relation should use the same implication operator in order to be coherent. The norm and implication operators used to execute the rules are instantiated via the centralized JefisDegreeFactory, which can easy be parametrized using a property file, parsed at factory startup. *Drools* uses a proprietary language, which might be mapped by a translator in the future. Recently, RuleML [11] has been proposed as a standard for rule representation and exchange. To support fuzzy and more general, imperfect extensions, the constructs introduced by the Fuzzy RuleML [2] module should be included.

4   Conclusions

This study had been done to demonstrate the power and feasibility of the extensions to the novel *Drools:Chance* development branch. The first tests are promising and show, that a fuzzy based reasoning within such an extendened classical rule engine is possible. With help of the factory build into the Jefis extender it is possible to define customizable fuzzy-logic based systems within the Drools Rule Language (DRL). The Jefis to Drools bridge is at present an experimental draft and some refactoring has to come, until a stable productive state is reached. It is developed as open-source under the Apache licence version 2.0 and accessible as a Maven2 project site at *http://www.lab4inf.fh-muenster.de/Lab4Jefis*.

References

1. Sottara, D., Mello, P., Proctor, M.: Adding uncertainty to a Rete-OO inference engine. In: Bassiliades, N., Governatori, G., Paschke, A. (eds.) RuleML 2008. LNCS, vol. 5321, pp. 104–118. Springer, Heidelberg (2008)
2. Damásio, C.V., Pan, J.Z., Stoilos, G., Straccia, U.: Representing Uncertainty in RuleML. Fundamenta Informaticae 82, 1–24 (2008)
3. Orchard, R.A., et al.: Fuzzy Extension to the CLIPS Expert System Shell (Fuzzy-CLIPS),
   http://wwwreno.nrc-cnrc.gc.ca/eng/projects/iit/fuzzy-reasoning.html
4. Pan, J., DeSouza, G.N., Kak, A.C.: FuzzyShell: A Large-Scale Expert System Shell using Fuzzy Logic for Uncertainty Reasoning. IEEE Trans. Fuzzy Syst. 6, 563–581 (1998)
5. Zadeh, L.A.: Fuzzy Sets. Information and Control 8, 338–353 (1965)
6. Gamma, E., Helm, R., Johnson, R., Vlissides, J.: Design Patterns – Elements of Resusable Object-Oriented Software. Addison-Wesley, Reading (1995)
7. Dubois, D., Prade, H.: Fuzzy Set and Systems: Theory and Applications. Academic Press, London (1980)
8. Dubois, D., Prade, H.: Fuzzy sets in approximate reasoning, Part I: Inference with possibility distributions. Fuzzy Sets and Systems 50, 143–202 (1991)
9. Mamdami, E.H., Assilian, S.: An Experiment in Linguistic Synthesis with a Fuzzy Logic Controller. International Journal of Man-Machine Studies, 1–13 (1985)
10. Larsen, P.M.: Industrial Applications of Fuzzy Logic Control. International Journal of Man-Machine Studies 12(1), 3–10 (1980)
11. Boley, H.: The rule-ml family of web rule languages. In: Alferes, J.J., Bailey, J., May, W., Schwertel, U. (eds.) PPSWR 2006. LNCS, vol. 4187, pp. 1–17. Springer, Heidelberg (2006)

Towards Modelling Defeasible Reasoning with Imperfection in Production Rule Systems

Davide Sottara[1], Paola Mello[1], and Mark Proctor[2]

[1] DEIS, Facolta di Ingegneria, Universita di Bologna
Viale Risorgimento 2, 40100 Bologna (BO) Italy
[2] JBoss, a division of Red Hat

Abstract. This paper introduces a novel extension to the object-oriented RETE algorithm, designed to create networks whose behaviour can be configured by plugging different modules in. The main feature is the possibility of asserting not just new objects as facts, but also information on how the facts satisfy the different constraints in the network. The underlying reasoning process has been created to process imperfect information, for example fuzzy or probabilistic, but the same framework can easily be adapted to reason with defeasible rules, both boolean and imperfect, by choosing the configuration modules appropriately.

1 Introduction

Production Rule Systems are Knowledge-Based systems using Rules, formal constructs here denoted by "**when** P **then** C" or $P \Rightarrow C$, stating that whenever some pre-conditions P are verified, then new information may be inferred and/or certain actions should be performed as a consequence C. Rules exploit the logical connection between the premise and the conclusion, denoted using the symbol $P \rightarrow C$. The core inferential engine of a Production Rule System is usually designed using a high performance forward chaining algorithm such as RETE [6]. Unfortunately, information on the state of the world may change, so it is not guaranteed that conclusions entailed at a certain time remain the same some time later, hence the knowledge inferred using rules may be not monotonic, but has to be revisioned and updated. The new information may refine the old, but may also conflict with what is know, leading to possbile inconsistencies. In particular, $C(t_1)$, true at time t_1, is updated if it is no longer guaranteed to hold at time t_2, when more information has been obtained; $C(t)$ is instead revisioned by $\neg C(t)$ if the acknowledgment of additional information changes the conclusion entailed so far. The latter category includes the acknowledgement of unexpected events, i.e. exceptions. Consider the classic example bird$(X) \Rightarrow$ flies(X). Since the rule is valid for *most* birds but not for *all*, there can be exceptions, from entire subclasses of birds (e.g. penguins) to individuals (e.g. Tweety). Even if incorrect, statements like this are simple and still cover the majority of cases, so it is convenient to consider them correct and possibly revise one's belief only when needed. To avoid inconsistencies, a rule should include additional checks in its

G. Governatori, J. Hall, and A. Paschke (Eds.): RuleML 2009, LNCS 5858, pp. 345–352, 2009.

preconditions (e.g. $(P \wedge \neg E) \Rightarrow C$, where E denotes an exceptional condition), but this is infeasible for several reasons, including the increased complexity and the necessity to modify a rule every time a new condition is found. To allow the revision of conclusions and the resolution of conflicts in logic reasoning, several proposals exist in literature, such as default reasoning frameworks [11] and defeasible logic [8]. This paper, in particular, will focus mainly on the latter, since it's more rule-oriented. The underlying idea is that some rules may be overridden by others under specific conditions. In particular, a rule $P \Rightarrow C$ may be countered either in its conclusion C, stating $\neg C$, or in its implication, stating $\neg(P \rightarrow C)$. The attacks may even be combined since the conflict arising from stating $\neg C$ has to be resolved, either assuming that one conclusion between C and $\neg C$ is stronger than the other, or by preventing the conflicting C from being entailed altogether. These principles are generally valid, but their applications depend on the choice of rule model and engine. For example, Governatori et al. have applied them to logic programming [1], where rules are Horn clauses and the engine is Prolog-compliant (i.e. backward, SLDNF resolution); other systems such as Dolores [7] use a forward chaining approach; this paper, instead, will discuss a possible implementation of non-monotonic reasoning using a RETE network. In particular, the architecture we are developing is a modification of the RETE-OO algorithm which builds networks capable of supporting reasoning when knowledge is stated with different degrees of imperfection (imperfection is a general term used to encompass both probabilistic and gradual approaches [12]), as described in [13]. This architecture is being developed in the DROOLS[1] rule engine. Moreover, imperfection and defeasibility can be mixed within the same rule: exceptions, for example, are possible but improbable, so they could be set in a probabilistic framework. Fuzzy exceptions could be considered as well: birds generally fly and penguins generally do not, but chickens, for example, do fly even if only to some limited degree. The paper is thus divided in two parts. First, the extended RETE architecture will be introduced; then it will be shown how this network, unlike its standard version, can process defeasible and defeater rules, both in the perfect and in the imperfect case.

2   Extended RETE Networks

The object-oriented version of the RETE algorithm matches objects stored in a working memory against a set of constraints specified in the rules, either as individual facts or as ordered tuples. The rules are compiled into a logic network which uses a combination of information sharing and caching to perform the required checks (for details see for example [6]).

Node types. The RETE algorithm compiles a rule base $R = \{r_j | r_j = P_j \Rightarrow C_j\}$ into a network which processes objects x or tuples $\mathbf{x}$, evaluating constraints σ on their fields.

---

[1] http://www.jboss.org/drools/

More specifically, the network is derived from the premises P_j, which in turn are logical formulas composed of patterns $P_{j,k}$. A pattern is a collection of constraints $\sigma_{j,k}^i$ on a single object. Each constraint, then, can be identified by its signature, also denoted by σ, and defined as: $\sigma = \texttt{Type.Field\# Op \# Args}$

Constraints are mapped to different nodes; in particular, two main types of nodes exist : α-**Nodes**, evaluating constraints on individual objects and β-**Nodes**, evaluating constraints between different objects joined in tuples.

Evaluations. When a generic tuple $\mathbf{X}$ passes through a constraint node, the local property σ is checked. This is conceptually similar to defining and evaluating the validity of a predicate $\pi =< \sigma, \mathbf{X} >$ stating that a generic tuple $\mathbf{X}$ satisfies σ. This evaluation is stored in a structure called Evaluation and denoted by ϵ, which records the degree at which a constraint is satisfied by a tuple, $\varepsilon^\pi(\mathbf{x})^2$. When dealing with imperfection, a constraint may be satisfied in a partial degree, resulting from the combination of different sources:

- ε_0^π : Prior information, provided as a fact.
- ε_σ^π : Direct evaluation, resulting from the evaluator embedded in the node.
- ε_i^π : As a consequence of one or more rules $r_{i \in I}$

Given an initial, off-line, lexical analysis of the rule base, it is possible to know the size of the potential index set $|I|$, i.e. how many pieces of information could theoretically concur to define the aggregate degree of an evaluation. Since not all contributions - especially logical consequences - could be available at runtime for every tuple, an information rate indicator ρ is defined as the ratio of the actual number of contributions and the number of possible contributions.

The overall behaviour is configurable with an appropriate set of strategies:

- The class $\mathbf{L}$ of degrees, generalizing the concepts of true (T), false (F) and unknown $(?)$ to be compatible with boolean and 3-valued logic (many logics use a real value in $[0,1]$). A degree can also have an associated confidence value $\chi \in [0,1]$, similarly to the ones used in FuzzyClips shell.
- A merge strategy $\cap : \mathbf{L}^{2+|I|} \mapsto \mathbf{L}$. Different degrees from different sources are combined into a single degree, so that $\varepsilon^\pi = \cap_{i \in \{0,\sigma\} \cup I} \varepsilon_i^\pi$.
- A null-handling strategy $\mathbf{S}_\emptyset : \emptyset \mapsto \mathbf{L}$. In case one of the sources is not available, its contribution is set to $\emptyset$ (null), which can be treated as if it were false (closed world assumption), or can be mapped to a degree representing lack of knowledge, e.g. the neutral element $?$ of $\cap$ such that $\forall d : d\cap? = d$.

Evaluations may be simple or composite: the latter model the logical operators used to create complex formulas in the premise of a rule. A composite evaluation is a tree node storing links to its operand evaluations, combined to compute its ε_σ^π contribution. Like the truth degrees, operators are pluggable to use different types of logic, as described (from a more theoretical point of view) in [4]. The evaluation tree, built progressively ars the constraints are evaluated, mirrors the

---

[2] When clear from the context, the arguments will be omitted.

structure of the premise being checked. When a degree in a node changes, due to the addition of a new contribution or to a change in rate ρ, the information is propagated to the root, where the global degree is stored.

Modus Ponens. A tuple $\mathbf{x}$ matching the patterns in a premise activates a rule in some degree (stored in a composite Evaluation) and the conclusions $\mathbf{y}$ are computed as function of $\mathbf{x}$; to obtain the conclusion degree, Modus Ponens $\Rightarrow$ has to be applied. Given a rule written in the form $P \Rightarrow C$ and its underlying logical implication $\rightarrow$, Modus Ponens has the generic form:

$$\frac{< P(\mathbf{x}), \varepsilon(\mathbf{x}) >, \forall \mathbf{X} :< P(\mathbf{X}) \rightarrow C(\mathbf{Y}(\mathbf{X})), \varepsilon(\rightarrow) >}{< C(\mathbf{y}(\mathbf{x})), \Rightarrow (\mathbf{x}, \varepsilon(\mathbf{x}), \varepsilon(\rightarrow)) >} \tag{1}$$

Usually rules are assumed true, and most definitions of $\Rightarrow$ just return $\varepsilon(\mathbf{x})$. In the proposed framework, instead, the implication $\rightarrow (\mathbf{x})$ itself is evaluated in a dedicated node before $\Rightarrow$ is applied to yield the consequence degree.

Logical Consequences. The conclusions of a rule can be divided in logical entailments and side effects. Ignoring the latter, in common OO production rule systems entailments consists in the assertion (or retraction) of one or more new objects in the WM. The proposed architecture, instead, also allows the entailment of the truth degree of a predicate, using the instruction **inject**$(\mathbf{y}, \sigma)$.

An injection consequence of rule r_i uses the degree computed from its premise P_i and its implication $\rightarrow_i$ by Modus Ponens: this value is used to set the contribution ε_i^π in the Evaluation of constraint σ for some other tuple $\mathbf{y}$, possibly derived from the activating tuple $\mathbf{x}$. Notably, evaluable constraints include implications: the truth degree of an implication is usually provided *de facto* stating $\varepsilon_0^\rightarrow = \texttt{true}$, but nothing prevents it from being evaluated by other sources. This feature allows a rule to condition the relationship between premise and conclusion of another rule.

Propagation. The injection of a degree poses a synchronization problem. If the tuple $\mathbf{y}$ hasn't been inserted in the working memory at the time of injection, the contribution is stored in a message box at the node responsible for evaluating σ, so that, should $\mathbf{y}$ be later evaluated at that node, the Evaluation will also include that contribution. If, on the other hand, $\mathbf{y}$ exists in the working memory and σ has already been evaluated on it, the relative Evaluation is retrieved and the new partial degree is added, propagating the update of the aggregate degree and the rate ρ. Moreover, the tuple propagation strategy has to be customized: in RETE, all formulas are conjunctions, so nodes propagate (resp. discard) an object or a tuple if it satisfies (resp. fails) the local constraint test. In presence of injections, which are not guaranteed to take place, a third option is possible: a tuple can be blocked while waiting for an injection. Nodes are customized with a propagation strategy S_f, which analyzes the current Evaluation tree, typically considering the aggregate values ε_π and ρ, and chooses between *Pass*, forwarding the tuple to the following node, *Hold*, waiting for injected degrees, and *Drop*, discarding the tuple. When *Hold* is chosen, the tuple is not propagated, but the current node monitors its Evaluation to apply the propagation criteria again should it change. A strategy requiring $\rho = 1$ to propagate a tuple can also be used to synchronize rules, but may be prone to deadlocks, unless all injecting rules are always guaranteed to fire.

3   Towards Defeasible Reasoning

The architecture of section 2 can support non-monotonic reasoning in a way that is similar to the fundamentals of defeasible logic, given some additional concepts.

Conflict. Two truth degrees ε_k and ε_v conflict iff $\varepsilon_k \cap \varepsilon_v = \emptyset$. Two values that can't be merged denote a logical inconsistency: this does not mean that the overall degree is unknown, even if setting conflicting merges to unknown may be a way to resolve them. Conflicts can be resolved using the confidence χ associated to the degrees: a strong evaluation, in fact, should be taken in greater account than an unreliable one. This approach has been widely used in literature, for example by Pollock [10], who calls it "justification degree", and by Bamber in his "scaled assertions"[2]. The merge strategy $\cap$ relies on a pluggable discount strategy, S_d, which applies a discount transformation $\delta : \varepsilon \times \chi \mapsto \varepsilon$. A degree with full confidence is never altered ($\delta(\varepsilon, 1) = \varepsilon$), while a totally unreliable one is transformed into the neutral element of $\cap$: ($\delta(\varepsilon, 0) = ?$). Hence, $S_\emptyset$ and S_d operate in pipeline, so that $\cap$ can always merge proper, discounted degrees, possibly avoiding conflicts arising from missing data or unreliable sources. We will denote by $\cap_\delta$ the confidence-aware merge strategy.

In other cases, instead, the user may want to programmatically state that some degrees should explicitly override any other, e.g. when writing an exception. So, the extended rule language provides defeaters with an "attacking"injection primitive, using the keyword **reject**($\mathbf{y}$,σ). A rejection works like an injection, but it uses the negated consequence degree of the invoking rule and sets an attack flag to true automatically. The attacking degrees are also handled by S_d: its default implementation alters the confidence of the attacked degrees in presence of attackers, before shaping the degrees for the merge according to its actual value. Defined $\chi_k = \max_j\{\chi_j : \varepsilon_j.\text{isAttacker} = \text{true }\}$, the confidence of all non-attacking degrees in the same Evaluation is lowered to $\max\{0, \chi - \chi_k\}$ before applying the discounting δ. This allows to take into account the confidence of the attacker, in addition to that of the attacked degree, limiting the effect of weak exceptions while preserving strong ones.

Rebutting defeat. A rule r_k is a rebutting defeater for a conclusion $C(\mathbf{y})$ of a rule r_v if both rules inject the same predicate $\pi :< \mathbf{y}, \sigma >$, there would be a conflict such that $\varepsilon_k \cap \varepsilon_v = \emptyset : \varepsilon_k$, but r_k attacks r_v with enough confidence, so that $\varepsilon_\pi = \varepsilon_k \cap_\delta \varepsilon_v = \varepsilon_k$. Notice that introducing an attacking injection among the actions of r_k is equivalent to defining a partial order relation $\succ$ between r_k and any rule r_v leading to the same conclusion.

Undercutting defeat. A rule r_k is an undercutting defeater for a conclusion $C(\mathbf{y}(\mathbf{x}))$ of a rule r_v if ε_k defeats $\varepsilon(\to)_v$, i.e. the implication constraint of r_v evaluated for $\mathbf{x}$. Since implications have a truth degree, they can be attacked: setting $\varepsilon(\to)$ to false will cause the Modus Ponens operator to return ?.

As a corollary, a rule r_s is *strict* iff the evaluation of $\to_{r_s}$ has $\chi = 1$ and its confidence can't be lowered. The conclusions of a strict rule can be refined, but not argued against since the deduced truth value will not be discounted.

3.1   Case Study: Rule Base with Defeasible Rules

A non-monotonic rule base includes attacking injections in the rules' conclusions. We will examine the "flying bird" problem, written in DROOLS' DRL language.

```
rule  "Bird(X) -> Flies(X)"   //r1
when      x : Bird( )
then      inject(x,"Bird.flier","==","true");

rule  "Chicken(X) -> neg Flies(X) , neg (Bird(X) -> Flies(X))"  //r2
when      x : Chicken( )
then      reject(x,"Bird(X) -> Flies(X)");
          reject(x,"Bird.flier","==","true");

rule  "Flies(X) -> ..."       //r3
when      x : Bird( flier == true )
then      ...
```

r_3 is the rule reacting to x being capable to fly: when the field `flier` is set to null, the constraint will have to be entailed logically. r_1 is the default rule, while r_2 models an exception. It acts both as rebutting (injecting the constraint in r_3) and as undercutting attacker (injecting the implication) for r_1, but, as will be discussed, one attack would be sufficient.

Boolean configuration. The network uses 3-valued logic with T, F and ? used to denote true, false and unknown respectively, while $\emptyset$ is used for conflict. Modus Ponens $\Rightarrow$ returns T when both premise and implication are true, ? otherwise; merging using $\cap$ has ? as neutral element and only returns $\emptyset$ if T and F are combined. Confidence can be either 0 or 1, so the discount δ returns ? or ε, respectively. $S_\emptyset$ always returns unknown (?) and, finally, S_f always chooses the *Pass* option. Initially, all implications are T; then, a Chicken (subtype of Bird) x is created and its field `flier` is set to `null`, so the direct evaluation of the constraint $\sigma^\star$: Bird.flier == true returns ?. After its insertion, r_1 and r_2 are T-activated, while r_3 is ?-activated. The agenda is non-deterministic, so suppose that the rules will activate in order r_3, r_1, r_2, causing the greatest number of revisions. The first activation of r_3 returns that it is unknown whether x flies or not. When r_1 fires, the combined information for $\sigma^\star$ evaluates to T, so r_3 entails that x flies. The eventual activation of r_2 revisions the belief: the attacking F it contributes discounts the existing evidence to ?, so the aggregate degree becomes F. It also sets the implication Bird$(x) \to$ Flies(x) to F, in turn changing the contribution of r_1 to ?. In either case, the merge at the constraint node is no longer conflictual. At the end, the consequence degree of r_3 is F (x does not fly), while the degree of r_2 is T; r_1 has a true premise, but its implication is false for x, as expected given the state of its conclusion, false, and its premise, true.

Imperfect configuration. The same network can be used with truth degrees other than booleans. Consider, for example, possibilistic logic: a degree has the form $[N, \Pi]_\chi$, where N and Π are respectively a lower bound of the necessity of a constraint and an upper bound of its possibility ([5]) and χ is the confidence degree. The filtering strategy remains the same (always *Pass*), but the other modules have to be adapted accordingly. Modus Ponens ($\Rightarrow$) takes the minimum ($\wedge$) of the premise and the implication's necessities, but then sets the possibility to 1; Negation ($\neg$) relies on the duality relation $N(x) = 1 - \Pi(\neg x)$; the merge operator $\sqcap$ has the effect of intersecting the intervals $[N_i, \Pi_i]$; null contributions are mapped by $S_\emptyset$ into $[0, 1]_0$; finally, the discount strategy uses confidence χ to broaden an interval into $[N \cdot \chi, 1 - \chi \cdot (1 - \Pi)]$. A full description of the properties of operators in possibilistic logic can be found in [5]. The same operators also combine the attached confidence degrees as in fuzzy CLIPS.

Suppose that the degree of $\rightarrow_1$ is set to $[.8, 1]_{.5}$, possibly because it has been learned by induction over a limited set of example birds (low confidence), not all of which were fliers (necessity < 1). Given a Bird x, rule r_1 returns $[.8, 1]_{.5}$. Since it does not trigger the exception, this degree is merged with the one resulting from direct evaluation at r_3. In particular, if x were known to fly, the evaluator would yield $[1, 1]_1$ so the merge would be true. If, instead, the field x.flies is null, the overall result is $[.4, 1]_{.5}$ since, due to the low confidence, N has been discounted; evaluation at r_3 returns

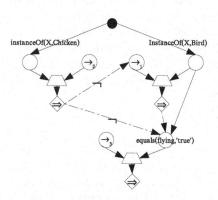

Fig. 1. RETE for rule base R

$[0, 1]_0$, while the premise of r_2 is false, so its activation degree is $[0, 1]_0$, which alters neither r_3 conclusion nor r_1 implication. When, instead, x is a Chicken, the system behaves like in the boolean case, overriding the contributions of $\rightarrow_1$ and r_3 to $[0, 0]_1$.

3.2   RuleML Encoding

Drools DRL is a proprietary language, but it is feasible to translate the rules into the standard, XML-based RuleML language. Given the object-oriented nature of Drools and its being a production rule system, and the customizable behaviour of operators and degrees, a translator would produce a combination of object oriented (OO, [3]), fuzzy [4] and Production rule (PR, [9]) RuleML. Being new, however, the `inject` logical action is not supported in any of the language modules, so a new Element should be introduced, with a minimal content model. Constraints and tuples can be serialized using Atom elements, while the attribute, false by default, distinguishes attacking from simple injections.

```
<!ELEMENT Inject Atom,  (Args) >
<!ATTLIST Inject attack \%boolean;  'false'>
<!ELEMENT Args (Atom+) >
```

4 Conclusions and Future Developments

We have introduced a novel extension of the RETE networks, stressing how its behaviour can be customized using a set of pluggable modules to affect the semantics of the rules it models. The architecture, implemented extending the DROOLS rule engine with imperfect reasoning as main goal, supports a primitive form of defeasible reasoning exploiting the explicit notion of truth degree and the evaluation of the implication part in the rules. Even if the engine is far from being a full framework supporting argumentations in defeasible logic, it supports the basic building blocks, allowing to write rules with exceptions, even in presence of imperfection or uncertainty in the data. However, the engine is very experimental and suffers some major drawbacks. No full scale tests have been carried out and a few real applications are still being developed. Moreover, a complete truth maintenance system has not been implemented for logical injections, so it is not possible to track the dependencies explicitly.

References

1. Antoniou, G., Billington, D., Governatori, G., Maher, M.J.: Embedding defeasible logic into logic programming. TPLP 6(6), 703–735 (2006)
2. Bamber, D., Goodman, I.R., Nguyen, H.T.: Robust reasoning with rules that have exceptions. Ann. Math. Artif. Intell. 45(1-2), 83–171 (2005)
3. Boley, H.: Object-oriented ruleml: User-level roles, uri-grounded clauses, and order-sorted terms. In: Schröder, M., Wagner, G. (eds.) RuleML 2003. LNCS, vol. 2876, pp. 1–16. Springer, Heidelberg (2003)
4. Damásio, C.V., Pan, J.Z., Stoilos, G., Straccia, U.: Representing uncertainty in ruleml. Fundam. Inf. 82(3), 265–288 (2008)
5. Dubois, D., Prade, H.: Possibilistic logic
6. Forgy, C.: Rete: A fast algorithm for the many patterns/many objects match problem. Artif. Intell. 19(1), 17–37 (1982)
7. Maher, M.J.: Propositional defeasible logic has linear complexity. In: Logic Programming, pp. 691–711 (2001)
8. Nute, D.: Defeasible logic. In: INAP, pp. 87–114 (2001)
9. Paschke, A., Kozlenkov, A., Boley, H.: A homogenous reaction rules language for complex event processing. In: International Workshop on Event Drive Architecture for Complex Event Process (2007)
10. Pollock, J.L.: Defeasible reasoning with variable degrees of justification. Artificial Intelligence 133(1-2), 233–282 (2001)
11. Reiter, R.: A logic for default reasoning. AI 13 (April 1980)
12. Smets, P.: Imperfect Information: Imprecision and Uncertainty, vol. 225, p. 254 (1996)
13. Sottara, D., Mello, P., Proctor, M.: Adding uncertainty to a rete-OO inference engine. In: Bassiliades, N., Governatori, G., Paschke, A. (eds.) RuleML 2008. LNCS, vol. 5321, pp. 104–118. Springer, Heidelberg (2008)

Author Index